Probability and Mathematical Statistics (Continued)

SEBER • Linear Regression Analysis
SEBER • Multivariate Observations
SEN • Sequential Nonparametrics: Invariance Principles and Statistical Inference
SERFLING • Approximation Theorems of Mathematical Statistics
TJUR • Probability Based on Radon Measures
WILLIAMS • Diffusions, Markov Processes, and Martingales, Volume I: Foundations
ZACKS • Theory of Statistical Inference

Applied Probability and Statistics

ABRAHAM and LEDOLTER • Statistical Methods for Forecasting
AGRESTI • Analysis of Ordinal Categorical Data
AICKIN • Linear Statistical Analysis of Discrete Data
ANDERSON, AUQUIER, HAUCK, OAKES, VANDAELE, and WEISBERG • Statistical Methods for Comparative Studies
ARTHANARI and DODGE • Mathematical Programming in Statistics
BAILEY • The Elements of Stochastic Processes with Applications to the Natural Sciences
BAILEY • Mathematics, Statistics and Systems for Health
BARNETT • Interpreting Multivariate Data
BARNETT and LEWIS • Outliers in Statistical Data
BARTHOLOMEW • Stochastic Models for Social Processes, *Third Edition*
BARTHOLOMEW and FORBES • Statistical Techniques for Manpower Planning
BECK and ARNOLD • Parameter Estimation in Engineering and Science
BELSLEY, KUH, and WELSCH • Regression Diagnostics: Identifying Influential Data and Sources of Collinearity
BHAT • Elements of Applied Stochastic Processes
BLOOMFIELD • Fourier Analysis of Time Series: An Introduction
BOX • R. A. Fisher, The Life of a Scientist
BOX and DRAPER • Evolutionary Operation: A Statistical Method for Process Improvement
BOX, HUNTER, and HUNTER • Statistics for Experimenters: An Introduction to Design, Data Analysis, and Model Building
BROWN and HOLLANDER • Statistics: A Biomedical Introduction
BROWNLEE • Statistical Theory and Methodology in Science and Engineering, *Second Edition*
CHAMBERS • Computational Methods for Data Analysis
CHATTERJEE and PRICE • Regression Analysis by Example
CHOW • Analysis and Control of Dynamic Economic Systems
CHOW • Econometric Analysis by Control Methods
COCHRAN • Sampling Techniques, *Third Edition*
COCHRAN and COX • Experimental Designs, *Second Edition*
CONOVER • Practical Nonparametric Statistics, *Second Edition*
CONOVER and IMAN • Introduction to Modern Business Statistics
CORNELL • Experiments with Mixtures: Designs, Models and The Analysis of Mixture Data
COX • Planning of Experiments
DANIEL • Biostatistics: A Foundation for Analysis in the Health Sciences, *Third Edition*
DANIEL • Applications of Statistics to Industrial Experimentation
DANIEL and WOOD • Fitting Equations to Data: Computer Analysis of Multifactor Data, *Second Edition*
DAVID • Order Statistics, *Second Edition*
DAVISON • Multidimensional Scaling
DEMING • Sample Design in Business Research
DILLON and GOLDSTEIN • Multivariate Analysis: Methods and Applications
DODGE and ROMIG • Sampling Inspection Tables, *Second Edition*
DOWDY and WEARDEN • Statistics for Research
DRAPER and SMITH • Applied Regression Analysis, *Second Edition*

continued on back

STATISTICAL COMPUTATION

STATISTICAL COMPUTATION

J. H. MAINDONALD
Applied Mathematics Division
Department of Scientific and Industrial Research
Auckland, New Zealand

JOHN WILEY & SONS
New York · Chichester · Brisbane · Toronto · Singapore

***Library of Congress Cataloging in Publication Data*:**

Maindonald, J. H. (John Hilary), 1937–
Statistical computation.

(Wiley series in probability and mathematical statistics)
Includes index.
1. Mathematical statistics—Data processing.
I. Title. II. Series.

QA276.4.M25 1984 519.5′028′54 83-21644
ISBN 0-471-86452-8

Printed in the United States of America

10 9 8 7 6 5 4 3 2 1

To Shireen and Evan, who share my fascination for computers

To my mother

Preface

This book is intended for the statistician or student of statistics who wants to learn about the computational techniques used in solving statistical problems. It should be useful also to those whose interest is in computation and wish to learn about the statistical calculations to which the techniques, mainly linear algebra and matrices, are relevant.

Any major collection of statistical computer programs will include a program for multiple regression. This form of analysis may be extended and generalized to take in analysis of variance and covariance, normal theory multivariate analysis, and loglinear analysis of multiway tables. These topics occupy the first seven chapters. Chapters 8 and 9 consider, but in less detail, splines, robust regression, time series, pseudorandom numbers, and several more minor topics of interest to the applied statistician. The final chapter reflects on the role of the computer in statistical analysis. In all cases I have been concerned to place methods of calculation in the context of real statistical problems.

Chapters 1 through 3 are largely devoted to the use of normal equation methods for linear least squares and correlation calculations, and lay the groundwork for the treatment in Chapter 4 of numerically superior methods based on the use of orthogonal reduction to upper triangular form. The advantages of starting with normal equation methods are that most statisticians are familiar with this approach and it is easy to demonstrate the calculations with data chosen to make the arithmetic simple. The matrix operations introduced in Chapters 1 through 4 are the basis for many of the calculations in Chapters 5 through 7. Simple devices are used to deal with linear dependence—there is no use of generalized matrix inverses as such.

In Chapter 6 the use of an algorithm for calculating eigenvalues and eigenvectors is assumed. Discussion of the alternative algorithms that are available is beyond the scope of this book. Nash (1979) is a good starting point for the reader who wishes to learn about algorithms for eigencalculations. Chapter 10 gives a BASIC program suitable for use with matrices of low order.

Except as an aid to learning, it is not expected that most readers will write their own computer programs. The aim is rather to encourage discriminating use of programs written by experts in statistical and numerical methods. Anyone who uses, or assists others in using statistical computer programs, will be aware of the need to understand in broad outline the algorithms used and to make simple checks as the calculations proceed. This will prevent the grosser forms of nonsense that unthinking users persuade statistical computer programs to provide. It is often possible for someone who understands a computer routine to turn it to a use different from what the program writer had immediately in mind.

The BASIC programs in Chapters 9 and 10 will interest those who wish to experiment with the use of small computers (including some of the handheld computers now available) for statistical calculations. It is intended to make these programs available in machine readable form through the Digital Equipment Users Society (DECUS) program library.

Unlike authors of calculus texts I do not have to justify yet another book on my chosen topic. There is little or nothing which, though giving a broad coverage of statistical computation, is suitable for the general reader who has a modest mathematical education. A further feature is that I have given attention to the statistical context of the calculations as well as to the calculations themselves.

The inclusion of a topic carries with it the implication that the calculations are, in appropriate circumstances, worth doing. Hence the attention given to questions of statistical practice, particularly where the models have only recently come into widespread use.

The descriptions of methods in Chapters 1 through 3, and the examples used to demonstrate them, should be accessible to anyone with a knowledge of matrix multiplication. Later chapters, and asterisked sections of Chapters 1 through 3, demand an understanding of elementary linear algebra. Chapter 7 assumes some familiarity with maximum likelihood methods. More difficult exercises, and those that involve extensive calculations, are marked with an asterisk.

J. H. MAINDONALD

Auckland, New Zealand
January 1984

Acknowledgments

I am grateful to Professor Vere-Jones and other former colleagues at Victoria University of Wellington (N.Z.) for their encouragement during the early stages of the writing of this manuscript, which was begun while I was on leave from the University in 1976. Dr. M. A. Saunders, while he was still with DSIR Applied Mathematics Division in Wellington, helped direct my ideas at some crucial points. Dr. R. B. Davies and other colleagues in DSIR have made many useful comments and suggestions. Special thanks are due to H. R. Henderson of Ruakura Research Centre (N.Z.) for his comments on draft versions of the manuscript. Others who have commented on some part of the manuscript include A. J. Miller (CSIRO, Australia), G. C. Arnold and A. Swift (Massey University, N.Z.) and J. R. Dale (DSIR). C. M. Triggs (DSIR) drew my attention to the method of Kuiper and Corsten, discussed in Section 5.5. G. W. Hill (now deceased) made several helpful suggestions in connection with the algorithm used in Fig. 9.7. None of these individuals can be held responsible for what I have made of their help! I am grateful to Maureen Cooper and Ann Koot for their assistance in typing Chapters 1 through 6. T. R. Cooper (DSIR) gave valued help with proofreading.

Where not otherwise acknowledged, the computer programs in Chapters 9 and 10 are included with the permission of the Director, Applied Mathematics Division, DSIR. I am grateful to J. Clearwater, E. Hewett, and S. Pennycook (all of DSIR) for permission to use their data. For permission to use material that appears as indicated, I am grateful to J. D. Beasley and S. G. Springer (Fig. 9.5), G. E. P. Box (quotation introducing Ch. 10), J. Nash (Fig. 10.9), R. P. Brent (Fig. 10.10), Prentice-Hall, Inc. (Fig. 10.10), T. R. Hopkins (Fig. 10.10), R. W. Cottle and E. Aguado (Appendix on Major Cholesky), H. S. Houthakker (Table 2.1), and the Royal Statistical Society (Table 2.1 and Fig. 9.5).

The writing of this text became a far larger project than I had intended when I began. I am grateful to my wife, Winifred, and to our children for their forbearance. Finally, I have to thank those, first at Victoria University and then at DSIR, whose statistical and computing problems have been a stimulus to my interest in statistical computation.

J.H.M.

Contents

Notation

The following conventions are chiefly relevant to Chapters 1 through 5:

$$\boldsymbol{X} = \begin{bmatrix} 1 & x_{11} & x_{12} & \cdots & x_{1p} \\ 1 & x_{21} & x_{22} & & x_{2p} \\ \vdots & \vdots & \vdots & & \vdots \\ 1 & x_{n1} & x_{n2} & \cdots & x_{np} \end{bmatrix}, \quad \boldsymbol{y} = \begin{bmatrix} y_1 \\ y_2 \\ \vdots \\ y_n \end{bmatrix}$$

$$= [\boldsymbol{x}_0, \boldsymbol{x}_1, \ldots, \boldsymbol{x}_p]$$

$$\mathbf{\mathsf{X}} = \begin{bmatrix} \mathsf{x}_{11} & \mathsf{x}_{12} & \cdots & \mathsf{x}_{1p} \\ \mathsf{x}_{21} & \mathsf{x}_{22} & & \mathsf{x}_{2p} \\ \vdots & \vdots & & \vdots \\ \mathsf{x}_{n1} & \mathsf{x}_{n2} & \cdots & \mathsf{x}_{np} \end{bmatrix}, \quad \mathbf{\mathsf{y}} = \begin{bmatrix} \mathsf{y}_1 \\ \mathsf{y}_2 \\ \vdots \\ \mathsf{y}_n \end{bmatrix}$$

$$= [\mathbf{\mathsf{x}}_1, \mathbf{\mathsf{x}}_2, \ldots, \mathbf{\mathsf{x}}_p]$$

where

$$\mathsf{x}_{ij} = x_{ij} - \bar{x}_j$$

$$\bar{x}_j = n^{-1} \sum_{i=1}^{n} x_{ij}$$

$$\mathsf{y}_i = y_i - \bar{y}$$

$$\bar{y} = n^{-1} \sum_{i=1}^{n} y_i$$

Then

$$\boldsymbol{S} = [\mathbf{X}, \boldsymbol{y}]'[\boldsymbol{X}, \boldsymbol{y}] \quad \text{is the SSP matrix}$$

$$\mathsf{S} = [\mathsf{X}, \mathsf{y}]'[\mathsf{X}, \mathsf{y}] \quad \text{is the CSSP matrix}$$

$\boldsymbol{T}$ is an upper triangular matrix such that $\boldsymbol{T}'\boldsymbol{T} = \boldsymbol{S}$

T is an upper triangular matrix such that $\mathsf{T}'\mathsf{T} = \mathsf{S}$

Take $q = p + 1$.

$$\boldsymbol{T} = \begin{bmatrix} \mathbf{t}_0' \\ \mathbf{t}_1' \\ \vdots \\ \mathbf{t}_q' \end{bmatrix} = \begin{bmatrix} \boldsymbol{T}_p & \boldsymbol{t}_q \\ \mathbf{0}' & t_{qq} \end{bmatrix}; \quad \boldsymbol{T}^{-1} = \begin{bmatrix} \boldsymbol{T}_p^{-1} & \boldsymbol{t}^q \\ \mathbf{0}' & t^{qq} \end{bmatrix}$$

Alternatively $\boldsymbol{t}_y = \boldsymbol{t}_q$; $t_{yy} = t_{qq}$; $\boldsymbol{t}^y = \boldsymbol{t}^q$; $t^{yy} = t^{qq}$. Below diagonal elements of a symmetric matrix are usually replaced by dots or omitted altogether. An upper triangular matrix has all of its below diagonal elements equal to zero. Below diagonal elements are usually printed as small zeros or omitted altogether.

STATISTICAL COMPUTATION

CHAPTER 1

Regression Calculations —Part I

Simple numerical examples, with numbers chosen to make the arithmetic easy, are used to illustrate the formation and solution of the least squares normal equations. The methods of solution used are based on algorithms for reducing any matrix of the form **X′X** to the form $T'T$, where T is an upper triangular matrix. (An upper triangular matrix is one where all below diagonal elements are zero.)

1.1. STRAIGHT-LINE REGRESSION

A line is to be fitted to the four data points:

	x_i	y_i
$i = 1$	-2	0
$i = 2$	-1	2
$i = 3$	2	5
$i = 4$	7	3

The number of points is $n = 4$. For use note the following:

$$n = 4 \quad \sum x_i = 6 \quad \sum y_i = 10$$

$$\sum x_i^2 = 58 \quad \sum x_i y_i = 29$$

$$\sum y_i^2 = 38$$

Figure 1.1 shows a line that has been drawn quite arbitrarily through the four points. The equation of such a line may be written $\hat{y} = a + bx$. To understand the use of the hat on $\hat{y}$, suppose that $x = x_i$ is the x-value for one of the four data points. The hat distinguishes the point $(x_i, \hat{y}_i)$ on the line from the data point (x_i, y_i).

Assume that y is the dependent variate and x the predictor or explanatory variate. Then the least squares method chooses a and b so that the sum of squares of deviations from the line in the y-direction is minimized. In order to give a formal mathematical description consider the equation

$$y_i = a + bx_i + e_i, \quad i = 1, 2, 3, 4$$

The e_i (positive for points above the line, and negative for points below the line) are represented by the heavy vertical lines. The parameter estimates a and b are required such that Σe_i^2 is minimized. It is shown in elementary statistics texts that the estimates of the parameters a and b may be obtained by solving the normal equations

$$na + \left(\sum x\right)b = \sum y \tag{1.1}$$

$$\left(\sum x\right)a + \left(\sum x^2\right)b = \sum xy \tag{1.2}$$

(Omission of the suffix i causes no ambiguity.)

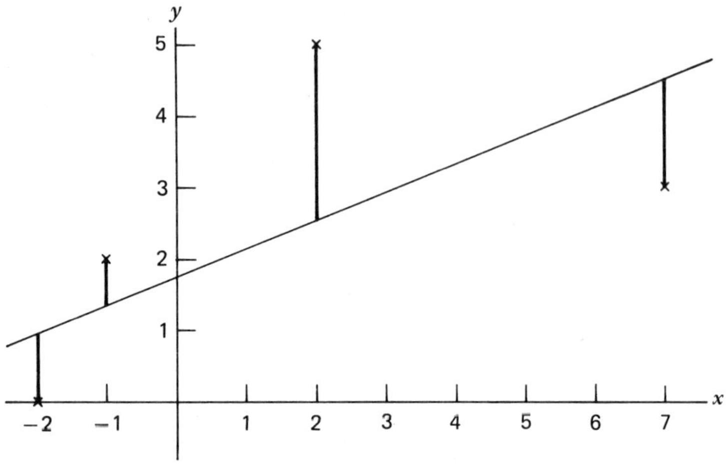

Figure 1.1. Straight-line regression.

In the present case

$$4a + 6b = 10 \tag{1.3}$$

$$6a + 58b = 29 \tag{1.4}$$

The parameter a may be eliminated by multiplying eq. (1.3) by $\frac{3}{2}$ and subtracting from eq. (1.4). This yields

$$4a + 6b = 10 \tag{1.3$'$}$$

$$49b = 14 \tag{1.4$'$}$$

The method now used to solve these is backward substitution. We solve first to obtain $b = \frac{2}{7}$ from eq. (1.4′) and then go back and substitute $b = \frac{2}{7}$ in eq. (1.3′), yielding

$$4a = 10 - 6 \times \tfrac{2}{7}, \quad a = \tfrac{29}{14}.$$

1.2. A MODIFIED SCHEME FOR SOLVING THE NORMAL EQUATIONS

The scheme we describe is the Cholesky or square root scheme. As before, we start with

$$4a + 6b = 10 \tag{2.1}$$

$$6a + 58b = 29 \tag{2.2}$$

Instead of eqs. (1.3′) and (1.4′) the Cholesky scheme gives

$$2a + 3b = 5 \tag{2.3}$$

$$7b = 2 \tag{2.4}$$

Equation (2.3) is obtained by dividing eq. (1.3′) through by the square root of the coefficient of a, that is, by $\sqrt{4}$. Equation (2.4) may be obtained from eq. (1.4′) by dividing through by $\sqrt{49}$, which is the square root of the coefficient of b in that equation.

The change serves at least two purposes:

1. The coefficient of b (i.e., of the second parameter) in eq. (2.3) tells what multiple of eq. (2.3) should be subtracted from eq. (2.2) (the second equation).

2. The terms on the right-hand side in eqs. (2.3) and (2.4) give us the information needed to set out the analysis of variance table:

	Sum of Squares
Due to fitting constant term [i.e., go from Σy^2 to $\Sigma(y - \bar{y})^2$]	5^2
Due to fitting x in addition [i.e., go to $\Sigma(y - a - bx)^2$]	2^2

A justification of this table follows. The residual sum of squares is calculated as $\Sigma y^2 - 5^2 - 2^2 = 9$.

Calculations proceed in this way (with the equation number retained unchanged whenever the equation is one that has appeared earlier):

(2.1)	$4a + 6b = 10$	Divide eq. (2.1) through by $\sqrt{4}$ →	$2a + 3b = 5$	(2.1′)
(2.2)	$6a + 58b = 29$		$6a + 58b = 29$	(2.2)

Multiply eq. (2.1′) by 3 (= coefficient of b in that equation) and subtract from eq. (2.2)

(2.1′)	$2a + 3b = 5$	Divide eq. (2.2′) through by $\sqrt{49}$ →	$2a + 3b = 5$	(2.1′)
(2.2′)	$49b = 14$		$7b = 2$	(2.2″)

As it is known that the coefficient of a in eq. (2.2′) will be zero, it need not be calculated.

In the general case, starting with

$$na + \left(\sum x\right)b = \sum y \tag{2.5}$$

$$\left(\sum x\right)a + \left(\sum x^2\right)b = \sum xy \tag{2.6}$$

the Cholesky scheme gives

$$n^{1/2}a + n^{-1/2}\left(\sum x\right)b = n^{-1/2}\sum y \tag{2.5′}$$

$$\left[\sum(x - \bar{x})^2\right]^{1/2}b = \left[\sum(x - \bar{x})^2\right]^{-1/2}\left[\sum(x - \bar{x})(y - \bar{y})\right] \tag{2.6′}$$

Equation (2.5′) is multiplied through by $n^{-1/2}\sum x$, that is, by the coefficient of b in that equation, before subtraction from eq. (2.6). To obtain eq. (2.6) in the form given, it is necessary to use the results

$$\sum x^2 - n^{-1}\left(\sum x\right)^2 = \sum (x - \bar{x})^2 \tag{2.7}$$

$$\sum xy - n^{-1}\left(\sum x\right)\left(\sum y\right) = \sum (x - \bar{x})(y - \bar{y}) \tag{2.8}$$

It is now easy to verify that the terms for the analysis of the foregoing variance table are correctly calculated. Squaring the term on the right-hand side of eq. (2.5′) gives

$$n^{-1}\left(\sum y\right)^2 = \sum y^2 - \sum (y - \bar{y})^2$$

(= reduction in sum of squares due to fitting constant term). Squaring the term on the right-hand side of eq. (2.6′) gives

$$\frac{\left[\sum (x - \bar{x})(y - \bar{y})\right]^2}{\sum (x - \bar{x})^2} = \sum (y - a - bx)^2 - \sum (y - \bar{y})^2$$

where a and b are given by eqs. (2.5) and (2.6). The algebraic details of this result may be found in elementary textbooks.

The coefficients a and b, and the equals signs, can with advantage be omitted in the written scheme of calculation. In addition it is useful, as will be seen in the next section, to include an extra row where information relevant to calculating the residual sum of squares can be placed.

1.3. SCHEMATIC REPRESENTATION

The information in the eqs. (2.3) and (2.4), that is,

$$2a + 3b = 5$$

$$7b = 2$$

is adequately represented by the scheme

a	b	
2	3	5
	7	2

A schematic representation will be used in all subsequent discussion.

We have an upper triangular array (see Section 1.4) with the element in the lower right-hand corner missing. It is appropriate to place the square root of the residual sum of squares in this position. For our example we then have

2	3	5	← Square this to get "due to constant term" sum of squares
	7	2	← Square this to get "due to x, given constant term" sum of squares
		3 (boxed)	← Square this to get residual sum of squares

This full array is obtained if the elimination scheme is applied to

$$\begin{array}{cc|c} a & b & \\ n & \sum x & \sum y \\ (\sum x) & \sum x^2 & \sum xy \\ \hline (\sum y) & (\sum xy) & \boxed{\sum y^2} \end{array}$$

that is,

$$\begin{array}{cc|c} a & b & \\ 4 & 6 & 10 \\ (6) & 58 & 29 \\ \hline (10) & (29) & \boxed{38} \end{array}$$

(Elements in below diagonal positions are bracketed because they need not be written down. They duplicate values in corresponding above diagonal positions.)

Calculations involving the initial two rows proceed as before, to give

2	3	5	←
(0)	7	2	←
(10)	(29)	38 (boxed)	← In the sequel this will be replaced by $(38 - 5^2 - 2^2)^{1/2}$.

Now continue and subtract

$$\begin{aligned}&(2, 3|5) \times 5\\ +&(0, 7|2) \times 2\end{aligned}$$

from the final row. The only element of the result that need concern us is the final element; the initial two elements will become zero. The effect is to replace the final 38 by

$$38 - 5^5 - 2^2 = 9.$$

The 9 is then replaced by its square root, to give

$$\begin{array}{cc|c} 2 & 3 & 5 \\ (0) & 7 & 2 \\ \hline (0) & (0) & \boxed{3} \end{array}$$

An array such as this, where all below diagonal elements are zero, is called an upper triangular array. It is again unnecessary to write down the below diagonal elements or to store them in a computer.

1.4. STORAGE OF SYMMETRIC OR UPPER TRIANGULAR ARRAYS

In this text it will be assumed that it is the upper triangle of any symmetric or upper triangular array that is stored. Where, as in the discussion of Section 1.3, the array corresponds to a regression equation which includes a constant term, rows and columns will be numbered $0, 1, \ldots, q$. Let $q = p + 1$. It will be assumed that variates $1, 2, \ldots, p$ are actual or potential explanatory variates. The final variate (variate q) will usually be taken as the dependent variate. The stored elements are those that occupy the storage locations:

$$\begin{array}{cccc} (0,0) & (0,1) & \cdots & (0,q) \\ & (1,1) & \cdots & (1,q) \\ & & \ddots & \vdots \\ & & & (q,q) \end{array}$$

(In Section 1.3, $q = 2$.)

Storage requirements will be kept to a minimum if these elements are stored in a one-dimensional array. A convenient ordering of the storage locations is

$$\begin{array}{cccc} 1 & 2 & \cdots & \frac{q(q+1)}{2}+1 \\ & 3 & \cdots & \frac{q(q+1)}{2}+2 \\ & & & \vdots \\ & & & \frac{(q+1)(q+2)}{2} \end{array}$$

The (i, j) element occupies storage location number $j(j+1)/2 + i + 1$.

1.5. MATRIX DECOMPOSITION ALGORITHMS

Sections 1.1 to 1.3 described a particular variant of the Cholesky algorithm for solving the least squares normal equations. In the subsequent discussion this algorithm will be known as the *complete Cholesky decomposition algorithm* (CCDA). Calculations involving the current row of the array are *completed* before making any changes to subsequent rows.

Another point of view on the calculations of Section 1.3 is that, starting with a matrix

$$\boldsymbol{S} = \begin{bmatrix} 4 & 6 & 10 \\ (6) & 58 & 29 \\ (10) & (29) & 38 \end{bmatrix}$$

the algorithm determined an upper triangle matrix $\boldsymbol{T}$ such that $\boldsymbol{T}'\boldsymbol{T} = \boldsymbol{S}$. This will be demonstrated by giving an alternative algorithm for forming such a matrix $\boldsymbol{T}$, and showing that this alternative algorithm carries out the same arithmetic operations as for the CCDA but in a different order. This alternative algorithm will, in the sequel and when reduced to its bare essentials, be known as the *sequential Cholesky decomposition algorithm* (SCDA).

Let $\boldsymbol{T}' = [\mathbf{t}_0, \mathbf{t}_1, \mathbf{t}_2]$. The first step involves finding $\mathbf{t}_0$ such that $\boldsymbol{S} - \mathbf{t}_0\mathbf{t}_0'$ has zeros in the first row and column. Now

$$\mathbf{t}_0\mathbf{t}_0' = \begin{bmatrix} t_{00} \\ t_{01} \\ t_{02} \end{bmatrix} [t_{00} \quad t_{01} \quad t_{02}] = \begin{bmatrix} t_{00}^2 & t_{00}t_{01} & t_{00}t_{02} \\ t_{01}t_{00} & & \\ t_{02}t_{00} & \text{etc.} & \end{bmatrix}$$

Then $t_{00}^2 = 4$ gives $t_{00} = 2$, $t_{00}t_{01} = 6$ gives $t_{01} = 3$, $t_{00}t_{02} = 10$ gives $t_{02} = 5$. Thus

$$\mathbf{t}_0\mathbf{t}_0' = \begin{bmatrix} 2 \\ 3 \\ 5 \end{bmatrix} [2 \quad 3 \quad 5] = \begin{bmatrix} 4 & 6 & 10 \\ 6 & 9 & 15 \\ 10 & 15 & 25 \end{bmatrix}$$

Thus

$$\boldsymbol{S} = \begin{bmatrix} 2 \\ 3 \\ 5 \end{bmatrix} [2 \quad 3 \quad 5] + \begin{bmatrix} 0 & 0 & 0 \\ 0 & 49 & 14 \\ 0 & 14 & 13 \end{bmatrix}$$

$$= \begin{bmatrix} 2 \\ 3 \\ 5 \end{bmatrix} [2 \quad 3 \quad 5] + \begin{bmatrix} 0 \\ 7 \\ 2 \end{bmatrix} [0 \quad 7 \quad 2] + \begin{bmatrix} 0 & 0 & 0 \\ 0 & 0 & 0 \\ 0 & 0 & 9 \end{bmatrix}$$

$$= \begin{bmatrix} 2 \\ 3 \\ 5 \end{bmatrix} [2 \quad 3 \quad 5] + \begin{bmatrix} 0 \\ 7 \\ 2 \end{bmatrix} [0 \quad 7 \quad 2] + \begin{bmatrix} 0 \\ 0 \\ 3 \end{bmatrix} [0 \quad 0 \quad 3]$$

$$= \mathbf{t}_0\mathbf{t}_0' + \mathbf{t}_1\mathbf{t}_1' + \mathbf{t}_2\mathbf{t}_2'$$

$$= [\mathbf{t}_0 \quad \mathbf{t}_1 \quad \mathbf{t}_2] \begin{bmatrix} \mathbf{t}_0' \\ \mathbf{t}_1' \\ \mathbf{t}_2' \end{bmatrix}$$

$$= \boldsymbol{T}'\boldsymbol{T}, \text{ where } \boldsymbol{T} = \begin{bmatrix} \mathbf{t}_0' \\ \mathbf{t}_1' \\ \mathbf{t}_2' \end{bmatrix} = \begin{bmatrix} 2 & 3 & 5 \\ 0 & 7 & 2 \\ 0 & 0 & 3 \end{bmatrix} \qquad (5.1)$$

The elimination procedure described in Section 1.3 reduces $\boldsymbol{S} = \boldsymbol{T}'\boldsymbol{T}$ to $\boldsymbol{T} = \boldsymbol{T}'^{-1}\boldsymbol{S}$. It is equivalent to premultiplication by $\boldsymbol{T}'^{-1}$. The matrix $\boldsymbol{T}$ is the *Cholesky* or *square root* decomposition of $\boldsymbol{S}$. Note furthermore that, writing $\boldsymbol{S}_1$ for the leading 2 by 2 submatrix of $\boldsymbol{S}$, and $\boldsymbol{T}_1$ for the leading 2 by 2 submatrix of $\boldsymbol{T}$,

$$\boldsymbol{S}_1 = \boldsymbol{T}_1'\boldsymbol{T}_1, \qquad \boldsymbol{T}_1 = \boldsymbol{T}_1'^{-1}\boldsymbol{S}_1$$

Alternative Algorithms

In the method just described the second row of $\boldsymbol{S}$ (identified with a 1) is modified by subtraction of $3\mathbf{t}_0' = t_{01}\mathbf{t}_0'$ prior to formation of $\mathbf{t}_1'$. This is just what happened in Section 1.3. With the third row (identified with a 2) the difference is that whereas in Section 1.3 the vector $5\mathbf{t}_0' + 2\mathbf{t}_1'$ was subtracted in one operation only after $\mathbf{t}_1'$ had been formed, here the order of operations is as follows:

1. Form $\mathbf{t}_0'$ and subtract the appropriate multiples from rows 2 and 3.
2. Form $\mathbf{t}_1'$ and subtract $2\mathbf{t}_1'$ from row 3.

The algorithm is a *sequential* one because the third row is modified sequentially in two steps, prior to the formation of $\mathbf{t}_2'$. A proof that either algorithm will always work is left until later.

1.6. CALCULATIONS BASED ON THE CSSP MATRIX

The CSSP matrix is the matrix of sums of squares and products about the means, also known as the matrix of *c*entered (or *c*orrected) *s*ums of *s*quares and *p*roducts. For the case where there are just two variates x and y, it is

$$\begin{bmatrix} \sum(x-\bar{x})^2 & \sum(x-\bar{x})(y-\bar{y}) \\ \sum(x-\bar{x})(y-\bar{y}) & \sum(y-\bar{y})^2 \end{bmatrix} = \begin{bmatrix} S_{xx} & S_{xy} \\ S_{xy} & S_{yy} \end{bmatrix}$$

The straight-line regression model (Section 1.1)

$$y_i = a + bx_i + e_i \quad (i = 1, 2, \ldots, n)$$

may equivalently be written

$$y_i - \bar{y} = a' + b(x_i - \bar{x}) + e_i \quad (i = 1, 2, \ldots, n)$$

where $a' = a - \bar{y} + b\bar{x}$. If in eqs. (1.1) and (1.2) we replace x_i by $x_i^* = x_i - \bar{x}$ and y_i by $y_i^* = y_i - \bar{y}$, then a is replaced by a' and b remains unchanged. As $\Sigma x_i^* = \Sigma y_i^* = 0$, the first equation reduces to $na' = 0$, whereas the second equation becomes

$$\left[\sum(x-\bar{x})^2\right]b = \sum(x-\bar{x})(y-\bar{y}) \tag{6.1}$$

Note that $\Sigma(x-\bar{x})^2$ and $\Sigma(x-\bar{x})(y-\bar{y})$ are the terms in the first row of the CSSP matrix. The quantities s_{xx}, s_{xy}, and s_{yy} may be calculated using the usual *centering* formulae:

$$s_{xx} = \sum (x-\bar{x})^2 = \sum x^2 - n\bar{x}^2 \tag{6.2}$$

$$s_{xy} = \sum (x-\bar{x})(y-\bar{y}) = \sum xy - n\bar{x}\bar{y}, \tag{6.2'}$$

and similarly for s_{yy}. Note here that these formulae should be used with some caution; this matter will be taken up in Section 1.7.

If we apply the Cholesky or square root scheme to the CSSP matrix we have

$$\left[\begin{array}{c|c} s_{xx} & s_{xy} \\ \hline s_{xy} & \boxed{s_{yy}} \end{array}\right]$$

$$\downarrow$$

$$\left[\begin{array}{c|c} s_{xx}^{1/2} & s_{xx}^{-1/2} s_{xy} \\ \hline 0 & \boxed{\left(s_{yy} - s_{xx}^{-1} s_{xy}^2\right)^{1/2}} \end{array}\right]$$

For example,

$$\left[\begin{array}{c|c} 49 & 14 \\ \hline 14 & \boxed{13} \end{array}\right]$$

$$\downarrow$$

$$\left[\begin{array}{c|c} 7 & 2 \\ \hline 0 & \boxed{3} \end{array}\right]$$

The *due to constant term* entry in the analysis of variance table now has to be calculated as $n\bar{y}^2$; other entries are

Due to x (having fitted constant term):	$s_{xx}^{-1} s_{xy}^2$
Residual:	$s_{yy} - s_{xx}^{-1} s_{xy}^2$.

For the case where there is more than one explanatory variate (these will be denoted $x_1, x_2, \ldots$), any examples of actual calculations will usually, unless the constant term is to be omitted from the regression, be based on the CSSP matrix. Before proceeding to the multiple regression (more than one explanatory variate) case, we pause to discuss the calculations of quantities such as $\mathbf{s}_{xx}$ and $\mathbf{s}_{xy}$.

1.7. STABLE CALCULATION OF CENTERED SUMS OF SQUARES AND PRODUCTS

If used carelessly the *centering* formulae (6.2) and (6.2′) may lead to serious loss of numerical precision. Consider an example. $\mathbf{s}_{xx} = \Sigma(x - \bar{x})^2$ is to be calculated for a variate which takes the three values:

$$2001 \qquad 2002 \qquad 2003$$

Then $\Sigma x^2 = 12{,}024{,}014$ and $n\bar{x}^2 = 12{,}024{,}012$. On a computer which stores the seven most significant decimal digits (or the equivalent in binary or octal or hexadecimal) in the result of any arithmetic operation, both numbers would be stored as 1.202401×10^7, possibly with some uncertainty about the final 1. The result would be incorrect in all digits.

Clearly the problem arises when the first several digits of Σx^2 and $n\bar{x}^2$ are the same, that is, when these numbers are nearly equal. The first h digits of Σx^2 and $n\bar{x}^2$ will be identical if the difference is about $10^{-h}\Sigma x^2$:

$$\mathbf{s}_{xx} \simeq 10^{-h}\sum x^2 \simeq 10^{-h} n\bar{x}^2$$

$$10^h \simeq \frac{n\bar{x}^2}{\mathbf{s}_{xx}}$$

Thus the number of decimal digits of precision lost is

$$h \simeq 2\log_{10}\frac{\bar{x}}{\mathbf{s}_x} \tag{7.1}$$

where $\mathbf{s}_x = \sqrt{n^{-1}\mathbf{s}_{xx}}$. For the example,

$$\bar{x} = 2002,\ \mathbf{s}_x = \sqrt{\tfrac{2}{3}} \quad \text{and} \quad h \simeq 6.8 \simeq 7$$

Stated thus, the problem is one of choosing a suitable origin of measurement, so that $\bar{x}$ is no more than three or four times larger than $\mathbf{s}_x$. A suitable recourse is to use the first observation, or the first few observations, to determine a working mean d. The quantities calculated are

$$\sum(x - d) \quad \text{and} \quad \sum(x - d)^2$$

Then

$$\mathbf{s}_{xx} = \sum(x-d)^2 - n^{-1}\left[\sum(x-d)\right]^2 \tag{7.2}$$

and so on. Similarly for calculation of $\mathbf{s}_{xy}$ and any other sums of products involving x.

The objection to the use of $d = \bar{x}$ is that this would require two passes through the data, one to calculate $\bar{x}$ and a second to evaluate $\Sigma(x - \bar{x})^2$. However a single pass formula is available, in which at each stage one determines the means of the data points so far, and the sum of squares and products about the current means.

Updating Formulae

Define $\bar{x}^{(k)} = k^{-1}\Sigma_{i=1}^{k} x_i$ and $\bar{y}^{(k)} = k^{-1}\Sigma_{i=1}^{k} y_i$ to be the means for the first k data points taken. Define

$$s_{xx}^{(k)} = \sum_{i=1}^{k}\left(x_i - \bar{x}^{(k)}\right)^2$$

$$s_{xy}^{(k)} = \sum_{i=1}^{k}\left(x_i - \bar{x}^{(k)}\right)\left(y_i - \bar{y}^{(k)}\right) \tag{7.3}$$

and similarly for $s_{yy}^{(k)}$. Then it is readily shown that

$$\bar{x}^{(k)} = \bar{x}^{(k-1)} + k^{-1}\left(x_k - \bar{x}^{(k-1)}\right)$$

$$s_{xx}^{(k)} = s_{xx}^{(k-1)} + \left(x_k - \bar{x}^{(k-1)}\right)\left(x_k - \bar{x}^{(k)}\right)$$

$$s_{xy}^{(k)} = s_{xy}^{(k-1)} + \left(x_k - \bar{x}^{(k-1)}\right)\left(y_k - \bar{y}^{(k)}\right) \tag{7.4}$$

The updating formulae are given in this form in Jennrich (1977); see also van Reeken (1968). Precision will be superior to that from use of eq. (7.2) and analogous formulae unless eq. (7.2) is able to use a working mean d which is close to the true mean $\bar{x}$. If d is close to $\bar{x}$, eq. (7.2) will give better precision because it involves less arithmetic.

1.8. MULTIPLE REGRESSION—MATRIX FORMULATION

Let $\boldsymbol{y}$ be an n by 1 vector of observed values, let $\boldsymbol{X}$ be an n by $p+1$ matrix whose rows are "observations," and let $\boldsymbol{b}$ be a $p+1$ by 1 vector of parameters, or regression coefficients. Then the least squares problem is that of finding $\boldsymbol{b}$ to minimize the sum of squares of elements of

$$\boldsymbol{e} = \boldsymbol{y} - \boldsymbol{X}\boldsymbol{b}$$

that is, to minimize

$$\boldsymbol{e}'\boldsymbol{e} = (\boldsymbol{y} - \boldsymbol{X}\boldsymbol{b})'(\boldsymbol{y} - \boldsymbol{X}\boldsymbol{b})$$

The minimum is given by the solution in $\boldsymbol{b}$ (or solutions if $\boldsymbol{b}$ is not uniquely determined) of the normal equations

$$\boldsymbol{X}'\boldsymbol{X}\boldsymbol{b} = \boldsymbol{X}'\boldsymbol{y} \tag{8.1}$$

This will be proved in Section 1.11.

For the straight-line regression case of Section 1.1

$$\boldsymbol{X} = \begin{bmatrix} 1 & x_1 \\ 1 & x_2 \\ \vdots & \vdots \\ 1 & x_n \end{bmatrix}, \quad \boldsymbol{y} = \begin{bmatrix} y_1 \\ y_2 \\ \vdots \\ y_n \end{bmatrix}, \quad \boldsymbol{b} = \begin{bmatrix} a \\ b \end{bmatrix}$$

Equation (8.1) then leads to the eqs. (1.1) and (1.2). (Alternatively, if each x_i is replaced by $x_i - \bar{x}$, and each y_i by $y_i - \bar{y}$, the equation $\boldsymbol{X}'\boldsymbol{X}\boldsymbol{b} = \boldsymbol{X}'\boldsymbol{y}$ then takes the form

$$\begin{bmatrix} n & 0 \\ 0 & \sum(x - \bar{x})^2 \end{bmatrix}\begin{bmatrix} a' \\ b \end{bmatrix} = \begin{bmatrix} 0 \\ \sum(x - \bar{x})(y - \bar{y}) \end{bmatrix}$$

which is another way of formulating the result of Section 1.6.)

More generally, suppose the model is

$$y_i = b_0 x_{i0} + b_1 x_{i1} + \cdots + b_p x_{ip} + e_i \quad (i = 1, 2, \ldots, n) \tag{8.2}$$

where x_{i0} is identically one. We have n values of each of y and the p explanatory variates $x_1, x_2, \ldots, x_p$. In this formulation

$$\boldsymbol{y} = \begin{bmatrix} y_1 \\ y_2 \\ \vdots \\ y_n \end{bmatrix}, \quad \boldsymbol{X} = \begin{bmatrix} 1 & x_{11} & \cdots & x_{1p} \\ 1 & x_{21} & \cdots & x_{2p} \\ \vdots & \vdots & & \vdots \\ 1 & x_{n1} & \cdots & x_{np} \end{bmatrix}, \quad \boldsymbol{b} = \begin{bmatrix} b_0 \\ b_1 \\ \vdots \\ b_p \end{bmatrix} \tag{8.3}$$

The quantities required for forming the normal equations may be obtained by omitting the final row from the sums of squares and products (SSP) matrix

$$[X, y]'[X, y] = \begin{bmatrix} X'X & X'y \\ y'X & y'y \end{bmatrix}$$

Calculations Based on the CSSP Matrix

The preferred formulation is

$$y_i - \bar{y} = b_0' + b_1(x_{i1} - \bar{x}_1) + \cdots + b_p(x_{ip} - \bar{x}_p) + e_i$$

$$\text{where } b_0' = b_0 - \bar{y} + \sum_{i=1}^{p} b_j \bar{x}_j \qquad (8.4)$$

For this replace X by $\begin{bmatrix} 1 & \\ \vdots & \mathbf{X} \\ 1 & \end{bmatrix}$ and y by $\mathbf{y}$, where

$$\mathbf{X} = \begin{bmatrix} x_{11} - \bar{x}_1 & \cdots & x_{1p} - \bar{x}_p \\ x_{21} - \bar{x}_1 & & x_{2p} - \bar{x}_p \\ \vdots & & \vdots \\ x_{n1} - \bar{x}_1 & & x_{np} - \bar{x}_p \end{bmatrix}, \quad \mathbf{y} = \begin{bmatrix} y_1 - \bar{y} \\ y_2 - \bar{y} \\ \vdots \\ y_n - \bar{y} \end{bmatrix}$$

Then the normal equations become

$$\begin{bmatrix} n & 0 & \cdots & 0 \\ 0 & & & \\ \vdots & & \mathbf{X}'\mathbf{X} & \\ 0 & & & \end{bmatrix} \begin{bmatrix} b_0' \\ b_1 \\ \vdots \\ b_p \end{bmatrix} = \begin{bmatrix} 0 \\ \mathbf{X}'\mathbf{y} \end{bmatrix}, \qquad (8.5)$$

that is, $nb_0' = 0$, and

$$\mathbf{X}'\mathbf{X}\mathbf{b} = \mathbf{X}'\mathbf{y} \qquad (8.6)$$

with

$$\mathbf{b} = \begin{bmatrix} b_1 \\ b_2 \\ \vdots \\ b_p \end{bmatrix}$$

Observe that the coefficients and right-hand side of the normal eq. (8.6) may be obtained by omitting the final row from the CSSP matrix for all variates, that is, from

$$[\mathbf{X}, \mathbf{y}]'[\mathbf{X}, \mathbf{y}] = \begin{bmatrix} \mathbf{X}'\mathbf{X} & \mathbf{X}'\mathbf{y} \\ \mathbf{y}'\mathbf{X} & \mathbf{y}'\mathbf{y} \end{bmatrix} \tag{8.7}$$

If b_0 is required to obtain the equation in the form (8.2), it may be calculated as

$$b_0 = \bar{y} - \sum_{j=1}^{p} b_j \bar{x}_j \tag{8.8}$$

The CSSP matrix, for all variates involved, provides a suitable starting point for various normal theory multivariate calculations. Such calculations have a multiple regression interpretation.

The Model Matrix

It is convenient to have a name for the matrix $\boldsymbol{X}$. The term *design matrix* is open to the objections discussed in Kempthorne (1980). Following Kempthorne's suggestion, we will call $\boldsymbol{X}$ the *model matrix*.

1.9. MULTIPLE REGRESSION—AN EXAMPLE

Four sets of values are provided for each of the variates x_1, x_2, and y. Values of $x_1 - \bar{x}_1$, $x_2 - \bar{x}_2$, and $y - \bar{y}$ are given on the right:

	x_1	x_2	y	$x_1 - \bar{x}_1$	$x_2 - \bar{x}_2$	$y - \bar{y}$
	−2	0	−3	−3.5	−2.5	−4.5
	−1	2	1	−2.5	−0.5	−0.5
	2	5	2	0.5	2.5	0.5
	7	3	6	5.5	0.5	4.5
Mean	1.5	2.5	1.5			

Using the notation of Section 1.8 the CSSP matrix is

$$[\mathbf{X}, \mathbf{y}]'[\mathbf{X}, \mathbf{y}] = \begin{array}{c} \begin{matrix} b_1 & b_2 & \end{matrix} \\ \left[\begin{array}{ccc} 49 & 14 & 42 \\ \cdot & 13 & 15 \\ \hline \cdot & \cdot & 41 \end{array}\right] \end{array}$$

(Below diagonal elements, whose values may be determined by inspecting the corresponding upper diagonal position, are replaced by a dot.)

Following the *complete* form (see Section 1.5) of the elimination scheme we have been using:

Subtract 2 × (row 1) from row 2

$$\begin{array}{cc|c} 49 & 14 & 42 \\ \cdot & 13 & 15 \\ \hline \cdot & \cdot & \boxed{41} \end{array} \quad \xrightarrow[\text{by } \sqrt{49}]{\text{Divide row 1}} \quad \begin{array}{cc|c} 7 & 2 & 6 \\ \cdot & 13 & 15 \\ \hline \cdot & \cdot & \boxed{41} \end{array} \leftarrow$$

Subtract 2 × (row 1) from row 2

$$\begin{array}{cc|c} 7 & 2 & 6 \\ 0 & 9 & 3 \\ \hline 0 & \cdot & \boxed{41} \end{array} \quad \xrightarrow[\text{by } \sqrt{9}]{\text{Divide row 2}} \quad \begin{array}{cc|c} 7 & 2 & 6 \\ 0 & 3 & 1 \\ \hline 0 & \cdot & \boxed{41} \end{array} \leftarrow$$

Subtract (6 × (row 1) + 1 × (row 2)) from row 3

$$\begin{array}{cc|c} 7 & 2 & 6 \\ 0 & 3 & 1 \\ \hline 0 & 0 & \boxed{4} \end{array} \quad \xrightarrow[\text{by } \sqrt{4}]{\text{Divide row 3}} \quad \begin{array}{cc|c} 7 & 2 & 6 \\ 0 & 3 & 1 \\ \hline 0 & 0 & \boxed{2} \end{array} \leftarrow$$

Thus the normal equations

$$\begin{bmatrix} 49 & 14 \\ 14 & 13 \end{bmatrix} \begin{bmatrix} b_1 \\ b_2 \end{bmatrix} = \begin{bmatrix} 42 \\ 15 \end{bmatrix}$$

have been reduced to

$$\begin{bmatrix} 7 & 2 \\ 0 & 3 \end{bmatrix} \begin{bmatrix} b_1 \\ b_2 \end{bmatrix} = \begin{bmatrix} 6 \\ 1 \end{bmatrix} \tag{9.1}$$

We have $b_2 = \frac{1}{3}$; $7b_1 + 2b_2 = 6$ so that $b_1 = \frac{16}{21}$. The constant term is

$$\begin{aligned} b_0 &= \bar{y} - b_1\bar{x}_1 - b_2\bar{x}_2 \\ &= 3 - \tfrac{16}{21} \times 1.5 - \tfrac{1}{3} \times 2.5 \\ &= -\tfrac{10}{21} \end{aligned}$$

As we are basing calculations on the CSSP matrix, the entry for placing in the *due to fitting constant term* row in the analysis of variance table is not available from the upper triangular array. Instead it must be calculated as

$$\begin{aligned} n\bar{y}^2 &= 4 \times 1.5^2 \\ &= 9 \end{aligned}$$

The entries for the remaining rows of the table are taken from the final column of the upper triangular array; one has

7	2	6	← Square this to get *due to* x_1, *given constant term* SS
0	3	1	← Square this to get *due to* x_2, *given* x_1, *constant term* SS
0	0	2	← Square this to get *residual* SS

(Note that SS = sum of squares.) These terms are the successive entries in a sequential analysis of the variance table. See Section 2.1.

Comparison with the Use of the SSP Matrix

For this example the SSP matrix $[X, y]'[X, y]$ is

b_0	b_1	b_2	
4	6	10	6
·	58	29	51
·	·	38	30
·	·	·	50

Using the Cholesky decomposition, this is reduced to

2	3	5	3	← Square this to get *due to constant term* SS
0	7	2	6	← Square this to get *due to* x_1, *given constant term* SS
0	0	3	1	← Square this to get *due to* x_2, *given constant*, x_1 SS
0	0	0	2	← Square this to get *residual* SS

The normal equations are now

$$\begin{bmatrix} 2 & 3 & 5 \\ 0 & 7 & 2 \\ 0 & 0 & 3 \end{bmatrix} \begin{bmatrix} b_0 \\ b_1 \\ b_2 \end{bmatrix} = \begin{bmatrix} 3 \\ 6 \\ 1 \end{bmatrix}$$

The only difference from the eq. (9.1), based on the CSSP matrix, is that we now have a first equation (which will be the last used) from which b_0 can be determined:

$$2b_0 + 3b_1 + 5b_2 = 3$$

This has the form

$$\sqrt{n}\, b_0 + (\sqrt{n}\, \bar{x}_1) b_1 + (\sqrt{n}\, \bar{x}_2) b_2 = \sqrt{n}\, \bar{y}$$

If one divides through by $\sqrt{n}$, the equation obtained is identical with eq. (8.8) of Section 1.8.

1.10. CALCULATIONS BASED ON THE CORRELATION MATRIX

This is a further possible refinement. The regression model may be written

$$\frac{y_i - \bar{y}}{\sqrt{S_{yy}}} = \sum_{j=1}^{p} b_j^* \frac{x_{ij} - \bar{x}_j}{\sqrt{S_{jj}}} + \frac{e_i}{\sqrt{S_{yy}}} \qquad (10.1)$$

that is,

$$y_i^* = \sum_{j=1}^{p} b_j^* x_{ij}^* + e_i^* \qquad (10.2)$$

where the b_j^* are related to the b_j that we have used earlier by

$$b_j = b_j^* \sqrt{S_{jj}^{-1} S_{yy}}$$

Minimization of the e_i^* is equivalent to minimization of the e_i. The b_j^* are known as *standardized* regression coefficients, or sometimes as *beta* coefficients.

The CSSP matrix that we require if we are to work with equation (10.2) is just the matrix of sums of squares and products for the x_{ij}^* and the y_i^*, that is, the correlation matrix. The attraction of working with the correlation matrix is that it provides a representation of the least squares problem that is independent of the origin and scale of measurement of the original data. Various quantities that appear in the course of the calculation, which we shall want to use later, are then readily interpretable as they stand.

The elements r_{ij} of the correlation matrix are readily obtained from the elements $\mathbf{s}_{ij}$ of the CSSP matrix for the original values:

$$r_{ij} = \frac{\mathbf{s}_{ij}}{\sqrt{\mathbf{s}_{ii}\mathbf{s}_{jj}}}$$

It is not necessary to evaluate the x_{ij}^* and the y_i^*. Readers should note that the standardized variate values, as this term is used in statistical contexts, are given by $\sqrt{n-1}\,x_{ij}^*$, $\sqrt{n-1}\,y_i^*$.

For the data of Section 1.9 the correlation matrix is (to three decimal places)

x_1	x_2	y	
1.0	0.555	0.937	$\sqrt{\mathbf{s}_{11}^{-1}\mathbf{s}_{yy}} = \sqrt{\frac{41}{49}} \simeq 0.915$
·	1.0	0.650	$\sqrt{\mathbf{s}_{22}^{-1}\mathbf{s}_{yy}} = \sqrt{\frac{41}{13}} \simeq 1.776$
·	·	1.0	

Following the elimination scheme we have been using, this becomes

1.0	0.555	0.937	← Square to get proportion of TSS which is *due to* x_1
0	0.832	0.156	← Square to get proportion of TSS which is *due to* x_2*, given* x_1
0	0	0.312	← Square to get proportion of TSS which remains

(Note that TSS = total sum of squares about the mean.)

The normal equations are thus

$$\begin{bmatrix} 1.0 & 0.555 \\ 0 & 0.832 \end{bmatrix} \begin{bmatrix} b_1^* \\ b_2^* \end{bmatrix} = \begin{bmatrix} 0.937 \\ 0.156 \end{bmatrix}$$

whence

$$b_2^* = 0.188 \qquad (\text{and } b_2 \simeq 0.33 \text{ as before})$$

$$b_1^* = 0.833 \qquad (\text{and } b_1 \simeq 0.76 \text{ as before})$$

The sequential analysis of variance table is

Due to fitting constant term	$n\bar{y}^2 = 9$
Due to x_1 *(having fitted constant)*	$0.937^2 \times s_{yy} = 36$
Due to x_2 *(having fitted constant,* x_1*)*	$0.156^2 \times s_{yy} = 1$
Residual	$0.312^2 \times s_{yy} = 4$

*1.11. THEORY—THE NORMAL EQUATIONS

The derivation of the normal equations will be based around the use of *Gram-Schmidt orthogonalization* to find an orthogonal basis for a Euclidean vector space.

Gram-Schmidt Orthogonalization

The subspace of n-dimensional space E_n spanned by columns $\boldsymbol{x}_0, \boldsymbol{x}_1, \ldots, \boldsymbol{x}_p$ of the matrix $\boldsymbol{X}$ is known as the column space of $\boldsymbol{X}$. Suppose initially that $\boldsymbol{x}_0, \boldsymbol{x}_1, \ldots, \boldsymbol{x}_p$ are linearly independent. (This restriction will be removed later). Then there is an orthogonal sequence of nonzero vectors $\boldsymbol{v}_0, \boldsymbol{v}_1, \ldots, \boldsymbol{v}_p$, which span the same subspace of E_n as the sequence $\boldsymbol{x}_0, \boldsymbol{x}_1, \ldots, \boldsymbol{x}_p$. The sequence will be chosen so that

$$\boldsymbol{v}_j = \boldsymbol{x}_j - \sum_{i<j} c_{ij}\boldsymbol{x}_i \quad (j = 0, 1, \ldots, p)$$

$$= \sum_{i \le j} h_{ij}\boldsymbol{x}_i, \tag{11.1}$$

where $h_{jj} = 1$; $h_{ij} = -c_{ij}$ for $i < j$.

To prove this, assume that orthogonal nonzero vectors $\boldsymbol{v}_0, \boldsymbol{v}_1, \ldots, \boldsymbol{v}_{k-1}$, of the form given by eq. (11.1), have been constructed. Thus $\boldsymbol{v}_0, \boldsymbol{v}_1, \ldots, \boldsymbol{v}_{k-1}$ span the same subspace as $\boldsymbol{x}_0, \boldsymbol{x}_1, \ldots, \boldsymbol{x}_{k-1}$. The device now used to construct a vector $\boldsymbol{v}_k$, also of the form given by eq. (11.1) and such that $\boldsymbol{v}_j'\boldsymbol{v}_k = 0$

for $j < k$, is known as *Gram-Schmidt orthogonalization*. For this set

$$\begin{aligned} \boldsymbol{v}_k &= \boldsymbol{x}_k - \sum_{j<k} \left(\frac{\boldsymbol{x}_k' \boldsymbol{v}_j}{\boldsymbol{v}_j' \boldsymbol{v}_j} \right) \boldsymbol{v}_j, \\ &= \boldsymbol{x}_k - \sum_{j<k} \left(\frac{\boldsymbol{x}_k' \boldsymbol{v}_j}{\boldsymbol{v}_j' \boldsymbol{v}_j} \right) \left[\sum_{i<j} h_{ij} \boldsymbol{x}_i \right] \\ &= \boldsymbol{x}_k - \sum_{i<j<k} \left[\left(\frac{\boldsymbol{x}_k' \boldsymbol{v}_j}{\boldsymbol{v}_j' \boldsymbol{v}_j} \right) h_{ij} \right] \boldsymbol{x}_i \end{aligned} \tag{11.2}$$

Thus $\boldsymbol{v}_k$ is of the form given in eq. (11.1). Furthermore $\boldsymbol{v}_k \neq \boldsymbol{0}$. Otherwise it would follow that $\boldsymbol{x}_k$ is a linear combination of $\boldsymbol{v}_0, \boldsymbol{v}_1, \ldots, \boldsymbol{v}_{k-1}$, and hence of $\boldsymbol{x}_0, \boldsymbol{x}_1, \ldots, \boldsymbol{x}_{k-1}$ contrary to the assumption that $\boldsymbol{x}_0, \boldsymbol{x}_1, \ldots, \boldsymbol{x}_k$ are linearly independent. The induction is completed by verifying that $\boldsymbol{v}_j' \boldsymbol{v}_k = 0$ for $j < k$.

If the vectors $\boldsymbol{x}_j$ $(j = 0, 1, \ldots, p)$ are not linearly independent, the only modification is that, in eq. (11.2), $\boldsymbol{v}_j$ will sometimes be zero. It is then not included as an element of the orthogonal basis, and the next nonzero vector $\boldsymbol{v}_{j'}$ becomes $\boldsymbol{v}_j$. This leads to an orthogonal basis $\boldsymbol{v}_0, \boldsymbol{v}_1, \ldots, \boldsymbol{v}_l$ which has $l < p$ elements. □

Gram-Schmidt Orthogonalization in Matrix Form

Let $\boldsymbol{V} = [\boldsymbol{v}_0, \boldsymbol{v}_1, \ldots, \boldsymbol{v}_l]$, where $l \leq p$. It has been proved above that $\boldsymbol{X}$ and $\boldsymbol{V}$ have the same column space. Thus $\boldsymbol{X} = \boldsymbol{VG}$, and $\boldsymbol{V} = \boldsymbol{XH}$, for suitable choices of the matrices $\boldsymbol{G}$ and $\boldsymbol{H}$. Notice that $\boldsymbol{x}_k$ is, by eq. (11.2), a linear combination of $\boldsymbol{v}_j$ for $j = 0, 1, \ldots, k$ only. Hence G is an upper triangular matrix.

Projection onto the Column Space of X

Let $\boldsymbol{y}$ be any vector in E_n. Let

$$\hat{\boldsymbol{y}} = \sum_{j<k} \left(\frac{\boldsymbol{y}' \boldsymbol{v}_j}{\boldsymbol{v}_j' \boldsymbol{v}_j} \right) \boldsymbol{v}_j \tag{11.3}$$

$$\begin{aligned} &= \boldsymbol{Vd} \quad \left(\text{where } \boldsymbol{d} \text{ has elements } d_j = \boldsymbol{y}' \boldsymbol{v}_j / \boldsymbol{v}_j' \boldsymbol{v}_j \right) \\ &= \boldsymbol{XHd} \\ &= \boldsymbol{X}\hat{\boldsymbol{b}} \quad (\text{where } \hat{\boldsymbol{b}} = \boldsymbol{Hd}) \end{aligned} \tag{11.4}$$

Then from eq. (11.3)

$$\hat{\boldsymbol{y}}'\boldsymbol{v}_i = \boldsymbol{y}'\boldsymbol{v}_i \quad (i = 0, 1, \ldots, l),$$

so that

$$\hat{\boldsymbol{y}}'\boldsymbol{V} = \boldsymbol{y}'\boldsymbol{V}$$

Thus

$$\hat{\boldsymbol{y}}'\boldsymbol{X} = \hat{\boldsymbol{y}}'\boldsymbol{V}\boldsymbol{G} = \boldsymbol{y}'\boldsymbol{V}\boldsymbol{G} = \boldsymbol{y}'\boldsymbol{X} \tag{11.5}$$

The Normal Equations

From eq. (11.4) $\hat{\boldsymbol{y}} = \boldsymbol{X}\hat{\boldsymbol{b}}$. Then

$$\begin{aligned} \boldsymbol{X}'\boldsymbol{X}\hat{\boldsymbol{b}} &= \boldsymbol{X}'\hat{\boldsymbol{y}} \\ &= \boldsymbol{X}'\boldsymbol{y} \quad \text{by eq. (11.5)} \end{aligned} \tag{11.6}$$

This proves the existence of a vector $\hat{\boldsymbol{b}}$ satisfying the normal eqs. (11.6). Write $\boldsymbol{e} = \boldsymbol{y} - \boldsymbol{X}\boldsymbol{b}$, and define

$$\|\boldsymbol{e}\| = \|\boldsymbol{e}\|_2 = \sqrt{(\boldsymbol{e}'\boldsymbol{e})}\,.$$

It will now be shown that $\|\boldsymbol{e}\|^2$ is minimized when $\boldsymbol{b}$ satisfies the normal equations $\boldsymbol{X}'\boldsymbol{X}\boldsymbol{b} = \boldsymbol{X}'\boldsymbol{y}$.

Suppose $\hat{\boldsymbol{b}}$ is such that $\boldsymbol{X}'\boldsymbol{X}\hat{\boldsymbol{b}} = \boldsymbol{X}'\boldsymbol{y}$, that is, $\boldsymbol{X}'(\boldsymbol{y} - \boldsymbol{X}\hat{\boldsymbol{b}}) = 0$. Write $\boldsymbol{e} = \boldsymbol{y} - \boldsymbol{X}\boldsymbol{b} = \boldsymbol{y} - \boldsymbol{X}\hat{\boldsymbol{b}} + \boldsymbol{X}\hat{\boldsymbol{b}} - \boldsymbol{X}\boldsymbol{b}$. Then as $\boldsymbol{X}'(\boldsymbol{y} - \boldsymbol{X}\hat{\boldsymbol{b}}) = \boldsymbol{0}$

$$\begin{aligned} \|\boldsymbol{e}\|^2 &= \|\boldsymbol{y} - \boldsymbol{X}\hat{\boldsymbol{b}}\|^2 + \|\boldsymbol{X}\hat{\boldsymbol{b}} - \boldsymbol{X}\boldsymbol{b}\|^2 \\ &\geq \|\boldsymbol{y} - \boldsymbol{X}\hat{\boldsymbol{b}}\|^2 \end{aligned}$$

with equality if and only if $\boldsymbol{X}\boldsymbol{b} = \boldsymbol{X}\hat{\boldsymbol{b}}$. Thus also $\boldsymbol{X}'\boldsymbol{X}\boldsymbol{b} = \boldsymbol{X}'\boldsymbol{y}$. □

An algorithm for determining $\boldsymbol{b}$ may be based on eqs. (11.3) and (11.4). However, rather than use these equations directly, numerical analysts have generally preferred the *modified Gram-Schmidt* (MGS) algorithm, which performs the operations in a different order. This algorithm will be discussed in Chapter 4.

The Existence of the Cholesky Decomposition of $X'X$

A proof of the existence of an upper triangular matrix T_p such that $T_p'T_p = X'X$ follows from the remark made, a few lines previous to eq. (11.3) that

$$X = VG$$

where the columns of V are mutually orthogonal and G is upper triangular. Thus

$$\begin{aligned} X'X &= G'V'VG \\ &= G'DG \\ &= (GD^{1/2})'GD^{1/2} \end{aligned}$$

where $D = V'V$ and $D^{1/2}$ are both diagonal matrices, and diagonal elements of $D^{1/2}$ are the square roots of those of D.

*1.12. THEORY—THE CHOLESKY DECOMPOSITION

Existence of the Cholesky Decomposition—A Second Proof

Let $\mathbf{X}$ be an real matrix, and let $S = \mathbf{X}'\mathbf{X}$. In the applications of this and of the next chapter $\mathbf{X} = [X, \mathrm{y}]$ or $\mathbf{X} = [\mathbf{X}, \mathbf{y}]$, where the notation is that of Section 1.8. It will be supposed that columns of $\mathbf{X}$, and hence rows and columns of S, are numbered from 0 to q.

The algorithm of Section 1.5 determines successively for $k = 0, 1, \ldots$ a matrix $S_{(k)}$ with zeros in rows and columns $0, 1, \ldots, k$ such that

$$\begin{aligned} S_{(k)} &= S - \sum_{i=1}^{k} \mathrm{t}_i \mathrm{t}_i' \\ &= S_{(k-1)} - \mathrm{t}_k \mathrm{t}_k' \end{aligned} \tag{12.1}$$

Now we may write

$$S_{(k-1)} = \left[\begin{array}{c|cc} \mathbf{0} & \multicolumn{2}{c}{\mathbf{0}} \\ \hline \mathbf{0} & s_{kk}^{(k-1)} & s_{k(k-1)}' \\ & s_{k(k-1)} & \tilde{S}_{(k-1)} \end{array}\right] \begin{array}{l} \} \text{ Rows 0 to } k-1 \\ \\ \\ \end{array}$$

Assume for purposes of a proof by induction that $\boldsymbol{S}_{(k-1)}$ is positive semidefinite, that is, $\boldsymbol{c}'\boldsymbol{S}_{(k-1)}\boldsymbol{c} \geq 0$ for every real vector $\boldsymbol{c}$. Then:

1. $s_{kk}^{(k-1)} \geq 0$. If $s_{kk}^{(k-1)} = 0$, then also $\boldsymbol{s}_{k(k-1)} = \mathbf{0}$. (See the proof provided next.)
2. If $s_{kk}^{(k-1)} > 0$, then form $\mathbf{t}_k$ by taking column k of $\boldsymbol{S}_{(k-1)}$ and dividing through by $t_{kk} = (s_{kk}^{(k-1)})^{1/2}$. Otherwise set $\mathbf{t}_k = \mathbf{0}$. Then $\boldsymbol{S}_{(k)} = \boldsymbol{S}_{(k-1)} - \mathbf{t}_k\mathbf{t}_k'$ has zeros in rows and columns 0 to k, and is such that $\boldsymbol{c}'\boldsymbol{S}_{(k)}\boldsymbol{c} \geq 0$ for every real vector $\boldsymbol{c}$. (Again, see the proof that follows.)

The induction is started by observing that $\boldsymbol{c}'\boldsymbol{S}\boldsymbol{c}$ is the sum of squares of elements of $\boldsymbol{c}'\mathbf{X}'$, and is therefore never less than zero.

To prove **1**, let $\boldsymbol{c}' = [0,\ldots,0, c_k, \boldsymbol{c}_2']$. Then

$$\boldsymbol{c}'\boldsymbol{S}_{(k-1)}\boldsymbol{c} = c_k^2 s_{kk}^{(k-1)} + 2c_k\boldsymbol{c}_2'\boldsymbol{s}_{k(k-1)} + \boldsymbol{c}_2'\tilde{\boldsymbol{S}}_{(k-1)}\boldsymbol{c}_2 \tag{12.2}$$

Setting $\boldsymbol{c}_2 = \mathbf{0}$, it immediately follows that $s_{kk}^{(k-1)} \geq 0$. Suppose $s_{kk}^{(k-1)} = 0$; then $\boldsymbol{s}_{k(k-1)} \neq \mathbf{0}$ would allow a choice of $\boldsymbol{c}_2$ such that $\boldsymbol{c}_2'\boldsymbol{s}_{k(k-1)} \neq 0$. Then

$$c_k = \left(2\boldsymbol{c}_2'\boldsymbol{s}_{k(k-1)}\right)^{-1}\left(-1 - \boldsymbol{c}_2'\tilde{\boldsymbol{S}}_{(k-1)}\boldsymbol{c}_2\right)$$

gives $\boldsymbol{c}'\boldsymbol{S}_{(k-1)}\boldsymbol{c} = -1$, contrary to the hypothesis.

The only part of the proof of **2** which remains is the demonstration that $\boldsymbol{c}'\boldsymbol{S}_{(k)}\boldsymbol{c} \geq 0$ for every real vector $\boldsymbol{c}$.

Suppose on the contrary that $\boldsymbol{c}'\boldsymbol{S}_{(k)}\boldsymbol{c} < 0$ for some particular choice of $\boldsymbol{c}$. Then from eq. (12.1),

$$\boldsymbol{c}'\boldsymbol{S}_{(k-1)}\boldsymbol{c} = \boldsymbol{c}'\boldsymbol{S}_{(k)}\boldsymbol{c} + \left(\boldsymbol{c}'\mathbf{t}_k\right)^2$$

If $t_{kk} = 0$, then by **1**, $\mathbf{t}_k = \mathbf{0}$, and $\boldsymbol{c}'\boldsymbol{S}_{(k-1)}\boldsymbol{c} < 0$, contrary to the hypothesis. If $t_{kk} \neq 0$, observe that $\boldsymbol{c}'\boldsymbol{S}_{(k)}\boldsymbol{c}$ is unaffected by the choice of c_k. Then

$$c_k = -t_{kk}^{-1} \sum_{j=k+1}^{q} c_j t_{kj}$$

gives $\boldsymbol{c}'\mathbf{t}_k = 0$, so

$$\boldsymbol{c}'\boldsymbol{X}_{(k-1)}\boldsymbol{c} = \boldsymbol{c}'\boldsymbol{S}_{(k)}\boldsymbol{c} < 0$$

contrary to hypothesis. □

A third proof may be found in Section 3.3. Or any of the algorithms for the orthogonal reduction of **X** to upper triangular form which are discussed in Chapter 4 may be made the basis of a proof of the existence of the Cholesky decomposition of $\mathbf{X'X}$.

Properties of the Cholesky Decomposition

Let $\boldsymbol{S} = \boldsymbol{T'T}$, where $\boldsymbol{T}$ is upper triangular, and suppose we partition

$$\boldsymbol{S} = \begin{bmatrix} \boldsymbol{S}_{11} & \boldsymbol{S}_{12} \\ \boldsymbol{S}'_{12} & \boldsymbol{S}_{22} \end{bmatrix} \qquad \boldsymbol{T} = \begin{bmatrix} \boldsymbol{T}_{11} & \boldsymbol{T}_{12} \\ 0 & \boldsymbol{T}_{22} \end{bmatrix}$$

Then by writing $\boldsymbol{T'T} = \boldsymbol{S}$ in terms of the submatrices we have

$$\boldsymbol{T}'_{11}\boldsymbol{T}_{11} = \boldsymbol{S}_{11} \tag{12.3}$$

$$\boldsymbol{T}'_{11}\boldsymbol{T}_{12} = \boldsymbol{S}_{12} \tag{12.4}$$

$$\boldsymbol{S}_{22} = \boldsymbol{T}'_{12}\boldsymbol{T}_{12} + \boldsymbol{T}'_{22}\boldsymbol{T}_{22} \tag{12.5}$$

Now take

$$\boldsymbol{S} = \begin{bmatrix} \boldsymbol{X'X} & \boldsymbol{X'y} \\ \boldsymbol{y'X} & \boldsymbol{y'y} \end{bmatrix}, \quad \boldsymbol{T} = \left[\begin{array}{c|c} \boldsymbol{T}_p & \boldsymbol{t}_y \\ \hline \boldsymbol{0}' & t_{yy} \end{array}\right]$$

Then eq. (12.3) implies that $\boldsymbol{T}_p'\boldsymbol{T}_p = \boldsymbol{X'X}$, and eq. (12.4) implies that $\boldsymbol{T}_p'\boldsymbol{t}_y = \boldsymbol{X'y}$. Hence, assuming that $\boldsymbol{T}_p$ is nonsingular

$$\begin{aligned} \boldsymbol{t}_y'\boldsymbol{t}_y &= \boldsymbol{y'X}\left(\boldsymbol{T}_p'\boldsymbol{T}_p\right)^{-1}\boldsymbol{X'y} \\ &= \boldsymbol{y'X}(\boldsymbol{X'X})^{-1}\boldsymbol{X'y} \quad \text{by (12.3)} \end{aligned} \tag{12.6}$$

From eq. (12.5)

$$\boldsymbol{y'y} = \boldsymbol{t}_y'\boldsymbol{t}_y + t_{yy}^2$$

that is,

$$\begin{aligned} t_{yy}^2 &= \boldsymbol{y'y} - \boldsymbol{t}_y'\boldsymbol{t}_y \\ &= \boldsymbol{y'y} - \boldsymbol{y'X}(\boldsymbol{X'X})^{-1}\boldsymbol{X'y} \quad \text{by (12.6)} \end{aligned} \tag{12.7}$$

This is the residual sum of squares obtained when $\boldsymbol{b}$ is replaced by its least squares estimate in $(y - Xb)'(y - Xb)$. See Exercise 6 at the end of this chapter for a proof that eq. (12.7) may be replaced by

$$t_{yy}^2 = \mathbf{y}'\mathbf{y} - \mathbf{y}'\mathbf{X}(\mathbf{X}'\mathbf{X})^{-1}\mathbf{X}'\mathbf{y} \tag{12.8}$$

where as in Section 1.8 values in each column of $[\mathbf{X}, \mathbf{y}]$ are measured from the mean for the column as origin.

1.13. THEORY—VARIANCES AND COVARIANCES OF THE ESTIMATED REGRESSION COEFFICIENTS

The practical details of these calculations will be treated in Chapter 2. Here we need to distinguish carefully the underlying model

$$y = X\boldsymbol{\beta} + \boldsymbol{\varepsilon} \tag{13.1}$$

where $\boldsymbol{\beta}$ is a vector of unknown parameters, from the fitted model

$$y = Xb + e \tag{13.2}$$

where b is the least squares estimate of $\boldsymbol{\beta}$. The assumption which we make here is that the elements $\varepsilon_1, \varepsilon_2, \ldots, \varepsilon_n$ of $\boldsymbol{\varepsilon}$ are independently and identically distributed as normal with mean 0 and variance σ^2. In other words

$$\operatorname{cov}(\varepsilon_i, \varepsilon_j) = \begin{cases} \sigma^2 & \text{if } i = j \\ 0 & \text{if } i \neq j \end{cases}$$

The same is true for $\operatorname{cov}(y_i, y_j)$. In the variance-covariance matrix $\operatorname{var}(y)$ the (i, j)th element is $\operatorname{cov}(y_i, y_j)$; thus we can write

$$\operatorname{var}(y) = \sigma^2 I_n$$

Using the result that

$$\operatorname{var}(Ay) = A \operatorname{var}(y) A' \tag{13.3}$$

it then follows, setting $A = (X'X)^{-1}X'$, that the variance-covariance matrix for the vector b determined by the normal equations $X'Xb = X'y$ is

$$\operatorname{var}(b) = (X'X)^{-1}\sigma^2 \tag{13.4}$$

(We assume that $\boldsymbol{X}$ is full column rank and so $\boldsymbol{X'X}$ is full rank.) Thus we have the variance-covariance table

	b_0	b_1	b_2	$\cdots$	b_p
b_0	The entries here are the elements of $(\boldsymbol{X'X})^{-1}\sigma^2$.				
b_1					
$\vdots$					
b_p					

Now assume the regression equation includes a constant term, so that $\boldsymbol{X}$ has an initial column of ones. Then the normal equations may alternatively be written

$$\mathbf{X'Xb} = \mathbf{X'y}$$

where in each column of $[\mathbf{X}, \mathbf{y}]$ values are measured from the column mean as origin. This may however be replaced by

$$\mathbf{X'Xb} = \mathbf{X'}y \quad (\text{NB: } y \text{ replaces } \mathbf{y}) \tag{13.5}$$

The reason is that the jth element of $\mathbf{X'}y$ differs from the jth element of $\mathbf{X'y}$ by an amount $\bar{y} \times$ (sum of elements in the jth column of $\mathbf{X}$) $= \bar{y} \times 0$. In this context we can replace $\mathbf{y}$, which has correlated elements, with y which has uncorrelated elements. From eq. (13.5) we then deduce, following the same pattern of argument as previously, that the variance-covariance matrix for $\boldsymbol{b}$ is

$$\text{var}(\mathbf{b}) = (\mathbf{X'X})^{-1}\sigma^2$$

Thus our table is

	b_1	b_2	$\cdots$	b_p
b_1	The entries here are the elements of $(\mathbf{X'X})^{-1}\sigma^2$.			
b_2				
$\vdots$				
b_p				

Comparison of this table with that following eq. (13.4) makes it evident that $(\boldsymbol{X'X})^{-1}$ may be obtained by deleting the first row and column from $(\mathbf{X'X})^{-1}$. This may be proved directly.

1.14. MULTIPLE AND ADJUSTED R-SQUARE

Using the notation of earlier sections, one has

$$t_{yy}^2 = \mathbf{y}'\mathbf{y} - \mathbf{y}'\mathbf{X}(\mathbf{X}'\mathbf{X})^{-1}\mathbf{X}'\mathbf{y} \tag{14.1}$$

as a version of the formula for the *residual sum of squares* (RSS) when y is regressed upon the columns of a matrix X which has an initial column of ones. The matrix $\mathbf{X}$ (with one less column than X) and the vector $\mathbf{y}$ are defined as in Section 1.8 and elsewhere; values in each column are measured from the mean for the column as origin. For a proof of eq. (14.1) see Exercise 6 at the end of this chapter. Define

$$\mathbf{y}'\mathbf{y} = \mathbf{s}_{yy} = \textit{Total sum of squares } (\text{TSS}) \textit{ about mean}$$

$$\mathbf{y}'\mathbf{X}(\mathbf{X}'\mathbf{X})^{-1}\mathbf{X}'\mathbf{y} = \textit{Due to regression sum of squares}$$

Then the squared multiple correlation for the above regression is defined as

$$R_{y(1\cdots p)}^2 = (\textit{Due to regression } \text{SS})/\text{TSS}$$

It may be viewed as the proportion of the *total sum of squares* which is *explained* by the regression. Then eq. (14.1) may be written

$$\begin{aligned} \text{RSS} &= \text{TSS} - \textit{Due to regression } \text{SS} \\ &= \text{TSS}\left[1 - R_{y(1\cdots p)}^2\right] \end{aligned}$$

Thus

$$t_{yy}^2 = \mathbf{s}_{yy}\left[1 - R_{y(1\cdots p)}^2\right] \tag{14.2}$$

Assessments of the strength of the relationship are, however, better based on the proportion of the mean sum of squares, which is explained. The motivation for this is the interpretation of mean sums of squares as variance estimates. If X has n rows, the mean sum of squares for y is $(n-1)^{-1}\mathbf{s}_{yy}$, whereas the residual mean square is $(n-p-1)^{-1}t_{yy}^2$. Thus the proportion of the mean TSS explained is the *adjusted* R^2 statistic

$$\overline{R}_{y(1\cdots p)}^2 = 1 - \frac{(n-p-1)^{-1}t_{yy}^2}{(n-1)^{-1}s_{yy}} = 1 - \frac{n-1}{n-p-1}\left[1 - R_{y(1\cdots p)}^2\right]$$

1.15. EXERCISES

1. Consider the set of data values

x_1	x_2	y
0	1	0
5	3	5
0	4	7
5	6	8

Form (a) the SSP matrix, (b) the CSSP matrix, and (c) the correlation matrix. For each of these matrices form the Cholesky decomposition. In each case carry through the calculations needed to determine the coefficients in the regression of y upon x_1 and x_2 (with constant term included).

2. Consider the set of data values

x_1	x_2	x_3	y
−1	0	1	0
3	0	1	0
2	−2	−2	−2
−2	−1	1	1
−1	1	−1	−1
3	3	1	3
2	2	2	4
−2	−1	−1	−2
2	1	1	3

The CSSP matrix $[\mathbf{X}, \mathbf{y}]'[\mathbf{X}, \mathbf{y}]$ is

$$\mathbf{S} = \begin{bmatrix} 36 & 12 & 6 & 18 \\ \cdot & 20 & 10 & 22 \\ \cdot & \cdot & 14 & 20 \\ \cdot & \cdot & \cdot & 40 \end{bmatrix}$$

Use the elimination scheme given in Section 1.9 to determine the upper triangle matrix $\mathbf{T}$ such that $\mathbf{T}'\mathbf{T} = \mathbf{S}$. Write down a variate by variate breakdown of the analysis of variance table. Determine the regression coefficients.

3. Calculations for the regression of y upon x_1 and x_2 are based on the correlation matrix, as in Section 1.10. Show that using the notation of

Section 1.10 the regression equation may be written

$$\hat{y}^* = b_1^* x_1^* + b_2^* x_2^*,$$

where, defining

$$r_{y1.2} = \frac{r_{y1} - r_{y2} r_{12}}{\sqrt{(1 - r_{y2}^2)(1 - r_{12}^2)}}$$

$$b_1^* = r_{y1.2} \sqrt{\frac{1 - r_{y2}^2}{1 - r_{12}^2}}$$

Write down the similar result for b_2^*.

4. (Continuation of Exercise 3)
As in Section 1.10 an analysis of variance table is set out in terms of the total sum of squares about the mean which is explained as each new variate is entered. Show that the fraction accounted for by x_1 is r_{1y}^2, whereas x_2 accounts for a further fraction of $(1 - r_{y1}^2) r_{y2.1}^2$. How do these fractions change when x_2 is taken first, then x_1? (*Note*: In chapter 3, $r_{y1.2}$ will be interpreted as the partial correlation between y and x_1 when x_2 is held constant.)

5. Let $[\boldsymbol{X}, \boldsymbol{y}]$ and $[\mathbf{X}, \mathbf{y}]$ be as in Section 1.8, and let $\boldsymbol{T}$ be upper triangular such that $\boldsymbol{T}'\boldsymbol{T} = [\boldsymbol{X}, \boldsymbol{y}]'[\boldsymbol{X}, \boldsymbol{y}]$. Show that the upper triangle matrix $\mathbf{T}$ such that $\mathbf{T}'\mathbf{T} = [\mathbf{X}, \mathbf{y}]'[\mathbf{X}, \mathbf{y}]$ may be obtained by deleting the first row and column from $\boldsymbol{T}$.

6. (Continuation of Exercise 5)
As in Section 1.8 let

$$\boldsymbol{b} = \begin{bmatrix} b_0 \\ \mathbf{b} \end{bmatrix}$$

be the vector of regression coefficients determined by the least squares equations. Show that

$$\begin{aligned} \boldsymbol{X}\boldsymbol{b} &= \mathbf{X}\mathbf{b} + \mathbf{1}\left[b_0 + (\bar{x}_1, \bar{x}_2, \ldots, \bar{x}_p)\mathbf{b}\right] \\ &= \mathbf{X}\mathbf{b} + \mathbf{1}\bar{y}, \end{aligned}$$

where $\mathbf{1}$ is an n by 1 vector of ones. [Refer to eqs. (8.4) and (8.8).]

Deduce that

$$y - Xb = \mathbf{y} - \mathbf{Xb}$$

and hence, using eq. (8.6), that the residual sum of squares may be written

$$\mathbf{y'y} - \mathbf{y'X(X'X)^{-1}X'y}$$

7. Show by writing elements of S in terms of elements of T that $T'T = S$ implies (using the suffix 0 to denote the first row or column):

$$t_{00} = \pm s_{00}^{1/2}, \quad t_{0j} = s_{0j} t_{00}^{-1} \quad (j > 0)$$

whereas for $i > 0$

$$t_{ii} = \pm \left(s_{ii} - \sum_{l=0}^{i-1} t_{li}^2 \right)^{1/2}$$

$$t_{ij} = \left(s_{ij} - \sum_{l=0}^{i-1} t_{li} t_{lj} \right) t_{ii}^{-1}, \quad j > i$$

(It is assumed that t_{ii} is never zero.)

If we insist that $t_{ii} > 0$, then T is unique. These are the formulae that describe the CCDA (Section 1.5).

8. Write $\mathbf{X} = [x_0, x_1, \ldots, x_q]$. Let $X_k = [x_0, x_1, \ldots, x_k]$. Let T_k be the leading k by k submatrix of the upper triangle matrix T such that $T'T = \mathbf{X'X}$. Let $P_k = X_k(X_k'X_k)^{-1}X_k$. (We assume that $X_k'X_k$ has an inverse.) Show that:
   (i) $T_k'T_k = X_k'X_k$.
   (ii) $P_k^2 = P_k$, $(I - P_k)^2 = I - P_k$.
   (iii) $t_{kk}^2 = x_k'(I - P_{k-1})x_k$, and is the sum of squares of elements of $(I - P_{k-1})x_k$.
   (iv) $t_{kk}^2 = 0$ if and only if $x_k = P_{k-1}x_k$. Deduce that $t_{kk}^2 = 0$ if and only if x_k is a linear combination of earlier columns of $\mathbf{X}$. [Given $x_k = X_{k-1}c$, it follows directly that $P_{k-1}x_k = x_k$. Conversely, $x_k = P_{k-1}x_k$ may be written $x_k = X_{k-1}c$, with $c = (X_{k-1}'X_{k-1})^{-1}X_{k-1}'x_k$.]

9. (Continuation of Exercise 8)
Let

$$T_k = \begin{bmatrix} T_{k-1} & t_k \\ 0 & t_{kk} \end{bmatrix}, \quad T^k = \begin{bmatrix} T^{k-1} & t^k \\ 0 & t^{kk} \end{bmatrix}$$

Prove that, if $T^{k-1} = T_{k-1}^{-1}$, then $T^k = T_k^{-1}$, provided:

(i) $t^{kk} = t_{kk}^{-1}$.

(ii) $t^k = -T^{k-1}t_k t^{kk}$.

Hence show that, given T with none of its diagonal elements zero, formation of T^{-1} is always possible. Give a practical scheme for the computation of T^{-1}, based on (i) and (ii). (See Section 2.3.)

10. (Continuation of Exercises 8 and 9)
Show that:

(i) $t_k = T_{k-1}'^{-1}s_k$, where $s_k = X_{k-1}'x_k$.

(ii) $t_{kk}^2 = s_{kk} - t_k't_k$, where $s_{kk} = x_k'x_k$.

Exercise 8(iv) implies that, provided no column of X_{k-1} is a linear combination of earlier columns and given T_{k-1} upper triangular such that $T_{k-1}'T_{k-1} = X_{k-1}'X_{k-1}$, then $t_{ii} \neq 0$ $(i = 1, 2, \ldots, k-1)$ and T_{k-1}^{-1} exists. Show that equations (i) and (ii) determine T_k such that $T_k'T_k = X_k'X_k$. Hence, assuming there are no linear dependencies among the columns of $\mathbf{X}$, prove that it is always possible to determine an upper triangular matrix T such that $T'T = \mathbf{X}'\mathbf{X}$. (Hence a third method, in addition to the CCDA and the SCDA, exists for the construction of T.)

11. Let S be a real symmetric positive definite matrix, with

$$S_k = \begin{bmatrix} S_{k-1} & s_k \\ s_k' & s_{kk} \end{bmatrix}$$

as its leading $k+1$ by $k+1$ submatrix. (This replaces the assumption made in Exercises 8 through 10, that $S = \mathbf{X}'\mathbf{X}$ for some real matrix $\mathbf{X}$.) Modify the argument of Exercise 9 to prove the existence of T such that $T'T = S$. [If $s_{kk} - t_k't_k \leq 0$, with $t_k = T_{k-1}'^{-1}s_k$, then for a suitable choice of h it would follow that $h'S_kh \leq 0$. Take $h = (S_{k-1}^{-1}s_k, h_k)$, etc.]

12. Prove that $\boldsymbol{S}$ is positive definite symmetric if and only if, for some $\mathbf{X}$, it can be written in the form $\mathbf{X}'\mathbf{X}$.

13. Suppose that $\boldsymbol{y} = \boldsymbol{X}\boldsymbol{b} + \boldsymbol{e}$, where $\boldsymbol{b}$ is chosen to minimize $\boldsymbol{e}'\boldsymbol{e}$. Prove that:

(i) $\boldsymbol{X}'\boldsymbol{e} = 0$.

(ii) $\hat{\boldsymbol{y}}'\boldsymbol{e} = 0$.

14. Let $\boldsymbol{X}^*$ be an n by q matrix such that $\boldsymbol{X}^{*\prime}\boldsymbol{X}^* = \Lambda^2$, where Λ is diagonal. Let $\mathbf{X} = \boldsymbol{X}^*\boldsymbol{U}$, where $\boldsymbol{U}$ is an upper triangular matrix. Show that the upper triangular matrix which results from the Cholesky decomposition of $\mathbf{X}'\mathbf{X}$ is, to within a change of sign of all elements in one or more rows, $\Lambda\boldsymbol{U}$. (This provides a convenient means for generating matrices $\mathbf{X}$ such that $\boldsymbol{U}$ is the Cholesky decomposition of $\mathbf{X}'\mathbf{X}$.)

CHAPTER 2

Regression Calculations —Part II

This chapter discusses some of the information, in addition to the regression coefficients, which is commonly wanted from regression calculations. Later sections discuss the problem of selecting explanatory variates in a regression equation, checks which computer programs ought to include, the effect of linear or near-linear relations between explanatory variates, alternative algorithms for handling calculations based on the normal equations, and weighted least squares.

2.1. DEPENDENCE ON THE FIRST k EXPLANATORY VARIATES ONLY

Columns of X will be written $x_0, x_1, \ldots, x_p$. Let

$$X_k = [x_0, x_1, \ldots, x_k].$$

The normal equations for studying the dependence of y on columns of $\mathbf{X}_k$ are

$$X_k' X_k b_k^0 = X_k' y \tag{1.1}$$

(As usual we may, if all elements of x_0 are one, replace X_k by the first k columns of $\mathbf{X}$, where values in each column are measured from the column mean as origin. The initial element of b_k^0 is then dropped.)

Now $X_k' X_k$ consists of rows and columns 0 to k of $X'X$, and $X_k' y$ consists of elements 0 to k of $X'y$. Hence in the process of reducing $[X'X, X'y]$ or the full array $[\boldsymbol{X}, \boldsymbol{y}]'[\boldsymbol{X}, \boldsymbol{y}]$ to upper triangular form, the array $[\boldsymbol{X}_k'\boldsymbol{X}_k, \boldsymbol{X}_k'\boldsymbol{y}]$ is incidentally reduced to upper triangular form. Thus

$$\begin{bmatrix} \boldsymbol{X}'\boldsymbol{X} & \boldsymbol{X}'\boldsymbol{y} \\ \boldsymbol{y}'\boldsymbol{X} & \boldsymbol{y}'\boldsymbol{y} \end{bmatrix} \qquad \text{carries with it} \qquad \left[\begin{array}{cc} \boldsymbol{X}_k'\boldsymbol{X}_k & \vdots \;\; \boldsymbol{X}_k'\boldsymbol{y} \\ \cdots & \vdots \\ \hline \cdots & \cdots \end{array}\right]$$

$$\downarrow \qquad\qquad\qquad\qquad \downarrow$$

$$\boldsymbol{T} = \begin{bmatrix} \boldsymbol{T}_p & \boldsymbol{t}_y \\ \boldsymbol{0}' & t_{yy} \end{bmatrix} \qquad\qquad \left[\begin{array}{cc} \boldsymbol{T}_k & \vdots \;\; \boldsymbol{t}_{y(k)} \\ \boldsymbol{0}' & \vdots \\ \hline \boldsymbol{0}' & \cdots \end{array}\right]$$

The normal eq. (1.1) has been reduced to

$$\boldsymbol{T}_k\boldsymbol{b}_k^0 = \boldsymbol{t}_{y(k)} \tag{1.2}$$

The solution to the eq. (1.1), for any k, is available from $\boldsymbol{T}$ simply by solving the requisite upper triangular set of equations.

The Sequential Analysis of Variance Table

We can now justify the way in which we have been writing down an analysis of variance table in Section 1.9 and subsequently. From the note accompanying eq. (12.7) in Chapter 1, but with $\boldsymbol{X}_k$ replacing $\boldsymbol{X}$, the residual sum of squares when $\boldsymbol{y}$ is regressed upon columns of $\boldsymbol{X}_k$ is

$$\boldsymbol{y}'\boldsymbol{y} - \boldsymbol{y}'\boldsymbol{X}_k\left(\boldsymbol{X}_k'\boldsymbol{X}_k\right)^{-1}\boldsymbol{X}_k'\boldsymbol{y} = \boldsymbol{y}'\boldsymbol{y} - \boldsymbol{t}_{y(k)}'\boldsymbol{t}_{y(k)}$$

For regression upon columns of $\boldsymbol{X}_{k+1}$ the vector $\boldsymbol{t}_{y(k)}$ must be replaced by $\boldsymbol{t}_{y(k+1)}$. The two vectors are identical except that $\boldsymbol{t}_{y(k+1)}$ has the additional element $t_{k+1,\,y}$. Thus the residual sum of squares is reduced by an amount $t_{k+1,\,y}^2$ when x_{k+1} is included as a further explanatory variate. This justifies denoting $t_{k+1,\,y}^2$ as the sum of squares which is *due to* x_{k+1}, *given* $x_1, x_2, \ldots, x_k$.

Use of a Variate Other than y as the Dependent Variate

Any use of normal equations

$$\boldsymbol{X}_k'\boldsymbol{X}_k\boldsymbol{b}_{q.(k)} = \boldsymbol{X}_k'\boldsymbol{x}_q, \quad q > k \tag{1.3}$$

for studying the dependence of $\boldsymbol{x}_q$ on columns of $\boldsymbol{X}_k$ is similarly catered for. We pick out from $\boldsymbol{T}$ the same submatrix $\boldsymbol{T}_k$ as before, together with $\boldsymbol{t}_{q(k)}$ consisting of elements 0 to k in column q of $\boldsymbol{T}$. The normal eq. (1.3) has been reduced to

$$\boldsymbol{T}_k \boldsymbol{b}_{q.(k)} = \boldsymbol{t}_{q(k)} \tag{1.4}$$

The case $q = k + 1$ will be of especial interest to us. If in eq. (12.7) of Chapter 1 we replace $\boldsymbol{X}$ by $\boldsymbol{X}_k$, $\boldsymbol{y}$ by $\boldsymbol{x}_{k+1}$, we then have

$$t_{k+1,k+1}^2 = \boldsymbol{x}_{k+1}'\boldsymbol{x}_{k+1} - \boldsymbol{x}_{k+1}'\boldsymbol{X}_k\left(\boldsymbol{X}_k'\boldsymbol{X}_k\right)^{-1}\boldsymbol{X}_k'\boldsymbol{x}_{k+1} \tag{1.5}$$

which is the residual sum of squares from the regression of $\boldsymbol{x}_{k+1}$ upon columns of $\boldsymbol{X}_k$.

Suppose that $\boldsymbol{X}$ has an initial column of ones. Then alternatively one may write

$$t_{k+1,k+1}^2 = \mathbf{x}_{k+1}'\mathbf{x}_{k+1} - \mathbf{x}_{k+1}'\mathbf{X}_k\left(\mathbf{X}_k'\mathbf{X}_k\right)^{-1}\mathbf{X}_k'\mathbf{x}_{k+1} \tag{1.6}$$

The notation is that of Section 1.8 and elsewhere; this result follows as in Exercise 6 of Chapter 1. Furthermore

$$t_{k+1,k+1}^2 = \mathsf{s}_{k+1,k+1}\left[1 - R_{k+1(1\cdots k)}^2\right] \tag{1.7}$$

$$R_{k+1(1\cdots k)}^2 = 1 - \mathsf{s}_{k+1,k+1}^{-1} t_{k+1,k+1}^2$$

(See Section 1.14 for the definition of R^2.)

Note further that $\mathsf{s}_{k+1,k+1}$ is the element in the $(k+1, k+1)$ position of $\mathbf{T}'\mathbf{T}$ (see Exercise 4, Chapter 1) and hence that

$$\mathsf{s}_{k+1,k+1} = \sum_{i=1}^{k+1} t_{i,k+1}^2$$

Thus

$$R_{k+1(1\cdots k)}^2 = 1 - t_{k+1,k+1}^2\left(\sum_{i=1}^{k+1} t_{i,k+1}^2\right)^{-1}$$

2.2. STANDARD ERRORS OF REGRESSION COEFFICIENTS

In this section we give the calculations for the example of Section 1.9. The estimated variance-covariance matrix for

$$\mathbf{b} = \begin{bmatrix} \frac{16}{21} \\ \frac{1}{3} \end{bmatrix}$$

is $(\mathbf{X}'\mathbf{X})^{-1}\sigma^2$, with σ^2 replaced by its estimate s^2. The ith diagonal element is the estimate of

$$\text{var}(b_i) = [\text{SE}(b_i)]^2$$

(Note SE = standard error.)

Estimation of σ^2

An unbiased estimate of σ^2 is

$$s^2 = (n - p - 1)^{-1} \times \textit{residual sum of squares}$$

The residual sum of squares is most simply obtained as t_{yy}^2, where t_{yy} is the final diagonal element of the Cholesky decomposition of $[X, y]'[X, y]$, or of $[\mathbf{X}, \mathbf{y}]'[\mathbf{X}, \mathbf{y}]$. Thus in the example of Section 1.9 $n - p - 1 = 4 - 3 = 1$, and $s^2 = 2^2 = 4$.

The Estimated Variance-Covariance Matrix

$$(\mathbf{X}'\mathbf{X})^{-1}s^2 = \begin{bmatrix} 49 & 14 \\ 14 & 13 \end{bmatrix}^{-1} \times 4$$

$$= \begin{bmatrix} \frac{52}{441} & -\frac{56}{441} \\ -\frac{56}{441} & \frac{196}{441} \end{bmatrix}$$

Thus the estimates we require are

$$\text{var}(b_1) = \tfrac{52}{441}, \quad \text{var}(b_2) = \tfrac{196}{441},$$

$\text{cov}(b_1, b_2) = -\frac{56}{441}$. The standard error (SE) of b_1 is the square root of $\text{var}(b_1)$, and similarly for SE of b_2.

If $\mathrm{var}(b_0)$ is required, it may be calculated as

$$\mathrm{var}(\bar{y} - b_1\bar{x}_1 - b_2\bar{x}_2) = \mathrm{var}(\bar{y}) + \bar{x}_1^2\mathrm{var}(b_1) + \bar{x}_2^2\mathrm{var}(b_2) + 2\bar{x}_1\bar{x}_2\mathrm{cov}(b_1, b_2)$$

The mean $\bar{y}$ is uncorrelated with b_1 and b_2. However b_0 is the predicted value corresponding to $x_1 = x_2 = 0$, and $\mathrm{var}(b_0)$ is best determined using the formula for the variance of a predicted value which will be given in Section 2.4.

Inversion of a Matrix of the form $X'X$—A General Method

Finding the inverse of a 2 by 2 matrix is not difficult; there is the explicit formula which was used in the preceding discussion:

$$\begin{bmatrix} c & h \\ g & d \end{bmatrix}^{-1} = (cd - gh)^{-1}\begin{bmatrix} d & -h \\ -g & c \end{bmatrix} \tag{2.1}$$

It will however be well to illustrate the method of the next section in this simple case. In Section 1.9 we found

$$\mathbf{X'X} = \begin{bmatrix} 49 & 14 \\ \cdot & 13 \end{bmatrix} \rightarrow \mathbf{T}_2 = \begin{bmatrix} 7 & 2 \\ 0 & 3 \end{bmatrix}$$

where $\mathbf{T}_2'\mathbf{T}_2 = \mathbf{X'X}$. Then using the method of the next section, or perhaps eq. (2.1) with $g = 0$, one has

$$\mathbf{T}_2^{-1} = \begin{bmatrix} \frac{1}{7} & -\frac{1}{7} \times 2 \times \frac{1}{3} \\ 0 & \frac{1}{3} \end{bmatrix} = \begin{bmatrix} \frac{1}{7} & -\frac{2}{21} \\ 0 & \frac{1}{3} \end{bmatrix} \tag{2.2}$$

Then

$$[\mathbf{X'X}]^{-1} = \mathbf{T}_2^{-1}\mathbf{T}_2'^{-1}$$

$$= \begin{bmatrix} (\frac{1}{7})^2 + (\frac{2}{21})^2 & -\frac{2}{21} \times \frac{1}{3} \\ \cdot & (\frac{1}{3})^2 \end{bmatrix} \tag{2.3}$$

We have $\mathrm{var}(b_1) = [(\frac{1}{7})^2 + (\frac{2}{21})^2]\sigma^2$, and so on.

An advantage of this way of writing the results is that one can see immediately how $\mathrm{var}(b_1)$ would change if we assume $\beta_2 = 0$ and set $b_2 = 0$. One would have $\mathrm{var}(b_1 | b_2 = 0) = (\frac{1}{7})^2\sigma^2$. Of course the estimate of σ^2 would change.

The *variance inflation factor* (VIF) provides information with a somewhat similar function to that provided by results in the form (2.3). It answers the question, How much is $\mathrm{var}(b_k)$ increased by comparison with $\mathrm{var}(b_k)$ for a model (with the same error variance σ^2) that has x_k as the only explanatory variate?

If we define $\mathbf{s}_{kk}$ to be the kth diagonal element of $\mathbf{X}'\mathbf{X}$, and s^{kk} to be the kth diagonal element of $(\mathbf{X}'\mathbf{X})^{-1}$, then

$$\mathrm{VIF}(b_k) = \mathbf{s}_{kk}s^{kk} \tag{2.4}$$

By Exercise 7 at the end of this chapter this equals

$$\left[1 - R^2_{k(1,\ldots,k-1,k+1,\ldots,p)}\right]^{-1}$$

2.3. THE INVERSE OF A MATRIX OF THE FORM $X'X$

For the calculations of the previous chapter matrix inversion was unnecessary. Its use in performing the calculations would have substantially increased the number of arithmetic operations, with some loss of precision. However in calculations such as those of Section 2.2, the inverse is required for its own sake.

Given S_p of the form $X'X$, we have to find S_p^{-1}. In the process of solving the normal equations

$$X'Xb = X'y \quad \text{or} \quad \mathbf{X}'\mathbf{X}\mathbf{b} = \mathbf{X}'\mathbf{y}$$

the array

$$S = \left[\begin{array}{c|c} X'X & X'y \\ \hline \cdot & y'y \end{array}\right]$$

(or the similar array with $\mathbf{X}$ in place of X, and $\mathbf{y}$ in place of y) has, we assume, been reduced to

$$T = \left[\begin{array}{c|c} T_p & t_y \\ \hline 0' & t_{yy} \end{array}\right]$$

where T is upper triangular, with zeros below the diagonal such that $T'T = S$. It is readily verified that $T_p'T_p = X'X$. We evaluate

$$(X'X)^{-1} = T_p^{-1}(T_p')^{-1} \tag{3.1}$$

by first finding T_p^{-1}, and then postmultiplying it by its transpose.

Assume that to date T_{k-1}^{-1} has been formed, where T_{k-1} is the leading $k-1$ by $k-1$ submatrix of T_p. Let

$$T_k = \begin{bmatrix} T_{k-1} & t_k \\ 0' & t_{kk} \end{bmatrix}, \quad T^k = \begin{bmatrix} T_{k-1}^{-1} & t^k \\ 0' & t^{kk} \end{bmatrix}$$

Then

$$T_kT^k = \begin{bmatrix} I_{k-1} & T_{k-1}t^k + t_kt^{kk} \\ 0' & t_{kk}t^{kk} \end{bmatrix}$$

This is an identity matrix, provided we take

$$t^{kk} = t_{kk}^{-1}, \quad t^k = -T_{k-1}^{-1}t_kt^{kk} \tag{3.2}$$

Thus, provided $t_{kk} \neq 0$, we can find $T^k = T_k^{-1}$. Computations are started with $T_0^{-1} = [t_{00}^{-1}]$.

Note that $-t_{kk}t^k$ is the vector of regression coefficients in the regression of x_k upon earlier columns of X.

An Interpretation for the Elements of T^{-1}

Consider the regression of x_k upon variates 1 to $k-1$. The vector of regression coefficients is

$$\begin{aligned} b_{k.(k-1)} &= T_{k-1}^{-1}t_k \\ &= -t^kt_{kk} \end{aligned}$$

by eq. (3.2). Thus element j ($j \leq k-1$) of $b_{k.(k-1)}$ is $-t^{jk}t_{kk} = -t^{jk}/t^{kk}$. The residual sum of squares in this regression is $t_{kk}^2 = (t^{kk})^{-2}$.

Example

$$S_2 = \begin{bmatrix} 4 & 6 & 10 \\ \cdot & 58 & 29 \\ \cdot & \cdot & 38 \end{bmatrix} \quad T_2 = \begin{bmatrix} 2 & 3 & 5 \\ 0 & 7 & 2 \\ 0 & 0 & 3 \end{bmatrix}$$

(S_2 is the SSP matrix $X'X$ for the data of Section 1.9.)

Notes

$$T_2 = \begin{bmatrix} 2 & 3 & 5 \\ 0 & 7 & 2 \\ 0 & 0 & 3 \end{bmatrix}$$

$$\downarrow$$

$$\rightarrow \begin{bmatrix} \frac{1}{2} & 3 & 5 \\ 0 & 7 & 2 \\ 0 & 0 & 3 \end{bmatrix} \qquad T_0^{-1} = [\tfrac{1}{2}]$$

$$\downarrow$$

$$\rightarrow \begin{bmatrix} \frac{1}{2} & -\frac{1}{2} \times 3 \times \frac{1}{7} & 5 \\ 0 & \frac{1}{7} & 2 \\ 0 & 0 & 3 \end{bmatrix} \qquad T_1^{-1} = \begin{bmatrix} \frac{1}{2} & -3 \times \frac{1}{14} \\ 0 & \frac{1}{7} \end{bmatrix}$$

$$T_2^{-1} = \begin{bmatrix} \frac{1}{2} & -3 \times \frac{1}{14} & -\frac{1}{3}(\frac{1}{2} \times 5 - 3 \times \frac{1}{14} \times 2) \\ 0 & \frac{1}{7} & -\frac{1}{3}(\frac{1}{7} \times 2) \\ 0 & 0 & \frac{1}{3} \end{bmatrix}$$

For forming elements 1 and 2 in column 3, $t^2 = \frac{1}{3} T_1^{-1} \begin{bmatrix} 5 \\ 2 \end{bmatrix}$.

$$= \begin{bmatrix} \frac{1}{2} & -\frac{3}{14} & -\frac{29}{42} \\ 0 & \frac{1}{7} & -\frac{2}{21} \\ 0 & 0 & \frac{1}{3} \end{bmatrix}$$

Thus

$$S_2^{-1} = \begin{bmatrix} (\frac{1}{2})^2 + (\frac{3}{14})^2 + (\frac{29}{42})^2 & -\frac{3}{14} \times \frac{1}{7} + \frac{29}{42} \times \frac{2}{21} & -\frac{29}{42} \times \frac{1}{3} \\ \cdot & (\frac{1}{7})^2 + (\frac{2}{21})^2 & -\frac{2}{21} \times \frac{1}{3} \\ \cdot & \cdot & (\frac{1}{3})^2 \end{bmatrix}$$

Use of this matrix for calculating variances and covariances in Section 2.2 would have allowed us to read off directly the coefficient for σ^2 in $\text{var}(b_0)$.

From the leading diagonal element we have

$$\text{var}(b_0) = \left[\left(\tfrac{1}{2}\right)^2 + \left(\tfrac{3}{14}\right)^2 + \left(\tfrac{29}{42}\right)^2\right]\sigma^2$$

If S_2^{-1} is written as shown here, it is easy to read off S_1^{-1} and S_0^{-1}. Thus S_1^{-1} is obtained from S_2^{-1} by deleting the final row and column and then omitting the last term in each of the remaining sums:

$$S_1^{-1} = \begin{bmatrix} \left(\tfrac{1}{2}\right)^2 + \left(\tfrac{3}{14}\right)^2 & -\tfrac{3}{14} \times \tfrac{1}{7} \\ \cdot & \left(\tfrac{1}{7}\right)^2 \end{bmatrix}$$

Notice that we do not need to form all elements of S_2^{-1} (or of S_1^{-1}). If our interest is in diagonal elements, we can with advantage refrain from calculating off-diagonal elements.

2.4. STANDARD ERRORS OF PREDICTED VALUES

The estimated regression coefficients $b_1, b_2, \ldots, b_p$ (and b_0) determine an estimated regression relation which may be written

$$\hat{y} = b_0 + b_1 x_1 + \cdots + b_p x_p \tag{4.1}$$

or

$$\hat{y} = \bar{y} + b_1(x_1 - \bar{x}_1) + \cdots + b_p(x_p - \bar{x}_p) \tag{4.2}$$

If $[x_1, x_2, \ldots, x_p] = [x_{i1}, x_{i2}, \ldots, x_{ip}]$ is one of the original data points, we have, if we use eq. (4.1),

$$\hat{y}_i = b_0 + b_1 x_{i1} + \cdots + b_p x_{ip}$$

The predicted value $\hat{y}_i$ is to be distinguished from the original observed y_i; in fact

$$e_i = y_i - \hat{y}_i$$

is the ith residual.

Calculations based on eq. (4.1) are straightforward. We have

$$\hat{y} = b_0 + b_1x_1 + \cdots + b_px_p$$
$$= \boldsymbol{l}'\boldsymbol{b}$$

where $\boldsymbol{l}' = (1, x_1, \ldots, x_p)$. Then by eq. (13.3) of Chapter 1

$$\text{var}(\hat{y}) = \boldsymbol{l}'(\boldsymbol{X}'\boldsymbol{X})^{-1}\boldsymbol{l}\sigma^2 \tag{4.3}$$

$$= \left(\boldsymbol{T}_p'^{-1}\boldsymbol{l}\right)'\boldsymbol{T}_p'^{-1}\boldsymbol{l}\sigma^2 \tag{4.3'}$$

$$= \boldsymbol{d}'\boldsymbol{d}\sigma^2$$

where $\boldsymbol{d} = \boldsymbol{T}_p'^{-1}\boldsymbol{l}$. The vector $\boldsymbol{d}$ may be formed either by postmultiplying $\boldsymbol{T}_p'^{-1}$ by $\boldsymbol{l}$, or by solving $\boldsymbol{T}_p'\boldsymbol{d} = \boldsymbol{l}$ for $\boldsymbol{d}$. As an example, the reader may care to verify that, if $\boldsymbol{l}' = [1, -2, 0]$, then $\boldsymbol{d}' = [\frac{1}{2}, -\frac{1}{2}, -\frac{1}{2}]$, and that, if σ^2 is replaced by $\hat{\sigma}^2 = 4$, eq. (4.3′) gives $\text{var}(\hat{y}) = 3.0$.

Notice also that $\text{var}(b_0)$ may be obtained by setting $\boldsymbol{l}' = [1, 0, 0]$, which gives $\boldsymbol{d}' = [\frac{1}{2}, -\frac{3}{14}, -\frac{29}{42}]$, and so on.

Calculations Based on Eq. (4.2)

In this case one has

$$\hat{y} = \bar{y} + \boldsymbol{\ell}'\mathbf{b}$$

where $\boldsymbol{\ell}' = [x_1 - \bar{x}_1, x_2 - \bar{x}_2, \ldots, x_p - \bar{x}_p]$. Now $\mathbf{b}$, because it is given in terms of $y_1 - \bar{y}, y_2 - \bar{y}, \ldots, y_n - \bar{y}$ (elements of $\boldsymbol{y}$) is independent of $\bar{y}$. Hence

$$\text{var}(\hat{y}) = \text{var}(\bar{y}) + \text{var}(\boldsymbol{\ell}'\boldsymbol{b})$$

$$= n^{-1}\sigma^2 + \boldsymbol{\ell}'(\mathbf{X}'\mathbf{X})^{-1}\boldsymbol{\ell}\sigma^2 \tag{4.4}$$

$$= n^{-1}\sigma^2 + \left(\mathbf{T}_p'^{-1}\boldsymbol{\ell}\right)'\left(\mathbf{T}_p'^{-1}\boldsymbol{\ell}\right)\sigma^2 \tag{4.4'}$$

where $\mathbf{T}_p$ is upper triangular such that $\mathbf{T}_p'\mathbf{T}_p = \mathbf{X}'\mathbf{X}$.

Example

As pointed out above, $\text{var}(b_0)$ may be calculated as $\text{var}(\hat{y})$ when $x_1 = x_2 = \cdots = 0$. For the data of Section 1.9, we had $\bar{x}_1 = 1.5$, $\bar{x}_2 = 2.5$,

$$\left(\mathbf{T}_2'\right)^{-1} = \begin{bmatrix} \frac{1}{7} & 0 \\ -\frac{2}{21} & \frac{1}{3} \end{bmatrix} \quad [(\text{eq. } 2.2)]$$

Then with $\ell = \begin{bmatrix} -1.5 \\ -2.5 \end{bmatrix}$,

$$(\mathbf{T}_2')^{-1}\ell = \begin{bmatrix} -\frac{3}{14} \\ \frac{3}{2} \times \frac{2}{21} - \frac{5}{6} \end{bmatrix} = \begin{bmatrix} -\frac{3}{14} \\ -\frac{29}{42} \end{bmatrix}$$

Thus $\ell'(X'X)^{-1}\ell = (\frac{3}{14})^2 + (\frac{29}{42})^2$, and using eq. (4.4), $\text{var}(\hat{y}) = [\frac{1}{4} + (\frac{3}{14})^2 + (\frac{29}{42})^2]\sigma^2 \simeq 0.7727\sigma^2$, which we estimate as $0.7727 \times 4 = 3.09$. Thus $\text{SE}(b_0) = \sqrt{3.09} = 1.76$.

Formation of $T_p'^{-1}l$ or $\mathbf{T}_p'^{-1}\ell$

In the preceding example we formed $T_2'^{-1}l$ by postmultiplying $T_2'^{-1}$ by l. For $p > 2$, and especially if $T_p'^{-1}$ is not already available, it is preferable to determine d by solving $T_p'd = l$. Solving the upper triangular equation requires only slightly more work than the postmultiplication.

Consider the earlier calculation based on the matrix $\mathbf{T}_2$. For this

$$\begin{bmatrix} 7 & 0 \\ 2 & 3 \end{bmatrix}\begin{bmatrix} d_1 \\ d_2 \end{bmatrix} = \begin{bmatrix} -1.5 \\ -2.5 \end{bmatrix}$$

Thus $7d_1 = -1.5$, $d_1 = -\frac{3}{14}$, and $2d_1 + 3d_2 = -2.5$, $d_2 = \frac{29}{42}$.

*2.5. OMISSION OF ANY EXPLANATORY VARIATE OR VARIATES

The notation used here is that of Sections 2.1 through 2.4. In Section 2.1 it was shown that omission of the final or pth explanatory variate increases the residual sum of squares by an amount $\theta_p = t_{p,y}^2$. As $b_p = t_{pp}^{-1}t_{p,y}$ it follows that

$$\theta_p = b_p^2 t_{pp}^2$$
$$= b_p^2 (s^{pp})^{-1}$$

where s^{pp} is the final diagonal element of S_p^{-1} or of $\mathbf{S}_p^{-1}$. (See Section 1.13.) Now s^{ii} and b_i will be the same whether the variate in question is taken as the final variate ($i = p$) or as an earlier variate ($i < p$). Hence, quite generally, the increase in the residual sum of squares due to omission of the ith variate (with variates $1, \ldots, i-1, i+1, \ldots, p$ still included) is

$$\theta_i = b_i^2 (s^{ii})^{-1} \tag{5.1}$$

Note incidentally that the t-statistic for testing the significance of b_i is $\hat{\sigma}^{-1}\theta_i^{1/2}$, where $\hat{\sigma}^2$ is the usual unbiased estimate of the error variance. In using eq. (5.1), s^{ii} may be calculated as demonstrated in Section 2.3.

Let θ_{ij} be the increase in the residual sum of squares from omitting the two variates x_i and x_j from the regression equation. Clearly,

$$\theta_{ij} \geq \theta_i, \quad \theta_{ij} \geq \theta_j \tag{5.2}$$

Given θ_i and θ_j, this is as much as one can say regarding θ_{ij}. For the example of Sections 1.9 and 2.3, $\theta_1 = 336/13$, $\theta_2 = 1$, and $\theta_{12} = 37$.

Omission of More Than One Explanatory Variate

Let $\boldsymbol{b}'_{(*)} = [b_i, b_j]$, and let $\boldsymbol{S}^{**}$ be the matrix that results from picking out from $\boldsymbol{S}^{-1}$ elements that belong to the ith or jth row and to the ith or jth column. Then it may be shown that

$$\theta_{ij} = \boldsymbol{b}'_{(*)}(\boldsymbol{S}^{**})^{-1}\boldsymbol{b}_{(*)} \tag{5.3}$$

Exercise 8 at the end of the chapter gives a result that may be used in proving this. Equation (5.3) generalizes in the obvious way when an arbitrary number of variates are omitted.

Application to All-Subsets Regression

If two variates are to be omitted from the regression equation, the pair x_i and x_j that give the smallest possible θ_{ij} will be chosen. Now, on the whole, small values of θ_{ij} will tend to be associated with small values of θ_i and θ_j. Hence the search should start by taking i and j such that θ_i and θ_j are the two smallest such values, and using eq. (5.3) to determine the corresponding θ_{ij}. Suppose now that for one or more suffixes ℓ, $\theta_\ell > \theta_{ij}$. It then follows that pairs that include variate ℓ can be excluded from consideration; this is a consequence of eq. (5.2).

Now consider the problem of finding, for an arbitrary value of k, the subset of k out of the p explanatory variates which leads to the minimum residual sum of squares; thus $k - p$ out of the p variates are to be excluded. Use of obvious extensions of preceding argument may substantially reduce the number of candidate subsets for which explicit calculation of the change in the residual sum of squares is necessary. For further discussion see the account in Seber (1977).

2.6. A COMPLETE SEQUENCE OF CALCULATIONS

This is a suitable point at which to demonstrate the use of the BASIC program described in Section 10.7. Calculations will be demonstrated for the example from Section 1.9. Figure 2.1 gives a record of program requests for user input, user responses, and computer output.

```
RUN

GIVE NO. OF VAR'S? 3

DO YOU WISH TO NAME THE VAR'S (Y OR N)? N

DO ALL ROWS HAVE EQUAL WEIGHT (Y OR N)? Y

ENTER VALUES ROW BY ROW, IN THE ORDER:
VAR. NO.  1          2          3
    NAME  V01        V02        V03

FINISH WITH:   EOD

? -2 0 -3
? -1 2 1
? 2 5 2
? 7 3 6
? EOD

* CSSP MATRIX  *
 1        49         14         42
 2                   13         15
 3                              41


VAR. NO.  1          2          3
    NAME  V01        V02        V03
MEANS:    1.5        2.5        1.5
S.D.'S:   4.0415     2.0817     3.6968

R-SQUARED: ----      .30769     .90244
(THIS MEASURES DEPENDENCE ON EARLIER VARIATES)


* CHOLESKY UPPER TRIANGULAR MATRIX  *
VAR. NO.  0          1          2          3
    NAME  CONST.     V01        V02        V03
 0        2          3          5          3
 1                   7          2          6
 2                              3          1
 3                                         2

ENTER  P  TO PROCEED WITH REGRESSION CALCULATIONS
OTHERWISE ENTER  D (TO DELETE A COLUMN)
             OR  T (TO TRANSPOSE A COLUMN)
             OR  L (TO LOOK AT DATA)
             OR  J (JUMP TO NEXT SET OF OPTIONS)
? P
```

Figure 2.1. Use of the BASIC program described in Section 10.7 to handle calculations for the example of Section 1.9.

```
* INVERSE OF UPPER TRIANGULAR MATRIX *
 0          .5        -.21429   -.69048    .2381
 1                     .14286   -.095238  -.38095
 2                               .33333   -.16667
 3                                         .5


* CALCULATE REGRESSION COEFFICIENTS *
GIVE DEP. VARIATE - TAKE VAR. NO.? 3
INCLUDE EXPLAN. VAR'S UP TO NO.? 2

DEPENDENT VAR. IS < V03 >.      ERROR VAR. = 4    (DF = 1 )

VAR. NO.   0          1         2
    NAME   CONST.     V01       V02
COEFFS.: -.47619     .7619     .33333
S.E.'S:    1.758     .34339    .66667


* VARIANCE-COVARIANCE MATRIX *
VAR. NO.   0          1         2
    NAME   CONST.     V01       V02
 0         3.0907    .14059   -.92064
 1                   .11791   -.12698
 2                              .44444


* TABLE OF RESIDUALS *
ROW NO.  OBSERVED  RESIDUAL   EXPECTED   SE(EXP)   WEIGHT
 1        -3       -1         -2         1.7321    1
 2         1        1.5714    -.57143    1.2372    1
 3         2       -.71429     2.7143    1.8681    1
 4         6        .14286     5.8571    1.9949    1

ENTER   F (FURTHER ANALYSIS ON THE SAME DATA)
OR      N (NEW PROBLEM)
OR      Q (QUIT)
? Q


STOP AT LINE 1000

READY
```

Figure 2.1 (*Continued*)

The matrix $\boldsymbol{T}$ which is the Cholesky decomposition of the SPP matrix may be formed by taking the Cholesky decomposition of the CSSP matrix and adjoining first an initial column of zeros and then an initial row $(\sqrt{n}, \sqrt{n}\,\bar{x}_1, \ldots)$. See Exercise 5, Chapter 1 (see also Section 3.2). This is the approach used for the calculations that follow. It gives the better precision which results from basing the formation of $\mathbf{T}$ on the CSSP matrix, and the simplification of certain of the calculations which results from working with $\boldsymbol{T}$ rather than with $\mathbf{T}$.

Observe that the inverse is given for the full Cholesky upper triangle matrix, including the final column which corresponds to the dependent variate. The information in the final column is not usually of immediate relevance in regression calculations.

2.7. VARIATE SELECTION AND RELATED TOPICS

In many applications of least squares regression an important part of the problem is to determine which of a number of candidate explanatory variates should be used. What calculations are relevant to variate selection problems?

A relatively straightforward case is that where there is a natural ordering of the variates, and the problem reduces to finding the cutoff point between variates that are included and those that are omitted. It is then possible to proceed as follows: the SSP or CSSP matrix is formed, with the variates taken in the prior order. For each $k = 0, 1, \ldots$ the residual sum of squares, when variates up as far as x_k have been fitted, is then determined by taking the final column of $\boldsymbol{T}$ and summing the squares of elements in row $k + 1$ and later rows. Division by the number of degrees of freedom (i.e., $n - k + 1$) then gives the error mean square $s^2(k)$. This is plotted against k (or equivalently, against $n - k + 1$). If at $k = k_0$, say, the graph begins to level out, this will indicate that inclusion of variates after x_k is of dubious benefit.

If there is no prior ordering of variates and only a few of the potential explanatory variates are candidates for exclusion, a suitable first line of attack begins by fitting all potential explanatory variates. For each regression coefficient the corresponding t-statistic is calculated by dividing the coefficient by its standard error. Variates with small t-statistics (e.g., less than 1.5 or 2) are then omitted, and the calculation repeated. For this the new set of rows and columns is picked out from the CSSP matrix, and the new Cholesky decomposition is formed. (An alternative method, which avoids going back to the CSSP matrix, is described in Chapter 4.) If one is fortunate, the mean square error (or equivalently the adjusted squared multiple correlation) will be similar to that when all variates are included. Scrutiny of the new t-statistics will then show whether any further variates can be omitted. In addition each of the variates omitted earlier should be tested in turn for possible reinclusion. If this approach works, it will provide a regression equation that is not far from optimal, in the sense that it leads to an error mean square that is near to the smallest possible. It will fail to work when some subset of variates, whose coefficients are individually nonsignificant when all variates are fitted, together give a significant reduction in the residual sum of squares. Other methods must then be used, such as *all-subsets regression* which will be discussed shortly, or some form of *stepwise regression*.

Various plots of residuals are a useful aid in deciding whether a given variate should be included or perhaps transformed before inclusion. (Other uses for residuals are discussed, briefly, at the end of this section.) Note the following:

1. The omission of variates that should appear in the model introduces a bias (see Exercise 2 at the end of this chapter). If the bias is small, exclusion of these variates may bring compensating advantages, such as a reduction in the mean square prediction error.

2. The standard error of the coefficient will be large for any variate that is strongly dependent on other explanatory variates (see further Sections 2.9, 3.4, and 3.12). In addition the magnitudes of the regression coefficients are likely to be inflated (see Exercise 9, Chapter 3).

Often one is justified in using only a few of the potential explanatory variates. Thus Kempton (1978) comments that in his experience of biological data "inclusion of more than three regressor variates is seldom if ever justified." The problem of assessing statistical meaningfulness will be eased if attention can in the first instance be limited to a small number of explanatory variates, chosen for theoretical reasons or because other investigators have found them to be important. One can then test the effect of including less likely possibilities one or two at a time.

All-Subsets Regression

With modern computing facilities it becomes possible, for modest values of p (perhaps less than 20), to contemplate methods that involve, potentially, consideration of all 2^{p-1} regression equations that use one or more of the p potential explanatory variates. Section 2.5 discussed devices that allow a substantial proportion of equations to be excluded without investigating them explicitly. Attention is drawn to the comments at the end of this section on the statistical consequences of selecting a regression equation from a large number of possible alternatives.

Forward Selection, Backward Deletion, and Stepwise Regression

Alternatives to all-subsets regression are *forward selection*, *backward deletion*, and a hybrid of these two usually called *stepwise regression*. In *forward selection* new explanatory variates are included one at a time. *Backward deletion* starts with the full set of explanatory variates, and deletes them one at a time. Any criterion used to decide which explanatory variate should next be included in *forward selection* corresponds in the obvious way to a criterion for deciding which variate should next be deleted in *backward deletion*. The criteria used in *forward selection* or in *stepwise regression* for choice of each new variate include:

1. The reduction in the residual sum of squares is a maximum.
2. The increase in the multiple correlation between y and the selected explanatory variates is a maximum.
3. The absolute value of the partial correlation with y, given the previously selected explanatory variates, is as large as possible.
4. The F-to-enter statistic (which equals the square of the t-statistic for that variate once it has been entered) is as large as possible.
5. The reduction in the mean residual sum of squares (obtained by dividing residual sum of squares by degrees of freedom) is the maximum possible.
6. The adjusted squared multiple squared correlation is as large as possible.

These criteria will all choose the same variate to be entered. Note however that Criteria 5 and 6 are far more satisfactory than 1 and 2 as measures of the adequacy of fit of a regression equation. The residual sum of squares (and the squared multiple correlation) must decrease or stay the same as further explanatory variates are taken. By contrast, the unwarranted inclusion of explanatory variates should make little difference to the mean sum of squares, and may even cause it to increase; similarly, for the adjusted squared multiple correlation. Ideally, one would like a criterion that attains its minimum when the choice of explanatory variates is in some sense optimum and then can be expected to increase as further explanatory variates are taken. Thus consider Mallows' statistic:

$$C_{\tilde{p}} = \frac{\text{RSS}}{s^2} - (n - 2\tilde{p} - 2)$$

Here RSS is the residual sum of squares from a model containing $\tilde{p}$ parameters, and s^2 is, ideally, a reliable unbiased estimate of the error variance. In practice, s^2 is usually taken as the mean residual sum of squares in a regression equation that includes all variates under consideration. Daniel and Wood (1980) demonstrate the use of Mallows' C_p.

In the usual form of stepwise regression each step involves (1) adding a variate from among those (if any exist) whose F-to-enter statistic exceeds some threshold; and (2) deleting a variate from among those (if any exist) whose F-to-delete statistic is less than some threshold. The choice of variate to add or delete is usually made with a view to obtaining a residual sum of squares that is as small as possible.

There is also a *backward* form of *stepwise regression*, which as in backward deletion starts from the full set of explanatory variates. None of these methods can be guaranteed to give the best subset of the finally chosen size; however, the subset that results will in many cases be near to optimal. Berk (1978) presents evidence on the comparison between all-subsets, forward selection, and backward deletion.

The Consequences of Selection

Attention has been drawn to the problem of checking a regression equation for statistical significance when its explanatory variates have been selected from among a large number of possibilities. Whenever possible, the regression equation should be checked (cross-validated) and coefficients re-estimated on a data set different from that used for the initial variate selection. Various elaborations of this technique have been suggested. Another promising technique is the *bootstrap*, which Efron has proposed for use in situations where the theoretical properties of a procedure are unknown or uncertain. Diaconis and Efron (1983) describe this technique from an elementary point of view. In the context of variate selection it involves the notion of an artificial infinite population in which each of the n rows of data (observations) appears with equal frequency. (For simplicity, independent observations, all with equal weights, are assumed.) Repeated random samples of n rows are drawn from this population. The selection procedure is repeated for each sample. The variation in the results so obtained is believed to give a good indication of the variation that might be expected if genuine replicate samples were available. Other approaches to checking for significance are discussed in Miller (1984).

Aitkin (1974) suggests an approach that may be used to identify subsets of explanatory variates which ought to be excluded from consideration. Miller (1982) comments on a method due to Spjötvoll (1972) that may be used to test whether one set of explanatory variates is significantly better than another.

Approaches to variate selection are discussed in Cox and Snell (1974), Draper and Smith (1981), Seber (1977, ch. 12), Hocking (1976 and 1983), Thompson (1978), Berk (1978), and Belsey et al. (1980).

The Uses of Residuals

Following any analysis one should routinely examine (1) a plot of residuals against fitted values and (2) a normal probability plot of residuals. This is a cumulative frequency plot of residuals, with the scale used for the cumulative frequencies transformed so that equal increments on the scale correspond to equal changes in the equivalent normal deviate.

Plot 1 will help in detecting cases where the y-variate should be transformed, or where the variance σ^2 changes systematically with the fitted y-value. It is undesirable to use observed in place of fitted values on the horizontal axis–the correlation between observed and fitted values induces a linear trend which complicates interpretation of the plot. Plot 2 will assist in detecting residuals associated with aberrant data points, or other types of non-normality in the distribution of values of the residuals.

Note, however, that the presence of two or more sizable outliers may so distort the fitted model that the calculated residuals are no longer a reliable guide in looking for aberrant data points. The robust regression methods discussed in Section 8.2 are designed to avoid this problem by fitting the model in such a way that data points with large residuals are given a reduced weight. Fits obtained using such robust regression methods provide a useful check on the results from standard linear least squares regression calculations. If the two regressions (robust and nonrobust) differ substantially, careful scrutiny of the data and/or the model used is called for.

Section 3.5 and 3.6 will discuss the use of *partial regression plots*. These may help decide whether a particular variate z should be included as an explanatory variate. For this determine the residuals from the regression of y on all explanatory variates except z, and plot these against the residuals from the regression of z on the same explanatory variates. A nonlinear relationship may be suggested, in which case one should look for some function $f(.)$ such that $f(z)$ enters into the model in a more nearly linear fashion. Nonlinearity will, however, show up better if deviations from linearity in the plot of partial residuals are plotted in the y-direction. Deviations from linearity in the plot of partial residuals are the residuals in the full model. It is good practice to examine such a plot for each explanatory variate finally included in the model. Sections 3.5 and 3.6 demonstrate the calculations.

Practical aspects of the use of plots of residuals are discussed in Cox and Snell (1974), Daniel and Wood (1980), Mosteller and Tukey (1977), Draper and Smith (1981), Atkinson (1982), and Allen and Cady (1982).

2.8. ELECTRICITY CONSUMPTION EXAMPLE

The data in Table 2.1 was obtained in a study of domestic electricity consumption in various British cities and towns over the period 1937–38. The CSSP matrix is given, together with variate means, in Table 2.2. The Cholesky decomposition and its inverse is given in Table 2.3. This will allow the reader to check through the separate steps of the calculations.

Table 2.1. Data for Electricity Consumption in Yorkshire, Cheshire, and Lancashire

	x_1	x_2	x_3	x_4	x_5	y
1	2.45	0.68	1.02	0.60	1.14	2.73
2	2.61	0.70	0.95	0.46	1.07	2.88
3	2.86	0.70	0.82	0.86	0.95	3.30
4	2.51	0.68	0.65	0.89	1.50	2.74
5	2.51	0.70	1.02	0.45	1.03	2.63
6	2.71	0.70	1.00	0.41	0.40	3.02
7	2.55	0.63	0.78	1.40	1.31	3.02
8	2.50	0.66	0.68	1.11	1.54	2.91
9	2.62	0.70	0.86	0.48	1.78	2.95
10	2.85	0.70	0.88	1.03	1.56	3.28
11	2.70	0.70	0.99	0.83	1.75	3.12
12	2.60	0.70	0.97	1.26	0.90	3.12
13	2.53	0.81	0.94	1.07	1.59	2.85
14	2.75	0.70	0.84	0.15	0.93	2.93
15	2.57	0.52	0.78	0.96	1.24	2.96
16	2.59	0.70	0.96	0.98	0.72	2.89
17	3.15	0.70	0.88	0.23	0.34	3.50

Source: From a larger dataset given in Houthakker (1951).
Note: Variates are used with permission.
x_1 = log(average annual household income, in pounds)
x_2 = log(marginal price of electricity, in pence per 10 kwh)
x_3 = log(marginal price of gas, in pence per therm)
x_4 = log(average holdings of heavy domestic electrical equipment, in tenths of a kilowatt)
x_5 = log(average number of consumers, in thousands)
y = log(average annual electricity consumption, in kilowatts)

Table 2.2. Variate Means and CSSP Matrix for the Data of Table 2.1

	x_1	x_2	x_3	x_4	x_5	y
Mean						
	2.651	0.6871	0.8835	0.7747	1.162	2.99
CSSP Matrix						
x_1	0.4913	0.02113	0.00996	−0.3990	−0.5442	0.5511
x_2		0.04895	0.04218	−0.05687	0.001188	0.00130
x_3			0.2070	−0.2007	−0.25431	−0.02320
x_4				2.1332	1.0614	0.02390
x_5					2.9998	−0.4258
y						0.7898

Table 2.3. Cholesky Upper Triangle Matrix and Its Inverse

Constant	x_1	x_2	x_3	x_4	x_5	y
Cholesky Upper Triangle Matrix						
4.123	10.929	2.833	3.643	3.194	4.790	12.328
	0.701	0.030	0.014	−0.569	−0.776	0.786
		0.219	0.190	−0.181	0.112	−0.102
			0.413	−0.383	−0.641	−0.036
				1.277	0.309	0.344
					1.371	0.049
						0.198
Inverse of Upper Triangle Matrix						
0.243	−3.782	−2.615	−0.804	−2.905	−2.497	4.078
	1.427	−0.196	0.041	0.621	0.704	−7.009
		4.562	−2.104	0.016	−1.361	2.277
			2.422	0.726	0.969	−1.058
				0.784	−0.179	−1.317
					0.730	−0.179
						5.050

Source: Derived from the CSSP matrix given in table 2.2.

When y is regressed on all of the variates $x_1, x_2, \ldots, x_5$, the coefficients, their standard errors, and the t-statistics (for testing the null hypothesis that the true coefficient is zero), are as follows:

	Coefficient	SE	t-statistic
constant	−0.808	0.361	−2.24
x_1	1.39	0.103	13.5
x_2	−0.451	0.311	−1.45
x_3	0.210	0.162	1.30
x_4	0.261	0.0479	5.44
x_5	0.0355	0.0436	0.82

The adjusted squared multiple correlation (or proportion of the estimated variance which is explained) is

$$1 - \frac{0.198^2/(17 - 5 - 1)}{0.7898/(17 - 1)} = 0.928$$

Table 2.4. Cholesky Upper Triangle Matrix and Its Inverse Without Rows and Columns 2, 3, and 5 of the CSSP Matrix in Table 2.2

Constant	x_1	x_4	y
Cholesky Upper Triangle Matrix			
4.123	10.929	3.194	12.328
	0.701	−0.569	0.786
		1.345	0.350
			0.221
Inverse of Upper Triangle Matrix			
0.243	−3.782	−2.176	3.377
	1.427	0.604	−6.037
		0.743	−1.180
			4.527

Three coefficients (those of x_2, x_3, and x_5) stand out as having small t-statistics (not even significant at the 10% level). It is worth trying the effect of omitting all three of these variates. For this rows 2, 3, and 5 must be omitted from the CSSP matrix. Table 2.4 shows the Cholesky decomposition and its inverse for this modified CSSP matrix. Coefficients, standard errors, and t-statistics are now:

	Coefficient	SE	t-statistic
constant	−0.746	0.258	−2.89
x_1	1.333	0.0915	14.6
x_4	0.261	0.0439	5.94

The adjusted squared multiple correlation is now

$$1 - \frac{0.221^2/(17 - 2 - 1)}{0.7898/(17 - 1)} = 0.929$$

Using the squared multiple correlation as a measure, this regression does as well as when all candidate explanatory variates are taken. It is thus near to the best possible. (Why?) Various plots of partial residuals may be used as discussed in Section 3.5 to check whether it is appropriate to enter the variates into the regression in linear form. (For example, it might be desirable to include x_1^2 as well as or instead of x_1.)

It is easy to find examples where the simple-minded approach used here will not work. It may be acceptable to omit one or two, but not all three, of the variates whose coefficients are not statistically significant (or otherwise judged important). Or the proper candidates for omission may be two or more of the variates whose coefficients are currently significant.

2.9. SOME CHECKS FOR COMPUTER PROGRAMS TO INCLUDE

Problems that may arise in computer programs for regression are best illustrated by reference to a particular set of data values:

Observation Number	x_1	x_2	x_3	x_4	x_5	y
1	0	−1	0	0.1	2.1	4
2	1	1	−4	3.9	2.1	1
3	5	3	−2	−2.1	2.1	6
4	4	1	2	2.0	2.1	8
5	3	2	−3	−3.1	2.1	3
6	3	0	3	3.0	2.1	4
7	4	0	5	4.9	2.1	7

The following points should be noted:

1. $x_5 = 2.1$.
2. $x_3 = 2x_1 - 3x_2 - 3$.
3. $x_4 \simeq 2x_1 - 3x_2 - 3$ (the estimated relationship is $\hat{x}_4 = 1.96x_1 - 3x_2 - 2.89$).

Features 1 and 2 are likely, with many of the programs that are in common use, to lead either to a run-time error or to nonsensical results.

A Constant Variate

It does not make sense to include $x_5 \equiv 2.1$ as an explanatory variate in a regression equation where the constant term is already catered for quite separately. Nevertheless, users of computer programs do this kind of thing from time to time, and it is desirable that computer programs should be robust against it. One has

$$y = b_0 + b_1x_1 + \cdots + b_4x_4 + b_5 + e$$

The coefficients b_0 and b_5 cannot be determined separately. A suitable recourse is to set $b_5 = 0$.

If calculations are based on the CSSP matrix, and if calculations were exact, one would have

$$\mathbf{s}_{55} = \sum_i (x_{i5} - \bar{x}_5)^2 = 0$$

$$\mathbf{s}_{5j} = \sum_i (x_{i5} - \bar{x}_5)(x_{ij} - \bar{x}_j) = 0 \qquad (9.1)$$

Following the notation we have been using, the corresponding diagonal element of the Cholesky upper triangle matrix **T** which results is written t_{55}. If all $\mathbf{s}_{5j}$ are zero, then t_{55} will be calculated as 0, leading to division by zero upon attempting to calculate t_{5y} in the usual way.

In practice, because of rounding error and the manner in which they are calculated, the quantities $\mathbf{s}_{55}$ and $\mathbf{s}_{5j}$ may not be zero. If the formula

$$\mathbf{s}_{55} = \sum x_{i5}^2 - n\bar{x}_5^2$$

is used, then $\mathbf{s}_{55}$, as calculated, may be a small (in magnitude) positive or negative number. In the absence of checks the consequence will be:

1. An attempt to calculate t_{55} as the square root of a small negative number, leading to program failure.
2. The calculation of t_{55} as a small nonsense number. If also $t_{5y} \neq 0$ this will lead to a nonsense value for b_5 and hence to nonsense values for earlier regression coefficients.

A suitable procedure is to determine maximum and minimum values for each variate and to check that the difference is in each case positive. It is desirable to have this information available for other reasons also; it may usefully be included with the details printed out in connection with each individual variate.

Linear Relations between the Explanatory Variates

To simplify the discussion, we omit variates x_4 and x_5 from consideration and take

$$\hat{y} = b_0 + b_1 x_1 + b_2 x_2 + b_3 x_3$$

Then also

$$\hat{y} = b_0 + b_1 x_1 + b_2 x_2 + b_3 x_3 + k(x_3 - 2x_1 + 3x_2 + 3)$$
$$= (b_0 + 3k) + (b_1 - 2k)x_1 + (b_2 + 3k)x_2 + (b_3 + k)x_3$$

where k is arbitrary. We can rewrite this:

$$\hat{y} = b_0^* + b_1^* x_1 + b_2^* x_2 + b_3^* x_3$$

Unless one of the b_0^*, b_1^*, b_2^*, and b_3^* is fixed, it is impossible to determine the others.

Suppose now that the CCDA (see Section 1.5) is used to reduce the CSSP matrix to upper triangular form **T**. Then

$$t_{kk} = \left(s_{kk}^{(k-1)}\right)^{1/2}$$

where, following the notation of Chapter 3,

$$s_{kk}^{(k-1)} = \mathbf{s}_{kk} - \sum_{i=1}^{k-1} t_{ik}^2 \tag{9.2}$$

From eq. (1.5), t_{kk}^2 is the residual sum of squares from the regression of x_k on earlier variates. Thus it is zero if and only if x_k is an exact linear combination of earlier variates. Thus in the preceding example $s_{33}^{(2)} = 0$, and hence $t_{33} = 0$. In practice, with calculations which are not exact, the possibilities are:

1. $s_{33}^{(2)}$ is calculated as a small (in magnitude) negative number; this will lead to an invalid square root argument.
2. $t_{33} = 0$; this will lead to a divide by zero in the attempt to calculate other elements in the same row of $\boldsymbol{T}$.
3. t_{33} is calculated as a small positive number. If also t_{3y} is nonzero, this will lead to nonsensical results.

A criterion is needed for deciding when a calculated t_{kk} is to all intents zero. Suppose first that the decision is made purely on grounds of numerical precision. Suppose the computer is one that retains the h most significant decimal digits, or the equivalent in another base, in the result of any arithmetic operation. Then any digit of $s_{kk}^{(k-1)}$ in eq. (9.2) with a place value of $10^{-h}\mathbf{s}_{kk}$ or less will certainly be meaningless. Unless $s_{kk}^{(k-1)}$ is somewhat larger than $10^{-h}\mathbf{s}_{kk}$ (e.g., at least 10 times as large), t_{kk} should be set to zero. (As noted later all other elements in that row of $\boldsymbol{T}$ will then also be set to zero.) For further discussion see Sections 3.4, 3.9, and 3.12.

Near-Linear Relations between Explanatory Variates

It is helpful to bear in mind the interpretation

$$s_{kk}^{-1}t_{kk}^2 = 1 - R_{k(1,\dots,k-1)}^2 \quad \text{by eq. (1.7)}$$

With data such as that of the example in Section 2.6, an $R_{k(1,\dots,k-1)}^2$ greater than about 0.99 would suggest that x_k is for all practical purposes a linear combination of earlier variates. The variate x_4 in the example in the present section, for which $R_{4(1,2)}^2 = 0.9998$, should clearly be regarded as a linear combination of x_1 and x_2. If x_1 and x_2 are to be retained, then x_4 should be omitted. It is unlikely that the small amount of its variation which cannot be explained in terms of x_1 and x_2 will be important for the regression, though this should be checked by examining its contribution to the analysis of variance table. It must be emphasized that it is not possible, given this data, to predict y when x_3 differs substantially from $2x_1 - 3x_2 - 3$.

The *variance inflation factor* (VIF) discussed in Section 2.2 provides another perspective on linear dependence. Observe that

$$R_{k(1,\dots,k-1,k+1,\dots,p)}^2 \geq R_{k(1,\dots,k-1)}^2$$

so that

$$\text{VIF}(b_k) \geq \left(1 - R_{k(1,\dots,k-1)}^2\right)^{-1}$$

Clearly, a variance inflation factor as large as, say, 1000 calls for scrutiny.

Near-linear dependence between explanatory variates leads also to a loss of numerical precision in the calculated regression coefficients, particularly when calculations are based on the normal equations. The discussion following eq. (9.2) shows how near-linear dependence, as evidenced by a value for $s_{kk}^{(k-1)}$ which is small relative to $\mathbf{s}_{kk}$, will mean that some (or in extreme cases all) digits in the calculated result are meaningless numerically.

Recall that

$$\mathbf{s}_{kk}^{-1}s_{kk}^{(k-1)} = 1 - R_{k(1,\dots,k-1)}^2$$

and that

$$\left(\mathbf{s}_{kk}s^{kk}\right)^{-1} = 1 - R_{k(1,\dots,k-1,k+1,\dots,p)}^2$$

Solving for b_i When $t_{ii=0}$

The solution of an upper triangular set of equations where one or more t_{ii} is zero can proceed without hitch if one sets $b_i = 0$, whenever $t_{ii} = 0$. A general form of solution, corresponding to the arbitrary choice of all such b_i, may then be obtained in the manner to be demonstrated in Section 3.9. Section 3.8 shows how to handle the calculation of variances and covariances when one or more of the b_i is arbitrarily set to zero.

How Linear Relations or Near-Linear Relations Arise

Consider a case where components with masses a, b, c, d, and e are available for examination only as part of larger pieces of equipment. As a result the following rough estimates of masses of combinations of these components are available:

$$a + b + c + d = 17.5 \quad \text{(g)}$$

$$b + c + d + e = 13$$

$$a + b = 12.4$$

$$b + c = 7.4$$

$$c + d = 5.7$$

$$d + e = 5.2$$

These equations give information on $a + b$, $b + c$, $c + d$, and $d + e$, but not on a, b, c, d and e individually. Here

$$\begin{array}{cc} & \begin{array}{ccccc} a & b & c & d & e \end{array} \\ \boldsymbol{X} = & \begin{bmatrix} 1 & 1 & 1 & 1 & 0 \\ 0 & 1 & 1 & 1 & 1 \\ 1 & 1 & 0 & 0 & 0 \\ 0 & 1 & 1 & 0 & 0 \\ 0 & 0 & 1 & 1 & 0 \\ 0 & 0 & 0 & 1 & 1 \end{bmatrix} \end{array}$$

If the columns of $\boldsymbol{X}$ are multiplied alternately by $+1$ and -1 and added, the result is a vector of zeros. This reflects the fact that there are more parameters than can in principle be estimated from the information provided.

It is instructive to carry through the calculations for the least squares estimation of a, b, c, d, and e. The general form of solution may be obtained as in the example in Section 3.9.

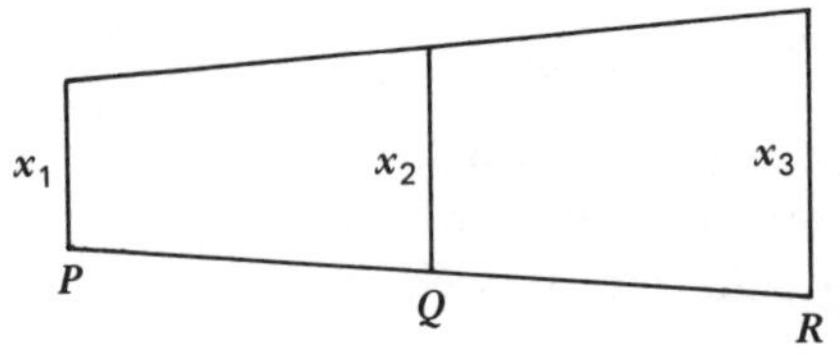

Figure 2.2. Linear dependence in biological data.

Similar problems arise when mistakes in design, or accidents, mean that an experiment does not yield all the information that was intended. An errant cow, an error in the application of treatments, or a hailstorm may destroy vital parts of the evidence in a field experiment so that information on some comparisons of interest is unavailable. Any analysis of the results should go as far as possible in estimating the quantities of interest and should identify those quantities that cannot be estimated individually.

The problem arises even where variates measure quantities that vary on a continuous scale. If, in order to describe differences of shape, a large number of measurements (variate values) are obtained on individuals from different species of shark, some variates are likely to be very close to a linear combination of others. As an extreme instance, suppose that in the species being studied distances through the body vary along the body length in the hypothetical manner shown in Fig. 2.2. If Q is halfway between P and R, then $x_2 = \frac{1}{2}(x_1 + x_3)$, whatever the ratio x_3/x_1.

Even if PQR and the similar line on the upper surface are curved rather than straight, it is still plausible that $x_2 \simeq c_1x_1 + c_3x_3$, for suitable constants c_1 and c_3. Because measurements are made to limited accuracy only, an underlying approximate relationship may become an exact relationship between the measured variate values. This is particularly likely if the number of data points is small.

*2.10. VARIANTS OF THE CHOLESKY DECOMPOSITION

For the example of Section 1.5

$$\boldsymbol{S} = \begin{bmatrix} 4 & 6 & 10 \\ \cdot & 58 & 29 \\ \cdot & \cdot & 38 \end{bmatrix}, \quad \boldsymbol{T} = \begin{bmatrix} 2 & 3 & 5 \\ 0 & 7 & 2 \\ 0 & 0 & 3 \end{bmatrix}$$

where $\boldsymbol{T}'\boldsymbol{T} = \boldsymbol{S}$. Let $\boldsymbol{D}^{1/2}$ be a diagonal matrix with its ith diagonal element equal to t_{ii}; thus

$$\boldsymbol{D}^{1/2} = \begin{bmatrix} 2 & 0 & 0 \\ 0 & 7 & 0 \\ 0 & 0 & 3 \end{bmatrix}, \quad \boldsymbol{D} = \begin{bmatrix} 4 & 0 & 0 \\ 0 & 49 & 0 \\ 0 & 0 & 9 \end{bmatrix}$$

Then we may write

$$S = LU = \begin{bmatrix} 1 & 0 & 0 \\ \frac{3}{2} & 1 & 0 \\ \frac{5}{2} & \frac{2}{7} & 1 \end{bmatrix} \begin{bmatrix} 4 & 6 & 10 \\ 0 & 49 & 14 \\ 0 & 0 & 9 \end{bmatrix}$$

Alternatively, $\boldsymbol{S} = \boldsymbol{LDL}'$, where $\boldsymbol{L} = \boldsymbol{D}^{-1/2}\boldsymbol{T}'$, $\boldsymbol{U} = \boldsymbol{D}^{1/2}\boldsymbol{T} = \boldsymbol{DL}'$. Either of the versions of the Cholesky algorithm which we have considered are readily modified to form $\boldsymbol{U}$ (with diagonal elements t_{ii}^2), or $\boldsymbol{L}'$ (with diagonal elements 1.0). Such versions of the Cholesky decomposition are *square root free*.

Used with a general square matrix $\boldsymbol{S}$, the Crout algorithm yields $\boldsymbol{S} = \boldsymbol{U}'\boldsymbol{L}'$. The Doolittle algorithm yields $\boldsymbol{S} = \boldsymbol{LU}$. (See Dahlquist and Bjorck, 1974, pp. 157–158.) When modified to take advantage of the symmetry of $\boldsymbol{S}$, these algorithms may in either case be regarded as square-root-free versions of the CCDA, the version of the Cholesky algorithm that forms each row of $\boldsymbol{T}$ in a single step.

Suppose $\boldsymbol{S} = \mathbf{X}'\mathbf{X}$. Then a variety of orthogonalization methods are available that form $\boldsymbol{T}$ directly from $\mathbf{X}$; brief details will be given in the next chapter. Again square-root-free versions are available. Such methods are more expensive computationally than methods that form $\boldsymbol{S}$ and work from it. However, in cases where later columns of $\mathbf{X}$ are strongly dependent (as measured by the squared multiple correlation) on earlier columns, they give better precision. Orthogonalization methods will be discussed in Chapter 4.

*2.11. THE GAUSS-JORDAN ALGORITHM

The Gauss-Jordan algorithm is so widely used in handling regression and related multivariate calculations that it is desirable to include an account of it here. Calculations are based on an SSP or CSSP matrix $\mathbf{X}'\mathbf{X}$, where as always $\mathbf{X}$ consists of an initial set of columns corresponding to explanatory variates, followed by one or more columns of dependent variates.

The successive steps in the calculation may be viewed as successive steps in the evaluation of $(\mathbf{X}'\mathbf{X})^{-1}$. However, in general, calculations will cease with the evaluation of the inverse of some submatrix $\boldsymbol{X}'\boldsymbol{X}$ of $\mathbf{X}'\mathbf{X}$. Details are then available of the regression coefficient for the regression of any of the remaining columns of $\mathbf{X}$ upon columns of $\boldsymbol{X}$, and of the matrix of sums of squares and products of residuals from these regressions.

Consider the partitioning $\mathbf{X} = [\boldsymbol{X}_{(1)}, \boldsymbol{X}_{(2)}]$ which corresponds in a manner that will shortly become clear to an intermediate stage of the calculations. It is not necessary that the columns of **X** be taken in their original order. Let

$$\boldsymbol{S} = \begin{bmatrix} \boldsymbol{S}_{11} & \boldsymbol{S}_{12} \\ \cdot & \boldsymbol{S}_{22} \end{bmatrix}$$

where $\boldsymbol{S}_{11} = \boldsymbol{X}'_{(1)}\boldsymbol{X}_{(1)}$, $\boldsymbol{S}_{12} = \boldsymbol{X}'_{(1)}\boldsymbol{X}_{(2)}$, and $\boldsymbol{S}_{22} = \boldsymbol{X}'_{(2)}\boldsymbol{X}_{(2)}$. Then at this stage of the calculations the following information is provided:

1. $\boldsymbol{S}_{11}^{-1}$, relevant to calculation of variances and covariances in the regression of a column of $\boldsymbol{X}_{(2)}$ upon columns of $\boldsymbol{X}_{(1)}$.
2. $\boldsymbol{S}_{11}^{-1}\boldsymbol{S}_{12}$. Each column of this matrix gives the regression coefficients for the regression of the corresponding column of $\boldsymbol{X}_{(2)}$ upon columns of $\boldsymbol{X}_{(1)}$.
3. $\boldsymbol{S}_{22.1} = \boldsymbol{S}_{22} - \boldsymbol{S}_{21}\boldsymbol{S}_{11}^{-1}\boldsymbol{S}_{12}$. This is the matrix of sums of squares and products of residuals from the preceding regressions.

Now set

$$\boldsymbol{G}^{(k)} = \begin{bmatrix} \boldsymbol{S}_{11}^{-1} & \boldsymbol{S}_{11}^{-1}\boldsymbol{S}_{12} \\ \cdot & \boldsymbol{S}_{22.1} \end{bmatrix}$$

where the k is designed to indicate that $\boldsymbol{X}_{(1)}$ has columns corresponding to variables denoted, at this point, $0,\ldots,k$. Let $g_{ij}^{(k)}$ be the (i, j)th element of $\boldsymbol{G}^{(k)}$. Then, supposing that it was the initial column of the previous version of $\boldsymbol{X}_{(2)}$ which was last appended to $\boldsymbol{X}_{(1)}$, the following is the rule by which $\boldsymbol{G}^{(k)}$ is formed from $\boldsymbol{G}^{(k-1)}$:

$$g_{ij}^{(k)} = g_{ij}^{(k-1)} + \frac{g_{ik}^{(k-1)}g_{jk}^{(k-1)}}{g_{kk}^{(k-1)}}, \qquad i < k,\ j < k$$

$$g_{ij}^{(k)} = g_{ij}^{(k-1)} - \frac{g_{ik}^{(k-1)}g_{kj}^{(k-1)}}{g_{kk}^{(k-1)}}, \qquad i < k,\ j > k$$

$$g_{ik}^{(k)} = -\frac{g_{ik}^{(k-1)}}{g_{kk}^{(k-1)}}, \qquad i < k \ (\text{and}\ j = k)$$

$$g_{kk}^{(k)} = \left(g_{kk}^{(k-1)}\right)^{-1},$$

$$g_{kj}^{(k)} = \frac{g_{kj}^{(k-1)}}{g_{kk}^{(k-1)}}, \qquad j > k \ (\text{and}\ i = k)$$

$$g_{ij}^{(k)} = g_{ij}^{(k-1)} - \frac{g_{ki}^{(k-1)}g_{kj}^{(k-1)}}{g_{kk}^{(k-1)}}, \qquad i > k,\ j > k.$$

A proof of this result is left to the reader (or see Jowett, 1963). It is sufficient to work with the elements of the upper triangle of $G^{(k)}$. The *pivotal* element $g_{kk}^{(k-1)}$ is algebraically identical to the quantity $s_{kk}^{(k-1)}$ which appears in Section 3.3. The same type of check that it is positive, and differs from zero by an amount greater than any numerical or statistical imprecision, is needed here also. The formulae stated here are summarized in Fig. 2.3.

Example

Calculations are for the CSSP matrix **S** in the example of Section 1.9:

$$\begin{array}{ccc} x_1 & x_2 & y \end{array}$$
$$\begin{bmatrix} 49 & 14 & 42 \\ \cdot & 13 & 15 \\ \cdot & \cdot & 41 \end{bmatrix}$$

$$\downarrow \text{Pivot} = g_{11}^{(0)} = 49$$

$$\begin{bmatrix} \dfrac{1}{49} & \dfrac{14}{49} & \dfrac{42}{49} \\ \cdot & 13 - \dfrac{14 \times 14}{49} & 15 - \dfrac{14 \times 42}{49} \\ \cdot & \cdot & 41 - \dfrac{42 \times 42}{49} \end{bmatrix} = \left[\begin{array}{c|cc} \dfrac{1}{49} & \dfrac{2}{7} & \dfrac{6}{7} \\ \hline \cdot & 9 & 3 \\ \cdot & \cdot & 5 \end{array}\right]$$

$$\downarrow \text{Pivot} = g_{22}^{(1)} = 9$$

$$\begin{bmatrix} \dfrac{1}{49} + \left(\dfrac{2}{7}\right)^2 \times \dfrac{1}{9} & -\dfrac{2}{7} \times \dfrac{1}{9} & \dfrac{6}{7} - \dfrac{2}{7} \times 3 \times \dfrac{1}{9} \\ \cdot & \dfrac{1}{9} & \dfrac{3}{9} \\ \cdot & \cdot & 5 - 3^2 \times \dfrac{1}{9} \end{bmatrix}$$

$$= \left[\begin{array}{cc|c} \dfrac{13}{441} & -\dfrac{2}{63} & \dfrac{16}{21} \\ \cdot & \dfrac{1}{9} & \dfrac{1}{3} \\ \hline \cdot & \cdot & 4 \end{array}\right]$$

Assuming that interest is in the regression of y upon x_1 and x_2, this is the point where the calculations stop. The elements in the final column are:

$b_1 = \frac{16}{21}$ (with variance $\frac{13}{441}\sigma^2$).
$b_2 = \frac{1}{3}$ (with variance $\frac{1}{9}\sigma^2$).

Residual sum of squares = 4.

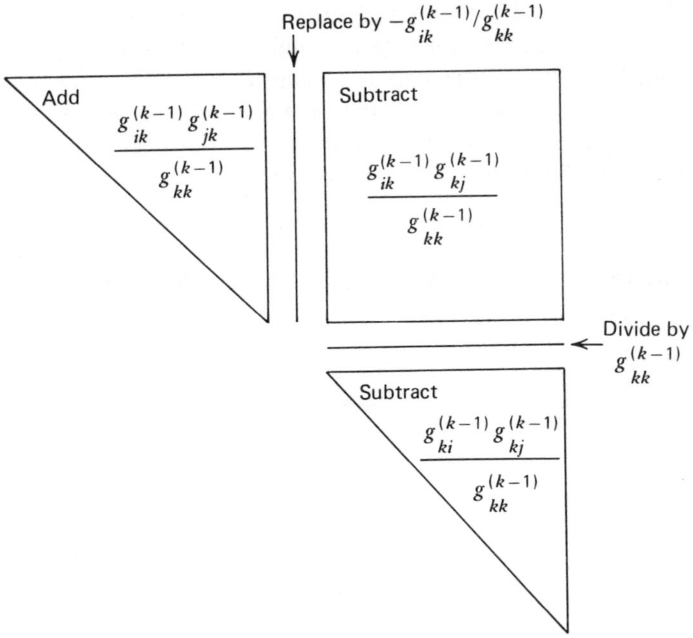

Figure 2.3. The Gauss-Jordan algorithm.

Stepping Explanatory Variates In and Out

Modification of the preceding formulae to handle the case where the variate to be taken next is x_ℓ rather than x_k ($\ell > k$) is straightforward. The pivot element is then $g_{\ell\ell}^{(k-1)}$. An obvious possibility is to shuffle elements of $\boldsymbol{G}^{(k-1)}$ around so that x_ℓ now corresponds to the kth row and column. Alternatively, the ordering of rows and columns may be left unchanged, and a record kept identifying the columns of $\boldsymbol{G}^{(k)}$ that correspond to columns of $\boldsymbol{X}_{(1)}$. Breaux (1968) suggests a possible approach.

The effect of removing a column from $\boldsymbol{X}_{(1)}$ and placing it back in $\boldsymbol{X}_{(2)}$ may be achieved by using in reverse the formulae given earlier. A careful check should be kept on cases where the associated subtractions may lead to a serious loss of precision in the elements of $\boldsymbol{S}_{11}^{-1}$.

The Gauss-Jordan algorithm gives what is easily the simplest available method for handling the stepping in and out of explanatory variates. Jennrich (1977) gives full computational details of its use in stepwise regression. There is a brief discussion of the Gauss-Jordan algorithm, comparing it with the sequential Cholesky decomposition algorithm and showing how that algorithm may be used for stepwise regression, in Maindonald (1977).

*2.12. WEIGHTED LEAST SQUARES

Simple Weighted Least Squares

Consider the model $\boldsymbol{y} = \boldsymbol{X\beta} + \boldsymbol{\varepsilon}$, where as before the ε_i are independently distributed with mean 0, but now $\mathrm{var}(\varepsilon_i) = \sigma_i^2$. It is assumed that σ_i^2 is known to within a constant multiplier. In order to apply least squares methods that assume constant variance, it is only necessary to transform the model equation to a form

$$\boldsymbol{y}^* = \boldsymbol{X}^*\boldsymbol{\beta} + \boldsymbol{\varepsilon}^* \tag{12.1}$$

where the elements ε_i^* of $\boldsymbol{\varepsilon}^*$ have constant variance c. This may be achieved by multiplying the ith row of the model equation through by $\sqrt{w_i}$, where for some constant c, $w_i = c\sigma_i^{-2}$. This replaces y_i by $y_i^* = \sqrt{w_i}\,y_i$, and ε_i by $\varepsilon_i^* = \sqrt{w_i}\,\varepsilon_i$; observe that

$$\mathrm{var}(y_i^*) = \mathrm{var}(\varepsilon_i^*) = c \tag{12.2}$$

In matrix notation

$$\boldsymbol{y}^* = \boldsymbol{W}^{1/2}\boldsymbol{y}, \quad \boldsymbol{X}^* = \boldsymbol{W}^{1/2}\boldsymbol{X}, \quad \boldsymbol{\varepsilon}^* = \boldsymbol{W}^{1/2}\boldsymbol{\varepsilon}$$

where $\boldsymbol{W}^{1/2}$ is a diagonal matrix with its ith diagonal element equal to $\sqrt{w_i}$.

Observe that in the model (12.1) the error variance estimate based on the residual sum of squares is an estimate of c.

Example

Consider the data of Section 1.9; however, now the ith row is taken with weight proportional to i^{-2}. Then

$$[\boldsymbol{X}, \boldsymbol{y}] = \begin{bmatrix} 1 & -2 & 0 & -3 \\ 1 & -1 & 2 & 1 \\ 1 & 2 & 5 & 2 \\ 1 & 7 & 3 & 6 \end{bmatrix}, \quad [\boldsymbol{X}^*, \boldsymbol{y}^*] = \begin{bmatrix} 1 & -2 & 0 & -3 \\ \frac{1}{2} & -\frac{1}{2} & 1 & \frac{1}{2} \\ \frac{1}{3} & \frac{2}{3} & \frac{5}{3} & \frac{2}{3} \\ \frac{1}{4} & \frac{7}{4} & \frac{3}{4} & \frac{3}{2} \end{bmatrix}$$

Reduction of $[\boldsymbol{X}^*, \boldsymbol{y}^*]$ to upper triangular form yields

$$\begin{bmatrix} \boldsymbol{T}_p, & \boldsymbol{t}_y \\ \boldsymbol{0}', & \boldsymbol{t}_{yy} \end{bmatrix} = \begin{bmatrix} 1.1932 & -1.3328 & 1.0418 & -1.8043 \\ 0 & 2.4455 & 1.3544 & 2.6230 \\ 0 & 0 & 1.1918 & 0.8921 \\ 0 & 0 & 0 & 1.0064 \end{bmatrix}$$

Solution of $\boldsymbol{T}_p \boldsymbol{b} = \boldsymbol{t}_y$ yields

$$b_0 = -1.4307, \quad b_1 = 0.6580, \quad b_2 = 0.7485$$

The analysis of variance table is:

Due to constant term	$1.8043^2 = 3.255$
Due to x_1 (given constant)	$2.6230^2 = 6.880$
Due to x_2 (given constant, x_1)	$0.8921^2 = 0.796$
Residual SS	$1.0064^2 = 1.013$ ($s^2 = 1.013/(4-3)$)

Thus the estimated variance for the ith data point y_i is $1.013i^{-2}$.

More General Types of Weighting

Let $\boldsymbol{W}$ be a positive definite symmetric matrix, and let $\boldsymbol{W}^{1/2}$ be any matrix such that $\boldsymbol{W}'^{1/2}\boldsymbol{W}^{1/2} = \boldsymbol{W}$. Then minimization of

$$(\boldsymbol{y} - \boldsymbol{X}\boldsymbol{b})'\boldsymbol{W}(\boldsymbol{y} - \boldsymbol{X}\boldsymbol{b})$$

is equivalent to minimization of

$$(\boldsymbol{y}^* - \boldsymbol{X}^*\boldsymbol{b})'(\boldsymbol{y}^* - \boldsymbol{X}^*\boldsymbol{b})$$

where $\boldsymbol{y}^* = \boldsymbol{W}^{1/2}\boldsymbol{y}$, $\boldsymbol{X}^* = \boldsymbol{W}^{1/2}\boldsymbol{X}$. The matrix $\boldsymbol{W}^{1/2}$ is conveniently taken to be the Cholesky decomposition of $\boldsymbol{W}$, that is, $\boldsymbol{W}^{1/2} = \boldsymbol{U}$, where $\boldsymbol{U}$ is upper triangular such that $\boldsymbol{U}'\boldsymbol{U} = \boldsymbol{W}$.

If $\text{var}(\boldsymbol{y}) = \text{var}(\boldsymbol{\varepsilon}) = c\boldsymbol{W}^{-1}$, it then follows that

$$\text{var}(\boldsymbol{y}^*) = \text{var}(\boldsymbol{\varepsilon}^*) = c\boldsymbol{I}_n$$

The foregoing method has the virtue of simplicity but is not a good one to use if $\boldsymbol{W}$ is near singular or singular. See Paige (1979) for an approach that will work in general.

*Weighted and Unweighted Least Squares with the Same Estimates

The condition for minimization of $(\boldsymbol{y} - \boldsymbol{X}\boldsymbol{b})'\boldsymbol{W}(\boldsymbol{y} - \boldsymbol{X}\boldsymbol{b})$ to give the same parameter estimates as minimization of $(\boldsymbol{y} - \boldsymbol{X}\boldsymbol{b})'(\boldsymbol{y} - \boldsymbol{X}\boldsymbol{b})$ is that, for some matrix $\boldsymbol{C}$, $\boldsymbol{W}\boldsymbol{X} = \boldsymbol{X}\boldsymbol{C}$. For example, this condition is satisfied if $\boldsymbol{X}$ has an initial column of ones and $\boldsymbol{W}^{-1}$ has elements $w^{ii} = \tau^2$, $w^{ij} = \rho\tau^2$ for $i \neq j$. Exercise 13 at the end of the chapter gives a more general form of matrices $\boldsymbol{W}$ such that $\boldsymbol{W}\boldsymbol{X} = \boldsymbol{X}\boldsymbol{C}$. Rao (1967) gives necessary and sufficient conditions that $\boldsymbol{W}\boldsymbol{X} = \boldsymbol{X}\boldsymbol{C}$.

If $\boldsymbol{W}$ and/or $\boldsymbol{X}$ is not of full rank, the preceding statement needs to be modified slightly. Suppose $\boldsymbol{W}\boldsymbol{X} = \boldsymbol{X}\boldsymbol{C}$. Then, quite generally,

$$\boldsymbol{X}'\boldsymbol{X}\boldsymbol{b} = \boldsymbol{X}'\boldsymbol{y} \quad \text{implies} \quad \boldsymbol{C}'\boldsymbol{X}'\boldsymbol{X}\boldsymbol{b} = \boldsymbol{C}'\boldsymbol{X}'\boldsymbol{y}$$

and hence that $\boldsymbol{X}'\boldsymbol{W}\boldsymbol{X}\boldsymbol{b} = \boldsymbol{X}'\boldsymbol{W}\boldsymbol{y}$. Thus solutions to the unweighted least squares problem are included among those to the weighted least squares problem. Conversely, suppose that every solution $\boldsymbol{b}$ of $\boldsymbol{X}'\boldsymbol{X}\boldsymbol{b} = \boldsymbol{X}'\boldsymbol{y}$ is also a solution of $\boldsymbol{X}'\boldsymbol{W}\boldsymbol{X}\boldsymbol{b} = \boldsymbol{X}'\boldsymbol{W}\boldsymbol{y}$. Then substituting

$$\boldsymbol{b} = (\boldsymbol{X}'\boldsymbol{X})^{-}\boldsymbol{X}'\boldsymbol{y} \quad \text{in} \quad \boldsymbol{X}'\boldsymbol{W}\boldsymbol{X}\boldsymbol{b} = \boldsymbol{X}'\boldsymbol{W}\boldsymbol{y}$$

where $(\boldsymbol{X}'\boldsymbol{X})^{-}$ is the generalized inverse that gives the solution $\boldsymbol{b}$, yields $\boldsymbol{C}'\boldsymbol{X}'\boldsymbol{y} = \boldsymbol{X}'\boldsymbol{W}\boldsymbol{y}$ with $\boldsymbol{C} = (\boldsymbol{X}'\boldsymbol{X})^{-}\boldsymbol{X}'\boldsymbol{W}\boldsymbol{X}$. As this is true for every $\boldsymbol{y}$, it follows that $\boldsymbol{W}\boldsymbol{X} = \boldsymbol{X}\boldsymbol{C}$. (See Section 3.8 and Exercises 13–15 in Section 3.16 for a brief account of generalized matrix inverses.)

*2.13. THEORY—ADDITION OF FURTHER TERMS TO THE MODEL

Let X $= [\boldsymbol{X}_{(1)}, \boldsymbol{X}_{(2)}]$ and $\boldsymbol{b} = [\boldsymbol{b}_{(1)}, \boldsymbol{b}_{(2)}]$. For fitting just the columns of $\boldsymbol{X}_{(1)}$ the regression equation is

$$\boldsymbol{X}'_{(1)}\boldsymbol{X}_{(1)}\boldsymbol{b}^0_{(1)} = \boldsymbol{X}'_{(1)}\boldsymbol{y} \tag{13.1}$$

If in addition the columns of $\boldsymbol{X}_{(2)}$ are fitted, the equation is

$$\boldsymbol{X}'\boldsymbol{X}\boldsymbol{b} = \boldsymbol{X}'\boldsymbol{y} \tag{13.2}$$

Now write

$$S = [X_{(1)}, X_{(2)}, y]'[X_{(1)}, X_{(2)}, y]$$

$$= \begin{bmatrix} S_{11} & S_{12} & X'_{(1)}y \\ \cdot & S_{22} & X'_{(2)}y \\ \cdot & \cdot & y'y \end{bmatrix}$$

(We do not bother to write down below diagonal submatrices.) The Cholesky decomposition reduces this to

$$T = \begin{bmatrix} T_{11} & T_{12} & t_{1y} \\ 0 & T_{22} & t_{2y} \\ 0' & 0' & t_{yy} \end{bmatrix} \tag{13.3}$$

If the variance-covariance matrix for y is $\sigma^2 I_n$, then t_{1y} is uncorrelated with t_{2y}. In fact all elements of t_y are uncorrelated, as will now be shown.

Elements of t_y Uncorrelated

Suppose that in $y = X\beta + \varepsilon$ the elements ε_i of ε are independent and identically distributed with variance σ^2, so that

$$\text{var}(y) = \text{var}(\varepsilon) = \sigma^2 I_n$$

Then

$$\begin{aligned} \text{var}(t_y) &= \text{var}(T'^{-1}X'y) \\ &= T'^{-1}X'\text{var}(y)XT^{-1} \\ &= T'^{-1}X'XT^{-1}\sigma^2 \\ &= I_{p+1}\sigma^2 \end{aligned}$$

Thus the elements of t_y are uncorrelated.

If the stronger assumption is made that the ε_i are independently and normally distributed with common variance σ^2, then it follows that the elements of t_y are independently and normally distributed with a common variance σ^2. Note that in deriving these results no assumption has been made regarding the model for the expectation of y.

Algebraic Identities Involving Submatrices of T

By writing $T'T = S$ in terms of the partitioned matrices,

$$T'_{11}T_{12} = S_{12} \tag{13.4}$$

$$T'_{11}t_{1y} = X'_{(1)}\,y \tag{13.5}$$

$$T'_{12}t_{1y} + T'_{22}t_{2y} = X'_{(2)}\,y \tag{13.6}$$

Then by eq. (13.6)

$$\begin{aligned} T'_{22}t_{2y} &= X'_{(2)}\,y - T'_{12}t_{1y} \\ &= X'_{(2)}\,y - T'_{12}T'^{-1}_{11}X'_{(1)}\,y, \quad \text{using eq. (13.5)} \end{aligned} \tag{13.7}$$

Independence of $b^0_{(1)}$ and $b_{(2)}$

In the process of reducing S to the upper triangular form T, eq. (13.1) has been reduced to

$$T_{11}b^0_{(1)} = t_{1y} \tag{13.8}$$

Equation (13.2) has been reduced to

$$\begin{bmatrix} T_{11} & T_{12} \\ 0 & T_{22} \end{bmatrix} \begin{bmatrix} b_{(1)} \\ b_{(2)} \end{bmatrix} = \begin{bmatrix} t_{1y} \\ t_{2y} \end{bmatrix} \tag{13.9}$$

Thus

$$T_{22}b_{(2)} = t_{2y}$$

so that

$$b_{(2)} = T^{-1}_{22}t_{2y} \tag{13.10}$$

As t_{1y} and t_{2y} are uncorrelated it follows that $b^0_{(1)}$ and $b_{(2)}$ are uncorrelated. Furthermore

$$T_{11}b_{(1)} + T_{12}b_{(2)} = t_{1y}$$

that is,

$$\begin{aligned} \boldsymbol{b}_{(1)} &= \boldsymbol{T}_{11}^{-1}\boldsymbol{t}_{1y} - \boldsymbol{T}_{11}^{-1}\boldsymbol{T}_{12}\boldsymbol{b}_{(2)} \\ &= \boldsymbol{b}_{(1)}^{0} - \boldsymbol{T}_{11}^{-1}\boldsymbol{T}_{12}\boldsymbol{b}_{(2)} \end{aligned}$$

This is to write

$$\boldsymbol{b}_{(1)} = \boldsymbol{b}_{(1)}^{0} + \Delta\boldsymbol{b}_{(1)}^{0} \tag{13.11}$$

where $\Delta\boldsymbol{b}_{(1)}^{0} = -\boldsymbol{T}_{11}^{-1}\boldsymbol{T}_{12}\boldsymbol{b}_{(2)}$ and is uncorrelated with $\boldsymbol{b}_{(1)}^{0}$.

Changes to Predicted Values

Now consider $\hat{y}^0 = \boldsymbol{l}'_{(1)}\boldsymbol{b}_{(1)}^{0}$, $\hat{y} = \boldsymbol{l}'\boldsymbol{b}$, where $\boldsymbol{l}' = [\boldsymbol{l}'_{(1)}, \boldsymbol{l}'_{(2)}]$. Then

$$\begin{aligned} \Delta\hat{y}^0 &= \hat{y} - \hat{y}^0 \\ &= \boldsymbol{l}'_{(1)}\boldsymbol{b}_{(1)} + \boldsymbol{l}'_{(2)}\boldsymbol{b}_{(2)} - \boldsymbol{l}'_{(1)}\boldsymbol{b}_{(1)}^{0} \\ &= \boldsymbol{l}'_{(1)}\,\Delta\boldsymbol{b}_{(1)}^{0} + \boldsymbol{l}'_{(2)}\boldsymbol{b}_{(2)} \end{aligned} \tag{13.12}$$

These results imply that $\Delta\hat{y}^0$ is uncorrelated with $\hat{y}^0$.

Such changes to predicted values can most readily be calculated by using the result now given. We have

$$\begin{aligned} \hat{y} &= \boldsymbol{l}'\boldsymbol{b} \\ &= \boldsymbol{l}'\boldsymbol{T}_p^{-1}\boldsymbol{t}_y \\ &= \left(\boldsymbol{T}_p'^{-1}\boldsymbol{l}\right)'\boldsymbol{t}_y \\ &= \boldsymbol{d}'_{(1)}\boldsymbol{t}_{1y} + \boldsymbol{d}'_{(2)}\boldsymbol{t}_{2y} \end{aligned} \tag{13.13}$$

where $(\boldsymbol{T}_p'^{-1}\boldsymbol{l})' = [\boldsymbol{d}'_{(1)}, \boldsymbol{d}'_{(2)}]$. By writing $\boldsymbol{T}_p'^{-1}\boldsymbol{l}$ in terms of submatrices of $\boldsymbol{T}_p^{-1}$, it is readily seen that $\boldsymbol{d}_{(1)} = \boldsymbol{T}_{11}'^{-1}\boldsymbol{l}_{(1)}$ and hence that $\boldsymbol{d}'_{(1)}\boldsymbol{t}_{1y} = \hat{y}^0$. It then follows from eq. (13.13) that

$$\begin{aligned} \Delta\hat{y}^0 &= \hat{y} - \hat{y}^0 \\ &= \boldsymbol{d}'_{(2)}\boldsymbol{t}_{2y} \end{aligned}$$

In Section 5.4 it will be shown how this result can be used in calculating estimates of the *effects* in the analysis of experimental designs with the orthogonality property.

*2.14. THEORY—REGRESSING RESIDUALS UPON RESIDUALS

By eq. (13.11)

$$\boldsymbol{y} - \boldsymbol{X}\boldsymbol{b} = \left[\boldsymbol{y} - \boldsymbol{X}_{(1)}\boldsymbol{b}^0_{(1)}\right] - \left[\boldsymbol{X}_{(2)} - \boldsymbol{X}_{(1)}\boldsymbol{B}_{12}\right]\boldsymbol{b}_{(2)} \qquad (14.1)$$

where $\boldsymbol{B}_{12}$ is such that $\boldsymbol{T}_{11}\boldsymbol{B}_{12} = \boldsymbol{T}_{12}$. Observe that $\boldsymbol{y} - \boldsymbol{X}_{(1)}\boldsymbol{b}^0_{(1)}$ is the vector of residuals when $\boldsymbol{y}$ is regressed upon columns of $\boldsymbol{X}_{(1)}$. Furthermore each column of $\boldsymbol{X}_{(2)} - \boldsymbol{X}_{(1)}\boldsymbol{B}_{12}$ is the vector of residuals from the regression of the corresponding column of $\boldsymbol{X}_{(2)}$ upon columns of $\boldsymbol{X}_{(1)}$. See eq. (1.4) for the argument.

Suppose now that $\boldsymbol{y} - \boldsymbol{X}_{(1)}\boldsymbol{b}^0_{(1)}$ is regressed on columns of $\boldsymbol{X}_{(2)} - \boldsymbol{X}_{(1)}\boldsymbol{B}_{12}$. Equation (14.1) implies that taking $\boldsymbol{b}_{(2)}$ as the vector of regression coefficients would give the same vector of residuals and the same residual sum of squares as when $\boldsymbol{y}$ is regressed on columns of $\boldsymbol{X}$. Furthermore no other choice (e.g., $\boldsymbol{b}^*_{(2)}$) can lead to a smaller residual sum of squares. For if it did, we could write

$$\begin{aligned}
&\left[\boldsymbol{y} - \boldsymbol{X}_{(1)}\boldsymbol{b}^0_{(1)}\right] - \left[\boldsymbol{X}_{(2)} - \boldsymbol{X}_{(1)}\boldsymbol{B}_{12}\right]\boldsymbol{b}^*_{(2)} \\
&\quad = \boldsymbol{y} - \left[\boldsymbol{X}_{(1)}, \boldsymbol{X}_{(2)}\right]\begin{bmatrix} \boldsymbol{b}^0_{(1)} - \boldsymbol{B}_{12}\boldsymbol{b}^*_{(2)} \\ \boldsymbol{b}^*_{(2)} \end{bmatrix} \\
&\quad = \boldsymbol{y} - \boldsymbol{X}\boldsymbol{b}^*
\end{aligned}$$

with a smaller sum of squares than $\boldsymbol{y} - \boldsymbol{X}\boldsymbol{b}$.

This result may be used to devise a method for handling multiple regression by means of a sequence of straight-line regressions. Details are given in a note at the end of Section 4.8. It becomes possible to check for linearity as each straight-line regression is performed. Anscombe (1967) comments on the large amount of information that this approach to the computations provides.

2.15. FURTHER READING AND REFERENCES

Graybill (1982) gives an elementary introduction to matrix algebra, with a view to statistical applications. Plackett (1960) has an account of the

Cholesky decomposition, and gives references to a number of earlier papers. See, in particular, Dwyer (1945) and Durand (1956). As explained in Section 2.10, there is a close connection with the earlier Doolittle method, first described in Doolittle (1878). See also Dwyer (1941). The names "complete Cholesky decomposition algorithm" and "sequential Cholesky decomposition algorithm" are due to Golub (1969). The first account of the method, ascribing it to Cholesky, appears in Benoit (1924). An appendix to the present book reproduces from the *Bullètin gèodèsique* for 1922 an account of Cholesky's life. The interpretation of elements of the inverse of the Cholesky triangle matrix as coefficients in appropriate regressions is discussed in Hawkins and Eplett (1982).

The LINPACK subroutine library is well suited for implementing the methods that have been described. The manual (Dongarra et al., 1979) contains much useful discussion of numerical linear algebra.

2.16. EXERCISES

1. Determine standard errors of regression coefficients for Exercise 1, Chapter 1. Investigate how the standard errors of b_1 and b_2 change if x_1 or x_2 alone are used as explanatory variates (but with the same estimate of σ^2 as before.)

2. Let

$$\boldsymbol{X} = \left[\boldsymbol{X}_{(1)}, \boldsymbol{X}_{(2)}\right], \quad \boldsymbol{\beta} = \begin{bmatrix} \boldsymbol{\beta}_{(1)} \\ \boldsymbol{\beta}_{(2)} \end{bmatrix},$$

where the model is $\boldsymbol{y} = \boldsymbol{X}\boldsymbol{\beta} + \boldsymbol{\varepsilon}$, with the expected value of $\boldsymbol{\varepsilon}$ equal to $\boldsymbol{0}$. Show that if the model fitted is $\hat{\boldsymbol{y}}^0 = \boldsymbol{X}_{(1)}\boldsymbol{b}^0_{(1)}$ [this assumes $\boldsymbol{y} = \boldsymbol{X}_{(1)}\boldsymbol{\beta}^0_{(1)} + \boldsymbol{\varepsilon}^0$], then the bias in the predicted value corresponding to $\boldsymbol{l} = [\boldsymbol{l}_{(1)}, \boldsymbol{l}_{(2)}]$ is

$$\mathrm{E}\left(\boldsymbol{l}'_{(1)}\boldsymbol{b}_{(1)} - \boldsymbol{l}'\boldsymbol{b}\right) = \left(\boldsymbol{l}'_{(1)}\left(\boldsymbol{X}'_{(1)}\boldsymbol{X}_{(1)}\right)^{-1}\boldsymbol{X}'_{(1)}\boldsymbol{X}_{(2)} - \boldsymbol{l}'_{(2)}\right)\boldsymbol{\beta}_{(2)}$$

3. Let $\boldsymbol{X}_k = [\boldsymbol{x}_0, \boldsymbol{x}_1, \ldots, \boldsymbol{x}_k]$ consist of the first $k+1$ columns of $\boldsymbol{X} = \boldsymbol{X}_q$. Let $\boldsymbol{T}$ be upper triangular such that $\boldsymbol{T}'\boldsymbol{T} = \boldsymbol{X}'\boldsymbol{X} = \boldsymbol{S}$, with $\boldsymbol{T}_k$ used to denote the submatrix consisting of rows and columns 0 to k of $\boldsymbol{T}$. Let $\boldsymbol{s}_{j(k)} = \boldsymbol{X}'_k\boldsymbol{x}_j$ be the vector consisting of elements 0 to k in column j of $\boldsymbol{S}$, and let $\boldsymbol{t}_{j(k)}$ similarly consist of elements 0 to k in column j of $\boldsymbol{T}$. By partitioning $\boldsymbol{T}$ and $\boldsymbol{S}$ suitably in $\boldsymbol{T}'\boldsymbol{T} = \boldsymbol{S}$:
   (i) Show that $\boldsymbol{T}'_{k-1}\boldsymbol{t}_{j(k)} = \boldsymbol{s}_{j(k)}, \ (j \geq k)$.
   (ii) $\boldsymbol{t}'_{k(k-1)}\boldsymbol{t}_{j(k-1)} + t_{kk}t_{kj} = s_{kj}, \ (j \geq k)$.

(iii) Deduce that, with $\boldsymbol{P}_{k-1} = \boldsymbol{X}_{k-1}(\boldsymbol{X}'_{k-1}\boldsymbol{X}_{k-1})^{-1}\boldsymbol{X}_{k-1}$,

$$t_{kk}t_{kj} = \boldsymbol{x}'_k(\boldsymbol{I} - \boldsymbol{P}_{k-1})\boldsymbol{x}_j$$

$$= (\boldsymbol{x}_k - \hat{\boldsymbol{x}}_{k(k-1)})'(\boldsymbol{x}_j - \hat{\boldsymbol{x}}_{j(k-1)})$$

where $\hat{\boldsymbol{x}}_{j(k-1)} = \boldsymbol{X}_{k-1}\boldsymbol{b}_{j.k-1}$ is the vector of predicted values in the regression of $\boldsymbol{x}_j$ upon columns of $\boldsymbol{X}_{k-1}$.

4. (Continuation of Exercise 3)
Let $\mathbf{t}'_k$ be row k of $\boldsymbol{T}$, and let $\boldsymbol{s}'_k$ be row k of $\boldsymbol{S}$. Show that:

(i) $t_{kk}\mathbf{t}'_k = \boldsymbol{s}'_k - \boldsymbol{s}'_{k(k-1)}[\boldsymbol{I}_{k-1}, \boldsymbol{T}'^{-1}_{11}\boldsymbol{T}_{12}]$, where $\boldsymbol{T}_{11} = \boldsymbol{T}_{k-1}$ and $\boldsymbol{T}_{12}$ is the matrix comprising the remaining columns in rows 0 to $k-1$ of $\boldsymbol{T}$.

(ii) $t_{kk}\mathbf{t}'_k = (\boldsymbol{x}_k - \hat{\boldsymbol{x}}_{k(k-1)})'\boldsymbol{X}$.

***5.** (Continuation of Exercise 3)
Let $s^{(k)}_{j\ell} = (\boldsymbol{x}_j - \hat{\boldsymbol{x}}_{j(k)})'(\boldsymbol{x}_\ell - \hat{\boldsymbol{x}}_{\ell(k)})$, $(j, \ell > k)$. Define

$$r_{j\ell.(1..k)} = \frac{s^{(k)}_{j\ell}}{\left(s^{(k)}_{jj}s^{(k)}_{\ell\ell}\right)^{1/2}}$$

to be the partial correlation between variates j and ℓ when variates $1, 2, \ldots, k$ are held constant. Show that:

(i) The preceding definition leads to the formula for $r_{y2.1}$ which results when variates 1 and 2 are interchanged in Exercise 3, Chapter 1.

(ii) $t_{kj}(s_{jj} - \sum_{i=0}^{k-1}t^2_{ij})^{-1/2} = r_{kj.\,(1..k-1)}$, $(j > k)$. [$s_{jj} - \sum_{i=0}^{k-1}t^2_{ij}$ is the residual sum of squares from the regression of x_j on variates $1, 2, \ldots, k-1$. See Section 2.1.]

6. Take $\boldsymbol{S}$ to be a nonsingular symmetric matrix and partition

$$\boldsymbol{S} = \begin{bmatrix} \boldsymbol{S}_{11} & \boldsymbol{S}_{12} \\ \boldsymbol{S}_{21} & \boldsymbol{S}_{22} \end{bmatrix}, \quad \boldsymbol{S}^{-1} = \begin{bmatrix} \boldsymbol{S}^{11} & \boldsymbol{S}^{12} \\ \boldsymbol{S}^{21} & \boldsymbol{S}^{22} \end{bmatrix}$$

Prove that:

(i) $(\boldsymbol{S}^{11})^{-1} = \boldsymbol{S}_{11} - \boldsymbol{S}_{12}\boldsymbol{S}_{22}^{-1}\boldsymbol{S}_{21}$, with a similar expression for $(\boldsymbol{S}^{22})^{-1}$.

(ii) $\boldsymbol{S}^{21} = \boldsymbol{S}_{22}^{-1}\boldsymbol{S}_{21}\boldsymbol{S}^{11}$.

7. (Continuation of Exercise 6, with S replaced by the CSSP matrix $\mathbf{S}$) Let

$$\mathbf{S}_p = \begin{bmatrix} \mathbf{S}_{p-1} & \mathbf{s}_{p(p-1)} \\ \mathbf{s}'_{p(p-1)} & s_{pp} \end{bmatrix} = \mathbf{X}'_p\mathbf{X}_p$$

$$\mathbf{S}^p = \begin{bmatrix} \mathbf{S}^{p-1} & \mathbf{s}^{p(p-1)} \\ \mathbf{s}^{p(p-1)'} & s^{pp} \end{bmatrix}$$

Show that $(s^{pp})^{-1} = s_{pp}(1 - R^2_{p(1..p-1)})$. Show that the result holds with p replaced by i $(i < p)$, and $R^2_{i(1..i-1,\,i+1..p)}$ in place of $R^2_{p(1..p-1)}$.

8. Given $S = \mathbf{X}'_q\mathbf{X}_q$, let T be an upper triangular matrix such that $T'T = S$. Corresponding to the partitioning of S in Exercise 6, let

$$T = \begin{bmatrix} T_{11} & T_{12} \\ \mathbf{0} & \mathbf{T}_{22} \end{bmatrix}, \quad T^{-1} = \begin{bmatrix} T^{11} & T^{12} \\ \mathbf{0} & \mathbf{T}^{22} \end{bmatrix}.$$

Prove that:
(i) $T'_{22}T_{22} = S_{22} - S_{21}S_{11}^{-1}S_{12} = (S^{22})^{-1}$;
(ii) $T^{11} = T_{11}^{-1}$, $\mathbf{T}^{22} = \mathbf{T}_{22}^{-1}$.
[The result for part (i) implies that $\mathbf{T}_{22}$ is the Cholesky decomposition of $(S^{22})^{-1}$.]

9. Suppose that $[X, y]$ and $[\mathbf{X}, \mathbf{y}]$ are as in Section 1.8. Using the notation of Exercise 8 take $\mathbf{X}_q = [X, y]$. Show that if $\mathbf{T}$ is upper triangular such that $\mathbf{T}'\mathbf{T} = [\mathbf{X}, \mathbf{y}]'[\mathbf{X}, \mathbf{y}]$, then $\mathbf{T}^{-1}$ may be obtained from T^{-1} by deleting the initial row and column.

10. Suppose that θ_i, θ_j, and θ_{ij} are defined as in Section 2.5. Show that

$$\theta_{ij} = \frac{\theta_i + \theta_j - 2\tau_{ij}\sqrt{\theta_i\theta_j}}{\left(1 - \tau_{ij}^2\right)}$$

where τ_{ij} is the correlation between b_i and b_j. (Note that, using the notation which will be introduced in Section 3.5, $\tau_{ij} = -r_{ij.\neg ij}$.)

11. Let W be a positive definite (or positive semidefinite) matrix. Suppose that in $y = Xb + e$, the vector b is chosen to minimize $e'We$. Let $\hat{y} = Xb$. Prove that:
(i) $X'We = \mathbf{0}$.
(ii) $\hat{y}'We = \mathbf{0}$.

12. Consider the model

$$y_i = X_i'\beta + \varepsilon_i \quad (i = 1,\ldots,m), \tag{16.1}$$

where X_i consists of n_i identical rows, for each i. Suppose that X is formed by taking a row from each of $X_1, X_2,\ldots,X_m$, and let $\bar{y}$ have as its ith element the mean $\bar{y}_i$ of elements of y_i.

Show that the least squares estimate of β in eq. (16.1) minimizing $\sum_{i=1}^{m}\varepsilon_i'\varepsilon_i$, is equal to the weighted least squares estimate of β in

$$\bar{y} = X\beta + \bar{\varepsilon} \tag{16.2}$$

where the weight for the ith row is n_i. Give an expression for the difference in the residual sum of squares in eqs. (16.1) and (16.2).

***13.** Let X be any n by $p+1$ matrix and Δ any $p+1$ by $p+1$ symmetric matrix. Show that if $V = X\Delta X' + \phi I_n$, then $VX = XK$ with $K = \Delta X'X + \phi I_{p+1}$. Deduce that if $W = V^{-1}$ exists, then the ordinary least squares equation $X'Xb = X'y$ has the same solutions b as the generalized least squares equation $X'WXb = X'Wy$. Show that if $b = (X'X)^{-1}X'y$ and $V = \text{var}(y)$, then $\text{var}(b) = K(X'X)^{-1}$.

***14.** Show that, if $\text{var}(y) = V$, with V as in Exercise 13, then

$$\text{E}\left[y'\left(I_n - X(X'X)^{-1}X'\right)y\right] = (n-p-1)\phi$$

(Hint: Set $y = V^{1/2}z$, and use Exercise 8, Chapter 3.)

***15.** Suppose V has elements $v_{ii} = \sigma^2$, $v_{ij} = \rho\sigma^2$ for $i \neq j$. If X is any n by $p+1$ matrix which has an initial column of ones, show that V has the form given in Exercise 13 with $\delta_{00} = \rho\sigma^2$, other elements of Δ equal to zero, and $\phi = (1-\rho)\sigma^2$.

***16.** Suppose that $V = \text{var}(y)$, W and X are as in Exercise 15. If $b = (X'X)^{-1}X'y$ show that

$$\text{var}(b) = \left[\Delta + (1-\rho)\sigma^2 I_{p+1}\right](X'X)^{-1}$$

Deduce that if s^2 is obtained by dividing the residual sum of squares by $n-p-1$, and $\text{var}(b)$ is estimated as $s^2(X'X)^{-1}$, the estimate will be unbiased for all elements of b except b_0.

***17.** Suppose that elements of the vector of observations y fall into k groups; elements are uncorrelated between groups and within any group have equal correlation ρ. Also $v_{ii} = \text{var}(y_i) = \sigma^2$.

The first k columns of $\boldsymbol{X}$ consist of indicator values that identify the group to which the observation belongs. Thus in column i $(1 \leq i \leq k)$ values are 1 for observations in the ith group, and 0 otherwise. Show that

$$\boldsymbol{V} = \boldsymbol{X}\boldsymbol{\Delta}\boldsymbol{X}' + \phi \boldsymbol{I}$$

where $\phi = (1 - \rho)\sigma^2$ and $\boldsymbol{\Delta}$ has elements

$$\delta_{ii} = \rho\sigma^2 \text{ if } 1 \leq i \leq k; \quad \text{while } \delta_{ii} = 0 \text{ if } i > k$$

$$\delta_{ij} = 0 \text{ if } i \neq j$$

Show that the generalized least squares estimate $\boldsymbol{b}$ of $\boldsymbol{\beta}$ in the model $\boldsymbol{y} = \boldsymbol{X}\boldsymbol{\beta} + \boldsymbol{\varepsilon}$ is the same as the ordinary least squares estimate. Determine the bias in var($\boldsymbol{b}$), and in its estimate, if the analysis is carried out assuming $\boldsymbol{V} = \boldsymbol{I}_n\sigma^2$.

CHAPTER 3

Matrix Manipulations for Regression and Correlation

The methods and ideas of Chapters 1 and 2 are considered in more detail and extended, emphasizing the matrix operations used and their numerical properties. Section 3.5 introduces the calculation of partial correlations.

3.1. PREAMBLE

Summary of Notation

The data matrix has one column of values for each of the variates $x_1, x_2, \ldots, x_p$, and y. In addition it is convenient to introduce an initial or zeroth column in which all values are 1.0. Values in this zeroth column are treated as values of a *dummy* variate x_0. They help define

$$\boldsymbol{X} = \begin{bmatrix} 1 & x_{11} & x_{12} & \cdots & x_{1p} \\ 1 & x_{21} & x_{22} & & x_{2p} \\ \vdots & \vdots & \vdots & & \vdots \\ 1 & x_{n1} & x_{n2} & & x_{np} \end{bmatrix}, \qquad \boldsymbol{y} = \begin{bmatrix} y_1 \\ y_2 \\ \vdots \\ y_n \end{bmatrix}$$

$$\mathbf{X} = \begin{bmatrix} x_{11} - \bar{x}_1 & x_{12} - \bar{x}_2 & \cdots & x_{1p} - \bar{x}_p \\ x_{21} - \bar{x}_1 & x_{22} - \bar{x}_2 & & x_{2p} - \bar{x}_p \\ \vdots & \vdots & & \vdots \\ x_{n1} - \bar{x}_1 & x_{n2} - \bar{x}_2 & & x_{np} - \bar{x}_p \end{bmatrix}, \quad \mathbf{y} = \begin{bmatrix} y_1 - \bar{y} \\ y_2 - \bar{y} \\ \vdots \\ y_n - \bar{y} \end{bmatrix}$$

Depending on the context $\mathbf{X}_q = [\boldsymbol{X}, \boldsymbol{y}]$, or $\mathsf{X}_q = [\mathsf{X}, \mathsf{y}]$. Define

$$\boldsymbol{S} = [\boldsymbol{X}, \boldsymbol{y}]'[\boldsymbol{X}, \boldsymbol{y}] \quad (\text{elements } s_{ij};\ i,\ j = 0,\ldots,q = p + 1)$$

to be the SSP (*sums of squares and products*) matrix, and define

$$\mathsf{S} = [\mathsf{X}, \mathsf{y}]'[\mathsf{X}, \mathsf{y}] \quad (\text{elements } \mathsf{s}_{ij};\ i,\ j = 1,\ldots,q)$$

to be the CSSP (*corrected sums of squares and products*) matrix.

The matrix $\boldsymbol{T}$ (elements t_{ij}; $i,\ j = 0,\ldots,q$) is upper triangular such that $\boldsymbol{T}'\boldsymbol{T} = \boldsymbol{S}$, whereas T (elements t_{ij}; $i,\ j = 1,\ldots,q$) is such that $\mathsf{T}'\mathsf{T} = \mathsf{S}$. Note that T may be obtained by deleting the first row and column of $\boldsymbol{T}$.

TOPICS TO BE CONSIDERED

This chapter will consider in more detail the matrix methods that were used for the calculations of Chapters 1 and 2. It will become apparent, here and in Chapter 5, that the methods are applicable to a very much wider class of problems than those so far discussed.

In the next chapter details will be given of several methods which form an upper triangular matrix $\boldsymbol{T}$ such that $\boldsymbol{T}'\boldsymbol{T} = \mathbf{X}_q'\mathbf{X}_q$ by applying orthogonal transformations to the columns of $\mathbf{X}_q$. Three alternative methods, based on three alternative forms of orthogonal transformation, will be considered. Such methods give better precision, at some extra computational cost, in cases where there are near-linear dependencies among the columns of $\mathbf{X}_q$. It will be instructive to consider circumstances where this additional precision is likely to be necessary and meaningful.

The methods of the last paragraph have other uses also, notably in modifying $\boldsymbol{T}$ in a way that corresponds to interchanging the columns of $\mathbf{X}$, or to adding or deleting one or more rows of $\mathbf{X}_q$. Applications of this type will be discussed in Chapter 4.

3.2. SUMMARY AND OPERATION COUNTS

Steps in the Formation of $\boldsymbol{T}$ or T

The calculations of Chapters 1 and 2 all started from columns 0 to q of the matrix $[\boldsymbol{X}, \boldsymbol{y}]$ of Section 3.1. The first step was the formation of the SSP matrix $\boldsymbol{S}$ or the CSSP matrix S. The preferred method for forming the CSSP matrix S uses the iterative updating formulae of Section 1.7.

Formation of $\boldsymbol{S}$ or of S requires slightly more than $\frac{1}{2}nq^2$ multiplications. For present purposes such a rough count of the number of multiplications, ignoring other operations, will provide an adequate guide to the amount of computation.

The next step was the formation of an upper triangular matrix $\boldsymbol{T}$ or $\mathbf{T}$; recall that

$$\boldsymbol{S} = \boldsymbol{T}'\boldsymbol{T}, \quad \mathbf{S} = \mathbf{T}'\mathbf{T}$$

The upper triangular matrix obtained is partitioned

$$\boldsymbol{T} = \begin{bmatrix} \boldsymbol{T}_p & \boldsymbol{t}_y \\ \mathbf{0}' & t_{yy} \end{bmatrix}, \quad \text{or} \quad \mathbf{T} = \begin{bmatrix} \mathbf{T}_p & \mathbf{t}_y \\ \mathbf{0}' & t_{yy} \end{bmatrix}$$

Note that the first row of $\boldsymbol{T}$ consists of the elements

$$\left[n^{1/2}, n^{1/2}\bar{x}_1, \ldots, n^{1/2}\bar{x}_p, n^{1/2}\bar{y}\right]$$

If this first row and the first column are deleted from $\boldsymbol{T}$, the matrix which results is $\mathbf{T}$ (Exercise 4, Chapter 1). It is thus easy to obtain $\boldsymbol{T}$ from $\mathbf{T}$, and vice versa. A good ploy is to work initially with the CSSP matrix, which will be formed using the iterative updating formulae of Section 1.7 and form $\mathbf{T}$. The matrix $\mathbf{T}$ is then augmented as shown earlier to give $\boldsymbol{T}$. This approach gives the precision that results from a carefully formed CSSP matrix, and the convenience of working with $\boldsymbol{T}$ rather than with $\mathbf{T}$.

Formation of $\boldsymbol{T}$ or $\mathbf{T}$ from $\boldsymbol{S}$ or $\mathbf{S}$ takes about $\frac{1}{6}q^3$ multiplications. (Again we do not bother about whether q should be replaced by $q + 1$.) This is much less than the work involved in forming $\boldsymbol{S}$ or $\mathbf{S}$ initially, especially if n is very much larger than q. Thus to a very rough approximation the number of multiplications required to form $\boldsymbol{T}$ or $\mathbf{T}$ from $[\boldsymbol{X}, \boldsymbol{y}]$ is $\frac{1}{2}nq^2$.

The following table gives the comparison in terms of Givens' (planar) rotations or Householder's reflections, as discussed in Chapter 4:

	Number of Multiplications
Form $\boldsymbol{S}$ or $\mathbf{S}$, then $\boldsymbol{T}$ or $\mathbf{T}$	$\frac{1}{2}nq^2$
Form $\boldsymbol{T}$ from $[\boldsymbol{X}, \boldsymbol{y}]$ using planar rotations	$2nq^2$ (discussed next)
Form $\boldsymbol{T}$ from $[\boldsymbol{X}, \boldsymbol{y}]$ using Householder	nq^2

An advantage of planar rotations over the method of Householder is that $[\boldsymbol{X}, \boldsymbol{y}]$ can be processed row by row, as in the usual normal equation method. This may be important on a machine with limited storage, if n is large by comparison with q. A modification of planar rotations is available which reduces the number of multiplications to a number closer to nq^2.

Further Calculations after Formation of T or $\mathbf{T}$

The regression coefficient estimates are obtained by solving for $\boldsymbol{b}$ in the equation

$$\boldsymbol{T}_p \boldsymbol{b} = \boldsymbol{t}_y \tag{2.1}$$

or for $\mathbf{b}$ in the equation

$$\mathbf{T}_p \mathbf{b} = \mathbf{t}_y \tag{2.2}$$

For the most part we assume eq. (2.1). In order to study the regression on the first k variates only, for $k < p$, $\boldsymbol{T}_p$ may be replaced by $\boldsymbol{T}_k$, $\boldsymbol{b}$ by $\boldsymbol{b}_k$, and $\boldsymbol{t}_y$ by $\boldsymbol{t}_{y(k)}$, as in Section 2.1. The calculations of eq. (2.1) or eq. (2.2), and subsequent calculations, can all be described in terms of the following matrix operations:

1. Solve an upper triangular system of equations, as in eq. (2.1) or (2.2).	$\frac{1}{2}p^2$ multiplications, approximately
2. Solve a lower triangular system of equations; thus to determine the standard error of $\hat{y} = \boldsymbol{l}'\boldsymbol{b}$, it was necessary to solve $\boldsymbol{T}_p'\boldsymbol{h} = \boldsymbol{l}$.	$\frac{1}{2}p^2$ multiplications
3. Determine the inverse of an upper triangular matrix $\boldsymbol{T}_p$.	$\frac{1}{6}p^3 + \frac{1}{2}p^2$ multiplications
4. Form $\boldsymbol{T}_p^{-1}\boldsymbol{T}_p'^{-1}$ from $\boldsymbol{T}_p^{-1}$.	$\frac{1}{6}p^3 + \frac{1}{2}p^2$

These operations are computationally cheap by comparison with the cost of forming $\boldsymbol{T}$ or $\mathbf{T}$ initially. The solution of upper or lower triangular systems of equations is particularly cheap and simple. The additional arithmetic needed to form $\boldsymbol{S}_p^{-1}$ once $\boldsymbol{T}$ (or $\mathbf{T}$) has been formed is unnecessary to the calculation of the regression coefficients and will introduce additional numerical error.

In the account given here all calculations turn on reducing $\boldsymbol{S}$ (or $\mathbf{S}$) to an upper triangular matrix $\boldsymbol{T}$, with $\boldsymbol{T}'\boldsymbol{T} = \boldsymbol{S}$. The matrix $\boldsymbol{T}$ provides a very convenient representation of $\boldsymbol{S}$ for purposes of all subsequent calculations. Calculations of $\boldsymbol{S}_p^{-1}$ from $\boldsymbol{S}_p$ involves about three times the amount of work required to form $\boldsymbol{T}$ from $\boldsymbol{S}$; in fact each of the stages

$$\boldsymbol{S} \to \boldsymbol{T}, \quad \boldsymbol{T}_p \to \boldsymbol{T}_p^{-1}, \quad \boldsymbol{T}_p^{-1} \to \boldsymbol{T}_p^{-1}\boldsymbol{T}_p'^{-1}$$

requires about $\frac{1}{6}p^3 + \frac{1}{2}p^2$ multiplications.

*3.3. EXISTENCE OF THE CHOLESKY DECOMPOSITION

In this section we outline a variation of the proof suggested in Exercises 8, 9, and 10 of Chapter 1, and extend it to handle the case where there are linear dependencies among the columns of $\mathbf{X}_q$.

Let $\boldsymbol{X}_k = [\boldsymbol{x}_0, \boldsymbol{x}_1, \ldots, \boldsymbol{x}_k]$ be the submatrix consisting of columns 0 to k of $\mathbf{X}_q$. Let $\boldsymbol{S}_k = \boldsymbol{X}_k'\boldsymbol{X}_k$, and let $\boldsymbol{s}_{j(k-1)} = \boldsymbol{X}_{k-1}'\boldsymbol{x}_j$ be the vector consisting of elements 0 to $k-1$ in column j of $\boldsymbol{S} = \mathbf{X}_q'\mathbf{X}_q$. Let $\boldsymbol{P}_{k-1} = \boldsymbol{X}_{k-1}(\boldsymbol{X}_{k-1}'\boldsymbol{X}_{k-1})^{-1}\boldsymbol{X}_{k-1}'$; we assume for the moment there are no linear dependencies among the columns of $\mathbf{X}_q$. The matrix $\boldsymbol{P}_{k-1}$ is a projection matrix or hat matrix; it projects onto the column space of $\boldsymbol{X}_{k-1}$.

Suppose that $\boldsymbol{T}_{k-1}$ and vectors $\boldsymbol{t}_{j(k-1)}$ $(j = k, k+1, \ldots, q)$ have been formed such that

$$\boldsymbol{T}_{k-1}'\boldsymbol{T}_{k-1} = \boldsymbol{S}_{k-1} \tag{3.1}$$

$$\boldsymbol{T}_{k-1}'\boldsymbol{t}_{j(k-1)} = \boldsymbol{s}_{j(k-1)} \quad (j \geq k) \tag{3.2}$$

In other words, rows 0 to $k-1$ of an intended matrix $\boldsymbol{T}$ such that $\boldsymbol{T}'\boldsymbol{T} = \boldsymbol{S}$ have been formed. Elements t_{kj} in row k of this intended matrix must satisfy

$$\begin{aligned} t_{kk}^2 &= s_{kk} - \boldsymbol{t}_{k(k-1)}'\boldsymbol{t}_{k(k-1)}' \\ &= \boldsymbol{x}_k'(\boldsymbol{I} - \boldsymbol{P}_{k-1})\boldsymbol{x}_k \end{aligned} \tag{3.3}$$

$$\begin{aligned} t_{kk}t_{kj} &= s_{kj} - \boldsymbol{t}_{k(k-1)}'\boldsymbol{t}_{j(k-1)} \\ &= \boldsymbol{x}_k'(\boldsymbol{I} - \boldsymbol{P}_{k-1})\boldsymbol{x}_j \quad (j > k) \end{aligned} \tag{3.4}$$

These follow by writing individual elements in row k of $\boldsymbol{S} = \boldsymbol{T}'\boldsymbol{T}$ as scalar products of the columns of $\boldsymbol{T}$. As we are assuming $\boldsymbol{x}_k$ is not a linear combination of columns of $\boldsymbol{X}_{k-1}$, it follows as in Exercise 8 of Chapter 1 that $\boldsymbol{x}_k'(\boldsymbol{I} - \boldsymbol{P}_{k-1})\boldsymbol{x}_k > 0$. Equation (3.3) then determines $t_{kk} \neq 0$; it is convenient to take $t_{kk} > 0$. Equation (3.4) determines t_{kj} for $j > k$. (For $j < k$ $t_{kj} = 0$.) Hence an inductive procedure allows us, if rows 0 to $k-1$ of T have been formed, to proceed to form row k. Elements in the zeroth row of $\boldsymbol{T}$ are formed as

$$t_{0j} = \frac{s_{0j}}{s_{00}^{1/2}} \quad (j = 0, 1, \ldots, q)$$

Now consider how the preceding argument may be modified when there are linear relations among the columns of $\boldsymbol{X}$. Suppose as before that rows 0 to $k-1$ of the intended matrix $\boldsymbol{T}$ have been formed, and that all elements t_{lj} in row l $(l < k)$ of $\boldsymbol{T}$ are zero whenever $t_{ll} = 0$. Let $\overline{\boldsymbol{T}}_{k-1}$ be identical with $\boldsymbol{T}_{k-1}$, except that any zero on the diagonal is replaced by a one. It remains true that eqs. (3.1) and (3.2) are satisfied; in addition

$$\overline{\boldsymbol{T}}'_{k-1}\boldsymbol{T}_{k-1} = \boldsymbol{S}_{k-1} \tag{3.1$'$}$$

$$\overline{\boldsymbol{T}}'_{k-1}\boldsymbol{t}_{j(k-1)} = \boldsymbol{s}_{j(k-1)} \quad (j \geq k) \tag{3.2$'$}$$

Whenever $t_{ll} = 0$ has been replaced by 1.0 in forming $\overline{\boldsymbol{T}}_{k-1}$, all elements in the lth row of $\boldsymbol{T}_{k-1}$ are zero. Hence the change has no effect on the eqs. (3.1) and (3.2).

Because $\overline{\boldsymbol{T}}_{k-1}$ is upper triangular with all diagonal elements nonzero, $\overline{\boldsymbol{T}}_{k-1}^{-1}$ exists, and we can write, using eq. (3.2′):

$$\begin{aligned} \boldsymbol{t}_{j(k-1)} &= \overline{\boldsymbol{T}}'^{-1}_{k-1}\boldsymbol{s}_{j(k-1)} \\ &= \overline{\boldsymbol{T}}'^{-1}_{k-1}\boldsymbol{X}'_{k-1}\boldsymbol{x}_j \end{aligned}$$

Now let

$$\boldsymbol{P}_{k-1} = \boldsymbol{X}_{k-1}\left(\overline{\boldsymbol{T}}'_{k-1}\overline{\boldsymbol{T}}_{k-1}\right)^{-1}\boldsymbol{X}'_{k-1} \tag{3.5}$$

It remains true (see Section 3.14) that

$$\boldsymbol{P}^2_{k-1} = \boldsymbol{P}_{k-1}; \quad (\boldsymbol{I} - \boldsymbol{P}_{k-1})^2 = \boldsymbol{I} - \boldsymbol{P}_{k-1}$$

that is, $\boldsymbol{P}_{k-1}$ is a projection matrix. Also $\boldsymbol{x}_k$ is a linear combination of earlier columns of $\boldsymbol{X}_q$ if and only if $\boldsymbol{x}_k = \boldsymbol{P}_{k-1}\boldsymbol{x}_k$. Again see Section 3.14 for the argument. With $\boldsymbol{P}_{k-1}$ defined as in eq. (3.5), t^2_{kk} is given as in eq. (3.3), that is, it is the sum of squares of elements of $(\boldsymbol{I} - \boldsymbol{P}_{k-1})\boldsymbol{x}_k$ and is zero if and only if $\boldsymbol{x}_k$ is a linear combination of earlier columns of $\boldsymbol{X}_q$.

If $t^2_{kk} \neq 0$, eqs. (3.3) and (3.4) then determine t_{kk} and t_{kj} $(j > k)$, as before. If $t_{kk} = 0$, it follows that $\boldsymbol{x}'_k(\boldsymbol{I} - \boldsymbol{P}_{k-1}) = 0$ so that eq. (3.4) is satisfied for any choice of the elements t_{kj} $(j > k)$. We choose $t_{kj} = 0$ for $j > k$ as well; it then remains true that, if a diagonal element is zero, then all remaining elements in the row are zero.

The preceding discussion provides us with a practical recourse when in the process of actual calculation it turns out that, to within machine precision, $t_{kk} = 0$. All elements in that row of $\boldsymbol{T}$ are set to zero; calculations then proceed as before.

Example

Consider

$$\mathbf{S} = \begin{bmatrix} 36 & 12 & 30 & 6 & 18 \\ \cdot & 20 & 2 & 10 & 22 \\ \cdot & \cdot & 29 & 1 & 7 \\ \cdot & \cdot & \cdot & 14 & 20 \\ \cdot & \cdot & \cdot & \cdot & 40 \end{bmatrix}$$

See Section 3.5 for details of a matrix $\mathbf{X}_q$ for which $\mathbf{S}$ is the CSSP matrix.

Setting out under each row the details of its modification when the CCDA is used, we have

	36	12	30	6	18	
Divide by $\sqrt{36}$	*6*	*2*	*5*	*1*	*3*	$= \mathbf{t}_1'$
		20	2	10	22	
Form $2\mathbf{t}_1'$		*4*	*10*	*2*	*6*	
Subtract		*16*	*−8*	*8*	*16*	
Divide by $\sqrt{16}$		*4*	*−2*	*2*	*4*	$= \mathbf{t}_2'$
			29	1	7	
Form $5\mathbf{t}_1' - 2\mathbf{t}_2'$			*29*	*1*	*7*	
Subtract			*0*	*0*	*0*	$= \mathbf{t}_3'$
				14	20	
Form $\mathbf{t}_1' + 2\mathbf{t}_2' + 0$				*5*	*11*	
Subtract				*9*	*9*	
Divide by $\sqrt{9}$				*3*	*3*	$= \mathbf{t}_4'$
					40	
Form $3\mathbf{t}_1' + 4\mathbf{t}_2' + 0 + 3\mathbf{t}_4'$					*34*	
Subtract					*6*	
Divide by $\sqrt{6}$					$\sqrt{6}$	$= \mathbf{t}_5'$

The upper triangle of the elements of $\mathbf{S}$ are in the standard typeface, intermediate quantities are printed in italics, and the finally calculated elements are given in italics and enclosed in rectangular boxes. Thus

$$\mathbf{T} = \begin{bmatrix} 6 & 2 & 5 & 1 & 3 \\ 0 & 4 & -2 & 2 & 4 \\ 0 & 0 & 0 & 0 & 0 \\ 0 & 0 & 0 & 3 & 3 \\ 0 & 0 & 0 & 0 & \sqrt{6} \end{bmatrix}$$

Observe that

$$\bar{\mathbf{T}} = \begin{bmatrix} 6 & 2 & 5 & 1 & 3 \\ 0 & 4 & -2 & 2 & 4 \\ 0 & 0 & 1 & 0 & 0 \\ 0 & 0 & 0 & 3 & 3 \\ 0 & 0 & 0 & 0 & \sqrt{6} \end{bmatrix}$$

*3.4. IMPLEMENTING THE CHOLESKY DECOMPOSITION

This section will round off a discussion that has given fairly exclusive attention to the methods which, starting from a version $\mathbf{X}_q$ of the data matrix, form $S = \mathbf{X}_q'\mathbf{X}_q$ and then from S an upper triangular matrix T such that $T'T = S$. Except to note that an algorithm designed to give T is easily modified to give one of the variant decompositions discussed in Section 2.10, the variants discussed in that section are not considered further. Attention is given to two algorithms for forming T—the *complete Cholesky decomposition algorithm* (CCDA), and the *sequential Cholesky decomposition algorithm* (SCDA). These algorithms provide a suitable context in which to develop a number of important practical and theoretical ideas. Chapter 4 will redress the balance in favor of algorithms that form T (or its equivalent) directly from X by an orthogonalization method.

Calculations will be described as though based on the SSP matrix S. Assuming $\mathbf{X}_q$ has an initial column of ones, it is desirable to ensure that any SSP matrix used in practice is for variate values that are approximately centred. (Variate means should be within one or two standard deviations of zero.) An alternative is to start with a carefully formed CSSP matrix $\mathbf{S}$, and to form $\mathbf{T}$ from it. Then the matrix T such that $T'T = S$, if it is required, can be formed by adjoining the appropriate initial row and column to $\mathbf{T}$.

It is assumed that rows and columns of S, and so rows and columns of T, are numbered from 0 to q. In order to aid discussion of the picture as it appears when rows 0 to $k - 1$ of T have been formed, consider the partitioning

$$T = \begin{bmatrix} T_{11} & T_{12} \\ 0 & T_{22} \end{bmatrix} \begin{matrix} \}\text{ Rows 0 to } k-1 \\ \}\text{ Rows } k \text{ to } q \end{matrix}$$

The symbol $\boldsymbol{T}_{11}$ is used as an abbreviation for $\boldsymbol{T}_{11(k-1)}$ (elsewhere written $\boldsymbol{T}_{k-1}$); its rows and columns are numbered from 0 to $k-1$. The matrix $\boldsymbol{T}_{12}$ $(= \boldsymbol{T}_{12(k-1)})$ consists of remaining columns in rows 0 to $k-1$ of $\boldsymbol{T}$. The matrix $\boldsymbol{S}$ is partitioned similarly.

The CCDA will be considered first.

The Complete Cholesky Decomposition Algorithm (CCDA)

It is convenient to suppose that elements of $\boldsymbol{T}$, as they are formed, overwrite elements of $\boldsymbol{S}$. At the stage when rows 0 to $k-1$ of $\boldsymbol{T}$ have been formed the picture is that shown in Fig. 3.1.

In order to form row k of $\boldsymbol{T}$, the following steps are repeated for each of $j = k, k+1, \ldots, q$:

1. Form $s_{kj}^{(k-1)} = s_{kj} - \sum_{i=0}^{k-1} t_{ik} t_{ij}$. [For $k = 0$, $s_{kj}^{(-1)} = s_{kj}$. For $j = k$ only, check whether, on appropriate statistical and/or numerical

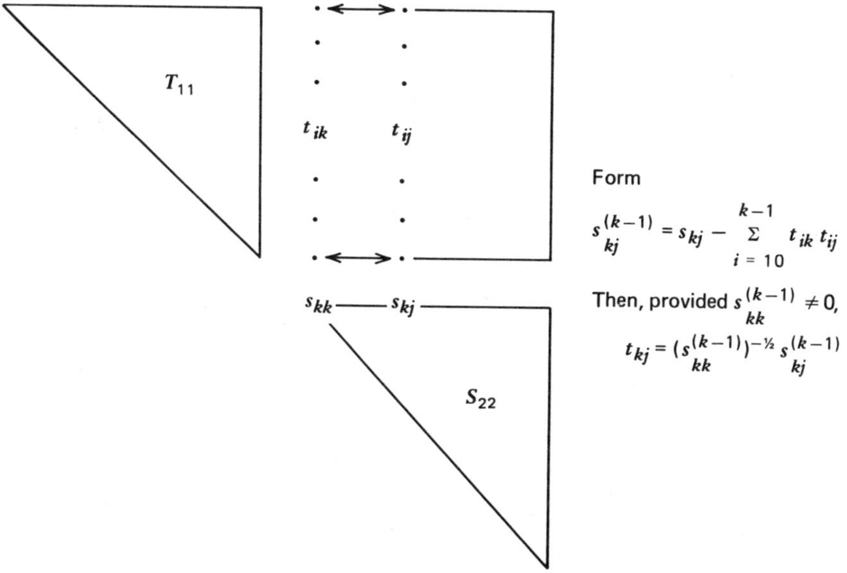

Figure 3.1. The complete Cholesky decomposition algorithm.

criteria, $s_{kk}^{(k-1)}$, should be treated as positive and nonzero. If it happens that $s_{kk}^{(k-1)} \simeq 0$, set $d_k = 0$. Otherwise, $d_k = (s_{kk}^{(k-1)})^{-1/2}$.]

2. Form $t_{kj} = d_k s_{kj}^{(k-1)}$ for $j \geq k$. The problem of detecting linear dependencies (one or more $d_k = 0$) will be considered following the discussion of the SCDA.

The Sequential Cholesky Decomposition Algorithm (SCDA)

With this algorithm the quantities s_{kj} in the kth row of $\boldsymbol{S}$ go through a sequence of modifications, to become

$$s_{kj}^{(0)}, \quad \text{then } s_{kj}^{(1)}, \ldots, \quad \text{then } s_{kj}^{(k-1)}$$

before finally being replaced by the elements t_{kj} $(j \geq k)$ in the kth row of $\boldsymbol{T}$. At the stage when rows 0 to $k-1$ of $\boldsymbol{T}$ have been formed, the picture is that shown in Fig. 3.2.

The steps associated with formation of row k of $\boldsymbol{T}$ are:

1. Decide whether, on appropriate statistical and/or numerical criteria, $s_{kk}^{(k-1)}$ should be treated as positive and nonzero. If $s_{kk}^{(k-1)} \simeq 0$, set $d_k = 0$. Otherwise, $d_k = (s_{kk}^{(k-1)})^{-1/2}$.
2. Form $t_{kj} = d_k s_{kj}^{(k-1)}$ for $j = k, k+1, \ldots, q$.
3. Form $s_{ij}^{(k)} = s_{ij}^{(k-1)} - t_{ki} t_{kj}$ for $i = k+1, \ldots, q$; $j \geq i$.

When Should $s_{kk}^{(k-1)}$ $(= t_{kk}^2)$ Be Treated as Zero?

This question has already been discussed, briefly, in Section 2.9. A variety of related issues will be considered in greater detail in Sections 3.10 through

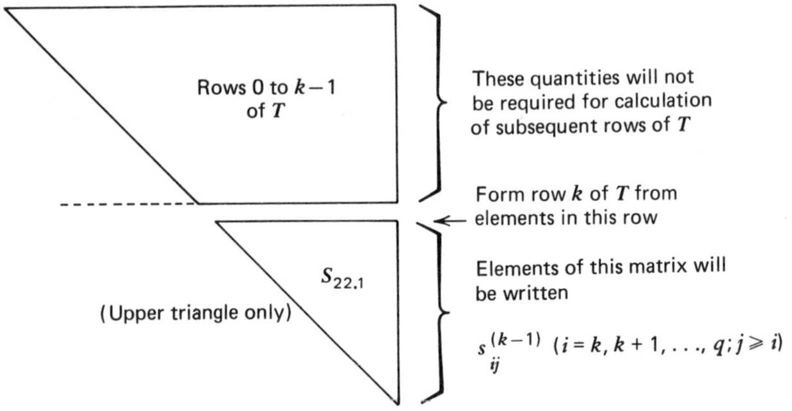

Figure 3.2. The sequential Cholesky decomposition algorithm.

3.13. The two main points which are relevant here are:

1. If $s_{kk}^{(k-1)}$ is very small, none of its digits will mean anything. Suppose, as in Section 2.9, that the computer used retains the h most significant decimal digits, or the equivalent in another base, in the result of any arithmetic operation. Then t_{kk} should certainly be set to zero if, as calculated, $s_{kk}^{(k-1)} < 10^{-h}s_{kk}$. In a more careful account of the matter, the machine constant ε will appear in place of 10^{-h}. Here we define ε to be the largest relative error that may occur in the rounding or truncation of the result of an individual arithmetic operation. For single precision arithmetic values of $h \simeq 7$ are common. Examples are provided by the IBM360/370 (6 hexadecimal digits, $\varepsilon \simeq 9.5 \times 10^{-7}$) and machines in the PDP11 range (24 binary digits, $\varepsilon \simeq 1.2 \times 10^{-7}$). Examples of machines that have much larger single precision values of h are the Burroughs B6700/7700 ($h \simeq 11.7$, $\varepsilon \simeq 7.3 \times 10^{-12}$) and the CDC6600 ($h \simeq 14.4$, $\varepsilon \simeq 7.1 \times 10^{-15}$).

2. Assume variate values are approximately centered, so that $s_{kk} = \mathbf{s}_{kk} + n\bar{x}_k^2$ is no more than about $5\mathbf{s}_{kk}$. Suppose, for example, $s_{kk}^{(k-1)} = 0.002s_{kk}$. Especially if h is much larger than 7, this would not lead us to worry unduly about numerical problems. However, from eq. (1.7) of Section 2.1, replacing t_{kk}^2 by $s_{kk}^{(k-1)}$,

$$R_{k(1,\ldots,k-1)}^2 = 1 - \mathbf{s}_{kk}^{-1}s_{kk}^{(k-1)}$$

$$\geq 0.99$$

Unless data values are known very precisely (and fit a linear model) this suggests that column k of $\boldsymbol{X}$ is to all intents and purposes a linear combination of earlier columns. In the program of which details are given in Section 10.7 the *tolerance* parameter T0 is set, by default, to 0.001. This means that t_{kk} will be set to 0 (and $t_{kj} = 0$ for $j > k$) if $\mathbf{s}_{kk}^{-1}s_{kk}^{(k-1)} < 0.001$. The user who believes that a smaller tolerance is justified in a particular case may change T0 appropriately, provided that the precision of the arithmetic used makes this safe. Otherwise, it will be necessary to change to one of the orthogonalization algorithms discussed in Chapter 4; these form $\boldsymbol{T}$ directly from $\mathbf{X}_q$. Alternatively, information on the nature of the near-linear relation between column k and earlier columns of $\mathbf{X}_q$ may be used to suggest an alternative parameterization that would avoid the problem.

Example of a Case Where $t_{kk} \simeq 0$

See Section 3.12 for details. In that example $x_3 \simeq \frac{1}{2} + x_1 - \frac{1}{2}x_2$. As a result $R_{3(1,2)}^2 \simeq 0.99997$ and $\mathrm{var}(b_3) = 1111\sigma^2$. If in that example it is thought important to retain x_3 or its equivalent, it would be advantageous to replace x_3 by $x_3^* = x_3 - x_1 + \frac{1}{2}x_2 - \frac{1}{2}$ in a rerun of the analysis.

*More Stringent Checks on Numerical Precision

Miller (1976) suggests a test for singularity which proceeds as follows: it is supposed that all elements in row i of $\boldsymbol{T}$ have a similar relative numerical error r_i which (except in the initial row) arises chiefly from the propagation of the relative error in a diagonal element t_{ii} through to other elements in the row which is consequent upon division by t_{ii}. Suppose that r_i has a variance of approximately γ_i^2. Then, as $[t_{ii}(1 + r_i)]^2 \simeq t_{ii}^2(1 + 2r_i)$, the absolute error in $s_{ii}^{(i-1)} = t_{ii}^2$ is

$$2r_i s_{ii}^{(i-1)} \simeq -2 \sum_{j=0}^{i-1} r_j t_{ji}^2$$

and

$$\gamma_i^2 = \text{var}(r_i) \simeq t_{ii}^{-4} \sum_{j=0}^{i-1} \gamma_j^2 t_{ji}^4$$

If the calculated value of $t_{ii}^2 = s_{ii}^{(i-1)}$ is less than say three times its standard error, that is, less than $6\gamma_i t_{ii}^2$, this will be taken as an indication of a possible singularity. The value of γ_0 in the initial row should be taken as at least ε, where ε is the machine constant defined earlier.

3.5. PARTIAL SUMS OF SQUARES AND PRODUCTS

The connection of the results now given with the SCDA is almost incidental; these are intermediate quantities when SCDA is used, and this is the natural place to introduce them. They will be central to the later discussion of classical multivariate calculations in Chapter 6.

At the stage when rows 0 to k of $\boldsymbol{T}$ have been formed using the SCDA, the setup is

$$\begin{bmatrix} \boldsymbol{T}_{11} & \boldsymbol{T}_{12} \\ 0 & \boldsymbol{S}_{22.1} \end{bmatrix} \begin{matrix} \}\ \text{Rows 0 to } k \text{ of } \boldsymbol{T} \\ \text{Intermediate quantities for} \\ \text{forming later rows of } \boldsymbol{T} \end{matrix}$$

Recall that $\boldsymbol{T}_{11}'\boldsymbol{T}_{11} = \boldsymbol{S}_{11}$, $\boldsymbol{T}_{11}'\boldsymbol{T}_{12} = \boldsymbol{S}_{12}$. Moreover

$$\boldsymbol{S}_{22.1} = \boldsymbol{T}_{22}'\boldsymbol{T}_{22} \tag{5.1}$$

$$= \boldsymbol{S}_{22} - \boldsymbol{T}_{12}'\boldsymbol{T}_{12} \tag{5.2}$$

Providing that S_{11} has an inverse, this equals $S_{22} - S_{21}S_{11}^{-1}S_{12}$. The matrix $S_{22.1}$ is important in a variety of multivariate calculations. As will be proved, it is the matrix of sums of squares and products of residuals:

$$x_j - \hat{x}_{j(k)} \quad (j = k+1, k+2, \ldots, q)$$

where $\hat{x}_{j(k)}$ is the vector of predicted values when x_j is regressed on columns of $X_k = [x_0, x_1, \ldots, x_k]$. In other words, the (j, l)th element $(j, l > k)$ of $S_{22.1}$ is

$$s_{jl}^{(k)} = (x_j - \hat{x}_{j(k)})'(x_l - \hat{x}_{l(k)}) \tag{5.3}$$

The matrix $S_{22.1}$ is a matrix of *partial sums of squares and products*; from it may be calculated *partial variances* and *covariances*, and *partial correlations*. (See Exercise 5, Chapter 2.)

Theoretical Derivations

Equation (5.1) holds because the steps by which T_{22} is formed are precisely those required for the Cholesky decomposition of $S_{22.1}$. Equation (5.2) is then a consequence of eq. (12.5) in Chapter 1. In order to derive eq. (5.3), observe that

$$s_{j\ell}^{(k)} = s_{j\ell}^{(k-1)} - t_{kj}t_{k\ell} \qquad (j, \ell > k)$$

(see Section 3.4). Expressing $s_{j\ell}^{(k-1)}$ similarly in terms of $s_{j\ell}^{(k-2)}$, and so on, one obtains, finally:

$$s_{j\ell}^{(k)} = s_{j\ell} - \sum_{i=0}^{k} t_{ij}t_{i\ell} \tag{5.4}$$

$$= s_{j\ell} - t_{j(k)}'t_{\ell(k)} \tag{5.4'}$$

Combining $s_{j\ell} = x_j'x_\ell$ and eq. (3.2) to write

$$t_{j(k)} = T_k'^{-1}s_{j(k)}$$

$$= T_k'^{-1}X_k'x_j$$

it follows [cf., eq. (3.4)] that

$$s_{j\ell}^{(k)} = x_j'\left(I - X_k\left(T_k'T_k\right)^{-1}X_k'\right)x_\ell$$

$$= x_j'(I - P_k)x_\ell$$

$$= x_j'(I - P_k)(I - P_k)x_\ell$$

$$= (x_j - \hat{x}_{j(k)})'(x_\ell - \hat{x}_{\ell(k)}) \tag{5.5}$$

The final line follows because, by taking $\boldsymbol{b}_{j.k}$ to be the vector of regression coefficients when $\boldsymbol{x}_j$ is regressed upon columns of $\boldsymbol{X}_k$,

$$\begin{aligned}\hat{\boldsymbol{x}}_j &= \boldsymbol{X}_k \boldsymbol{b}_{j.k} \\ &= \boldsymbol{X}_k\left(\boldsymbol{X}_k'\boldsymbol{X}_k\right)^{-1}\boldsymbol{X}_k'\boldsymbol{x}_j \\ &= \boldsymbol{P}_k \boldsymbol{x}_j\end{aligned}$$

For details of the manner in which the argument needs to be modified when $\boldsymbol{X}_k$ is not of full rank, see Section 3.14. It should be emphasized that eqs. (5.1) and (5.2) allow us to form $\boldsymbol{S}_{22.1}$ whether or not $\boldsymbol{S}_{11}$ has an inverse.

Note incidentally that the CSSP matrix is $\boldsymbol{S}_{22.1(k)}$ with $k = 0$.

Alternative Methods for Forming $\boldsymbol{S}_{22.1}$ and Associated Partial Correlations

If the SCDA is used, all that is needed is to cease calculations following formation of rows 0 to k of $\boldsymbol{T}$. Rows and columns $k + 1$ to q then contain the matrix $\boldsymbol{S}_{22.1}$. If the more usual CCDA is used, formation of rows 0 to k of $\boldsymbol{T}$ is all that is needed. Then from eq. (5.2)

$$\boldsymbol{S}_{22.1} = \boldsymbol{S}_{22} - \boldsymbol{T}_{12}'\boldsymbol{T}_{12} \tag{5.6}$$

If $\boldsymbol{T}_{11}$ is available and $\boldsymbol{T}_{12}$ has still to be formed, this is achieved either by solving $\boldsymbol{T}_{11}'\boldsymbol{T}_{12} = \boldsymbol{S}_{12}$ for $\boldsymbol{T}_{12}$ or by setting $\boldsymbol{T}_{12} = \boldsymbol{T}_{11}'^{-1}\boldsymbol{S}_{12}$. If $\boldsymbol{T}_{11}$ is singular, then replace $\boldsymbol{T}_{11}'^{-1}$ by $\overline{\boldsymbol{T}}_{11}'^{-1}$, where $\overline{\boldsymbol{T}}_{11}$ is defined as in Section 3.3.

If $\boldsymbol{T}$ has been formed using one of the orthogonalization algorithms of the next chapter, then the calculation proceeds from eq. (5.1) by using

$$\boldsymbol{S}_{22.1} = \boldsymbol{T}_{22}'\boldsymbol{T}_{22} \tag{5.7}$$

The use of the standard formula $\boldsymbol{S}_{22.1} = \boldsymbol{S}_{22} - \boldsymbol{S}_{21}\boldsymbol{S}_{11}^{-1}\boldsymbol{S}_{12}$ is to be discouraged. It involves a great deal of unnecessary arithmetic and requires $\boldsymbol{S}_{11}^{-1}$ to be replaced by a generalized inverse if $\boldsymbol{X}_k$ is not of full column rank.

Once $\boldsymbol{S}_{22.1}$ has been formed, the partial correlation $r_{j\ell}^{(k)}$ measuring the strength of linear relationship between x_j and $x_\ell\,(j, \ell > k)$ when variates 1 to k are *held constant* may be calculated as

$$r_{j\ell}^{(k)} = \frac{s_{j\ell}^{(k)}}{\left(s_{jj}^{(k)} s_{\ell\ell}^{(k)}\right)^{1/2}} \tag{5.8}$$

The next subsection shows how partial sums of squares and products, and partial correlations, may in general be calculated.

The Use of S^{-1} in Calculating Partial Sums of Squares and Products

The matrix $S_{22.1}$ of partial sums of squares and products when variates 1 to k inclusive are held constant may be written

$$S_{22.1} = (S^{22})^{-1}$$

where S^{22} consists of rows and columns $(k+1)$ and on from S^{-1} or from $\mathbf{S}^{-1}$. See Exercise 6(i), Chapter 2.

Now let S^{**} be the matrix obtained by taking an arbitrary sequence of rows and corresponding columns from S^{-1}. Then clearly (imagine the corresponding variates renumbered to follow other variates in sequence) the matrix $(S^{**})^{-1}$ is the matrix of partial sums of squares and products when variates corresponding to other rows and columns (those not included in forming S^{**}) are *held constant.*

In particular, consider the case where all variates except the ith and the jth are *held constant.* Then

$$S^{**} = \begin{bmatrix} s^{ii} & s^{ij} \\ s^{ji} & s^{jj} \end{bmatrix}$$

$$S^{**-1} = \frac{1}{s^{ii}s^{jj} - (s^{ij})^2} \begin{bmatrix} s^{jj} & -s^{ij} \\ -s^{ji} & s^{ii} \end{bmatrix} \tag{5.9}$$

The partial correlation between x_i and x_j, with all other variates *held constant*, is then

$$-\frac{s^{ij}}{(s^{ii}s^{jj})^{1/2}} \tag{5.10}$$

The quantities s^{ii}, s^{ij}, and s^{jj} may be obtained by taking, respectively: the sum of squares of elements in the ith row of T^{-1}, the sum of products of elements from the ith row with corresponding elements from the jth row, and the sum of squares of elements in the jth row. If T is singular, then one uses $\bar{T}^{-1}$.

Correlations and Partial Correlations Measure Linear Relationship

The calculated partial correlation will not reflect the nonlinear component in any relationship. As a check on possible nonlinearity, it is desirable to supplement calculation of $r_{j\ell}^{(k)}$ with examination of a plot of $x_j - \hat{x}_{j(k)}$ against $x_\ell - \hat{x}_{\ell(k)}$. In the context of regression calculations (e.g., with x_j as the dependent variate, variates $1,\ldots,k$ included as predictors, and variate ℓ as a candidate for inclusion as a predictor) this would be called a *partial regression plot*. If the modified Gram-Schmidt algorithm discussed in Section 4.8 is used to reduce $\boldsymbol{X}$ to upper triangular form, the vectors $\boldsymbol{x}_j - \hat{\boldsymbol{x}}_{j(k)}$ and $\boldsymbol{x}_\ell - \hat{\boldsymbol{x}}_{\ell(k)}$ appear in their respective columns (j and ℓ) at the stage when calculations on columns 0 to k of $\mathbf{X}_q$ are complete. Otherwise, they must be specially calculated, as shown in the following example.

Consider now the fitted equation

$$\hat{y} = b_0 + b_1x_1 + b_2x_2 + \cdots + b_px_p \tag{5.11}$$

The discussion of Section 2.16 implies that b_j may be obtained by taking the residuals $y - \hat{y}_{.\neg j}$ from the regression of y upon all x_i's except x_j. These are regressed upon the residuals $x_j - \hat{x}_{j.\neg j}$ from the regression of x_j upon all other explanatory variates x_i. It follows that

$$b_j = \frac{s_{yj.\neg j}}{s_{jj.\neg j}} = r_{yj.\neg j}\frac{\left(s_{yy.\neg j}\right)^{1/2}}{\left(s_{jj.\neg j}\right)^{1/2}}$$

where the symbol $.\neg j$ is used to mean "holding constant all explanatory variates except x_j." The residuals from this regression are the residuals $y - \hat{y}$ from the regression (5.11).

Example—The Calculation of Partial Correlations

Take as data matrix

$$\begin{array}{ccccc} x_1 & x_2 & x_3 & x_4 & x_5 \end{array}$$
$$\begin{bmatrix} -1 & 0 & -\frac{1}{2} & 1 & 0 \\ 3 & 0 & 3\frac{1}{2} & 1 & 0 \\ 2 & -2 & 3\frac{1}{2} & -2 & -2 \\ -2 & -1 & -1 & 1 & 1 \\ -1 & 1 & -1 & -1 & -1 \\ 3 & 3 & 2 & 1 & 3 \\ 2 & 2 & 1\frac{1}{2} & 2 & 4 \\ -2 & -1 & -1 & -1 & -2 \\ 2 & 1 & 2 & 1 & 3 \end{bmatrix}$$

Note that $x_3 = x_1 - \frac{1}{2}x_2 + \frac{1}{2}$; this will cause no difficulty for the calculations.

The SSP matrix is

$$
S = \begin{bmatrix} 9 & 6 & 3 & 9 & 3 & 6 \\ \cdot & 40 & 14 & 36 & 8 & 22 \\ \cdot & \cdot & 21 & 5 & 11 & 24 \\ \cdot & \cdot & \cdot & 38 & 4 & 13 \\ \cdot & \cdot & \cdot & \cdot & 15 & 22 \\ \cdot & \cdot & \cdot & \cdot & \cdot & 44 \end{bmatrix} \begin{matrix} x_0 \equiv 1 \\ x_1 \\ x_2 \\ x_3 \\ x_4 \\ x_5 \end{matrix}
$$

with Cholesky decomposition

$$
T = \begin{bmatrix} T_{11} & T_{12} \\ \mathbf{0} & T_{22} \end{bmatrix} = \left[\begin{array}{cc|cccc} 3 & 2 & 1 & 3 & 1 & 2 \\ 0 & 6 & 2 & 5 & 1 & 3 \\ \hline 0 & 0 & 4 & -2 & 2 & 4 \\ 0 & 0 & 0 & 0 & 0 & 0 \\ 0 & 0 & 0 & 0 & 3 & 3 \\ 0 & 0 & 0 & 0 & 0 & \sqrt{6} \end{array}\right]
$$

If T has been formed directly from $\mathbf{X}_q$ by orthogonal reduction to upper triangular form (see Chapter 4), the simplest way to form the matrix of partial sums of squares and products given x_1 is by taking

$$
S_{22.1} = T_{22}'T_{22} = \begin{bmatrix} 16 & -8 & 8 & 16 \\ \cdot & 4 & -4 & -8 \\ \cdot & \cdot & 13 & 17 \\ \cdot & \cdot & \cdot & 31 \end{bmatrix}
$$

Alternatively, $S_{22.1}$ may be obtained by carrying through two steps of the SCDA (Section 3.4) to yield

$$
\begin{bmatrix} T_{11} & T_{12} \\ 0 & S_{22.1} \end{bmatrix} = \left[\begin{array}{cc|cccc} 3 & 2 & 1 & 3 & 1 & 2 \\ 0 & 6 & 2 & 5 & 1 & 3 \\ \hline 0 & 0 & 16 & -8 & 8 & 16 \\ 0 & 0 & \cdot & 4 & -4 & -8 \\ 0 & 0 & \cdot & \cdot & 13 & 17 \\ 0 & 0 & \cdot & \cdot & \cdot & 31 \end{array}\right] \begin{matrix} \\ x_1 \\ x_2 \\ x_3 \\ x_4 \\ x_5 \end{matrix}
$$

If calculations had been based on the CSSP matrix, the first row and column would not appear.

Finally, $S_{22.1}$ may be formed as $S_{22} - T_{12}'T_{12}$:

$$S_{22.1} = \begin{bmatrix} 21 & 5 & 11 & 24 \\ \cdot & 38 & 4 & 13 \\ \cdot & \cdot & 15 & 22 \\ \cdot & \cdot & \cdot & 44 \end{bmatrix} - \begin{bmatrix} 1 & 2 \\ 3 & 5 \\ 1 & 1 \\ 2 & 3 \end{bmatrix} \begin{bmatrix} 1 & 3 & 1 & 2 \\ 2 & 5 & 1 & 3 \end{bmatrix}$$

$$= \begin{bmatrix} 16 & -8 & 8 & 16 \\ \cdot & 4 & -4 & -8 \\ \cdot & \cdot & 13 & 17 \\ \cdot & \cdot & \cdot & 31 \end{bmatrix}$$

The preceding matrix $S_{22.1}$ may be used in calculating any partial correlation in which x_1 is held constant. Thus

$$r_{35.1} = \frac{-8}{\sqrt{(4 \times 31)}} \simeq -0.72$$

This is the correlation coefficient between residuals from the regression of x_3 on x_1, and those from the regression of x_5 on x_1. A high (positive or negative) partial correlation indicates a strong linear relationship between the two sets of residuals. The term *partial residuals* is used for residuals that enter into the calculation of a partial correlation.

Plots of Partial Residuals

As has just been pointed out, the partial correlation is a measure of linear relationship between two sets of residuals. The linearity of the relationship can be checked by examining a plot of one set of residuals against the other. The two sets of residuals relevant to the correlation $r_{35.1}$ are:

$x_5 - \hat{x}_{5.1}$	$\frac{1}{6}$	$\frac{-11}{6}$	$\frac{-20}{6}$	$\frac{10}{6}$	$\frac{-5}{6}$	$\frac{7}{6}$	$\frac{16}{6}$	$\frac{-8}{6}$	$\frac{10}{6}$
$x_3 - \hat{x}_{3.1}$	$\frac{-2}{9}$	$\frac{10}{9}$	$\frac{25}{9}$	$\frac{4}{9}$	$\frac{-11}{9}$	$\frac{-17}{9}$	$\frac{-11}{9}$	$\frac{2}{9}$	$\frac{-2}{9}$

This plot is known as a *partial regression* plot. The regression in mind is the regression of x_5 on x_1 and x_3.

3.6. PARTIAL REGRESSION PLOTS

In the example of Section 3.5 consider the regression of x_5 on earlier variates. Setting $b_3 = 0$ (recall that x_3 is a linear combination of x_1 and x_2), the fitted equation is

$$\hat{x}_5 = \tfrac{1}{18} + \tfrac{1}{6}x_1 + \tfrac{1}{2}x_2 + x_4$$

As a check whether the linear form of relationship which this equation assumes really is appropriate, one examines the partial regression plots:

$$x_5 - \hat{x}_{5.24} \quad \text{vs.} \quad x_1 - \hat{x}_{1.24}$$

$$x_5 - \hat{x}_{5.14} \quad \text{vs.} \quad x_2 - \hat{x}_{2.14}$$

$$x_5 - \hat{x}_{5.12} \quad \text{vs.} \quad x_4 - \hat{x}_{4.12}$$

Consider now the last of these sets of plots. The residuals (perhaps calculated as indicated below) are:

$x_5 - \hat{x}_{5.12}$	$\frac{-1}{18}$	$\frac{-13}{18}$	$\frac{-10}{18}$	$\frac{38}{18}$	$\frac{-37}{18}$	$\frac{-13}{18}$	$\frac{26}{18}$	$\frac{-16}{18}$	$\frac{26}{18}$
$x_4 - \hat{x}_{4.12}$	$\frac{5}{6}$	$\frac{5}{6}$	$\frac{-7}{6}$	$\frac{8}{6}$	$\frac{-10}{6}$	$\frac{-4}{6}$	$\frac{5}{6}$	$\frac{-4}{6}$	$\frac{2}{6}$

The residual in the regression of $x_5 - \hat{x}_{5.12}$ upon $x_4 - \hat{x}_{4.12}$ is the same as the residual from the regression of x_5 upon x_1, x_2 and x_4, and the slope is b_4. Thus

$$x_5 - \hat{x}_{5.124} = x_5 - \hat{x}_{5.12} - b_4(x_4 - \hat{x}_{4.12})$$

The residual $x_4 - \hat{x}_{4.12}$ may thus be calculated as $[(x_5 - \hat{x}_{5.12}) - (x_5 - \hat{x}_{5.124})]/b_4$. (In this instance $b_4 = 1$.) The plot appears as Fig. 3.3.

Simultaneous Calculation of all Sets of Partial Residuals

A simple algorithm is available which calculates all p sets of residuals from the regression of column j ($j = 1,\ldots,p$) of $\boldsymbol{X}$ upon other columns. Let $\boldsymbol{X} = [\boldsymbol{X}_{(1)}, \boldsymbol{x}_p]$, and suppose that the SSP matrix $\boldsymbol{S}_p$ and its inverse $\boldsymbol{S}_p^{-1}$ are partitioned correspondingly as

$$\boldsymbol{S}_p = \begin{bmatrix} \boldsymbol{S}_{11} & \boldsymbol{s}_{1p} \\ \cdot & s_{pp} \end{bmatrix}, \quad \boldsymbol{S}_p^{-1} = \begin{bmatrix} \boldsymbol{S}^{11} & \boldsymbol{s}^{1p} \\ \cdot & s^{pp} \end{bmatrix}$$

Consider

$$\boldsymbol{X}(\boldsymbol{X}'\boldsymbol{X})^{-1} = \boldsymbol{X}\begin{bmatrix} \boldsymbol{S}^{11} & \boldsymbol{s}^{1p} \\ \cdot & s^{pp} \end{bmatrix}$$

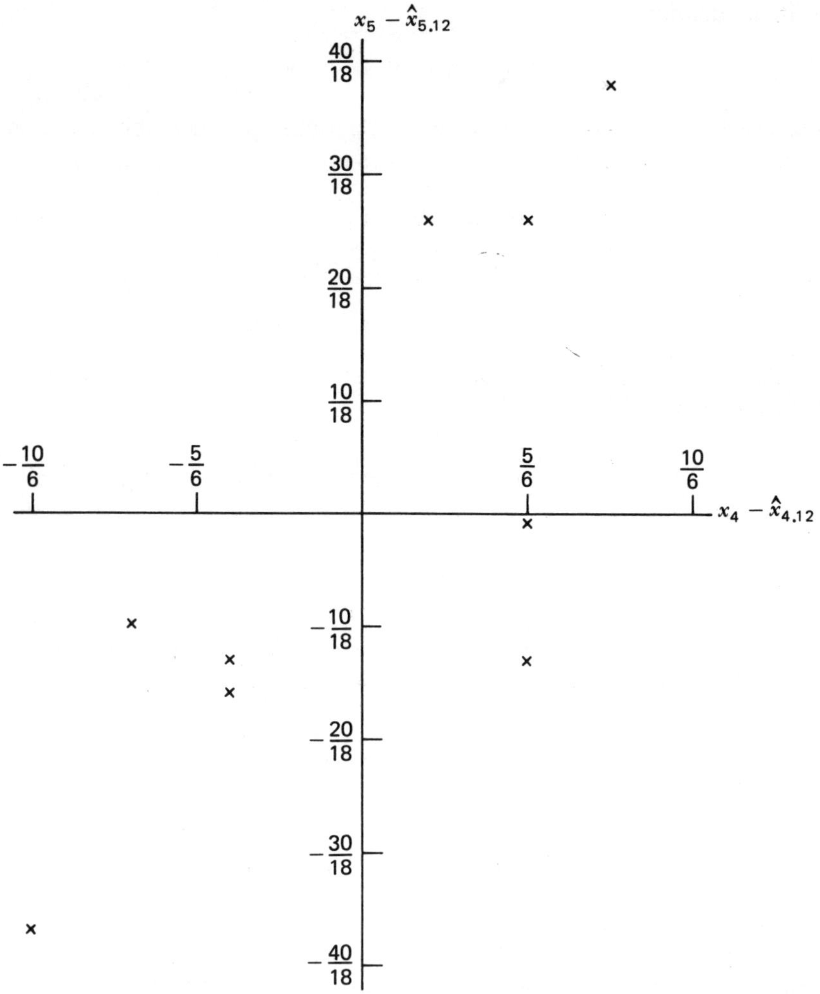

Figure 3.3. Partial regression plot: $x_5 - \hat{x}_{5.12}$ against $x_4 - \hat{x}_{4.12}$.

The final column is

$$\begin{aligned} \boldsymbol{X}_{(1)}\boldsymbol{s}^{1p} + \boldsymbol{x}_p s^{pp} &= -\boldsymbol{X}_{(1)}\boldsymbol{S}_{11}^{-1}\boldsymbol{s}_{1p}s^{pp} + \boldsymbol{x}_p s^{pp} \\ &= s^{pp}\left[\boldsymbol{x}_p - \boldsymbol{X}_{(1)}\boldsymbol{b}_{p.\neg p}\right] \\ &= s^{pp}[\boldsymbol{x}_p - \hat{\boldsymbol{x}}_{p.\neg p}] \end{aligned}$$

By moving column j of $\boldsymbol{X}$ and row and column j of $\boldsymbol{S}$ into the final row and column position and repeating the preceding algebra, it follows that column j of $\boldsymbol{X}(\boldsymbol{X}'\boldsymbol{X})^{-1}$ is $s^{jj}[\boldsymbol{x}_j - \hat{\boldsymbol{x}}_{j.\neg j}]$. This algorithm is due to Velleman and Welch (1981).

Note the use of

$$\boldsymbol{y} - \hat{\boldsymbol{y}}_{.\neg j} = \boldsymbol{y} - \hat{\boldsymbol{y}} + b_j(\boldsymbol{x}_j - \hat{\boldsymbol{x}}_{j.\neg j})$$

to calculate the quantities required for plotting in the y-direction.

Component Plus Residual Plots

Suppose that b_j is the coefficient of x_j in the regression of y on $x_1, x_2, \ldots, x_p$. Let

$$e = y - \hat{y} = y - b_0 - b_1 x_1 - \cdots - b_p x_p$$

The partial regression plot for x_j plots

$$y - \hat{y}_{.\neg j} = e + b_j(x_j - \hat{x}_{j.\neg j}) \quad \text{against} \quad (x_j - \hat{x}_{j.\neg j})$$

Ezekial (1924) proposed instead to plot $e + b_j x_j$ against x_j. Ezekial's plot is often known as a *partial residual* plot. However, it may be better to call it a *component plus residual* plot, using the terminology of Wood (1973). Several of the participants in the discussion following Atkinson (1982) comment on the comparison between these two types of plot. It is not yet clear what their relative merits are. The least squares regression line has the same slope b_j for both types of plot.

3.7. UPPER AND LOWER TRIANGULAR SYSTEMS OF EQUATIONS

Upper Triangular Systems (Backward Substitution)

Consider $\boldsymbol{T}_p \boldsymbol{b} = \boldsymbol{g}$, where in the applications given earlier [e.g., eq. (2.1)], $\boldsymbol{g} = \boldsymbol{t}_{y(p)}$. The elements of $\boldsymbol{b}$ are given by

$$b_p = t_{pp}^{-1} g_p$$

and for $i = p - 1, p - 2, \ldots, 0$ (i moves backward!),

$$b_i = t_{ii}^{-1}\left(g_i - \sum_{j=i+1}^{p} b_j t_{ij}\right) \tag{7.1}$$

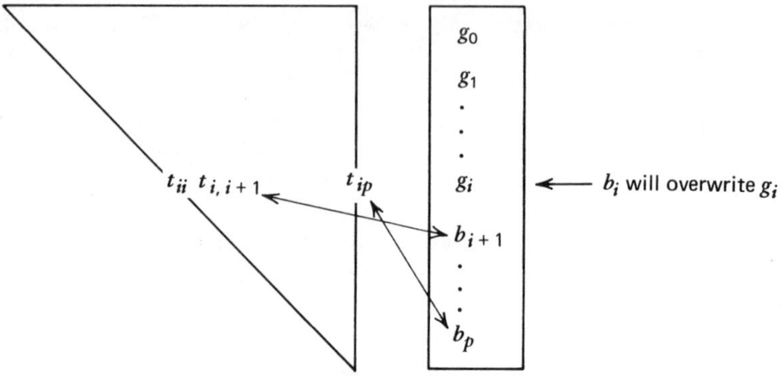

Figure 3.4. The solution of an upper triangular system of equations.

provided $t_{ii} \neq 0$. Figure 3.4 gives a diagrammatic representation. It is convenient to allow the elements of $\boldsymbol{b}$, as they are formed, to overwrite corresponding elements of $\boldsymbol{g}$.

If $t_{ii} = 0$, then according to the convention of Section 3.3 $t_{ij} = 0$ for $j > i$ as well. This implies that $g_i = 0$, as will be the case if $\boldsymbol{g} = \boldsymbol{t}_{y(p)}$. The element b_i may then be chosen arbitrarily; usually the choice will be $b_i = 0$. This is equivalent to replacing $\boldsymbol{T}_p\boldsymbol{b} = \boldsymbol{g}$ by $\overline{\boldsymbol{T}}_p\boldsymbol{b} = \boldsymbol{g}$, where $\overline{\boldsymbol{T}}_p$ is as defined in Section 3.3.

Lower Triangular Systems (Forward Substitution)

In the present development the lower triangular matrix of coefficients will have the form $\boldsymbol{T}_p'$, where $\boldsymbol{T}_p$ is upper triangular. Thus the system of equations to be solved is

$$\boldsymbol{T}_p'\boldsymbol{h} = \boldsymbol{l}$$

The elements of $\boldsymbol{h}$ are given by

$$h_0 = t_{00}^{-1} l_0$$

and for $i = 1, 2, \ldots, p$ (i moves forwards):

$$h_i = t_{ii}^{-1}\left(l_i - \sum_{j=0}^{i-1} h_j t_{ij}\right) \tag{7.2}$$

provided $t_{ii} \neq 0$. Figure 3.5 gives a diagrammatic representation.

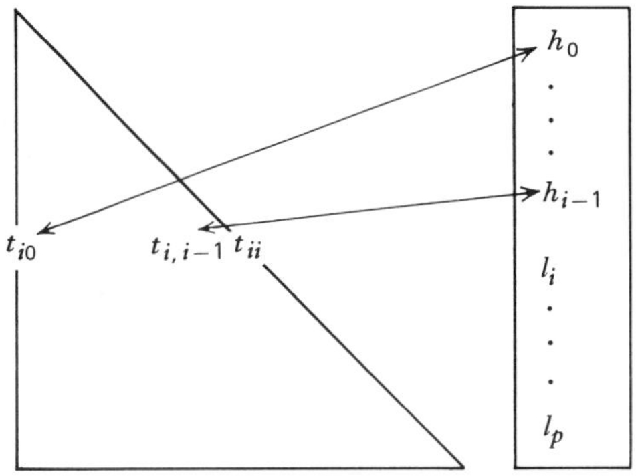

Figure 3.5. The solution of a lower triangular system of equations.

If $t_{ii} = 0$, it will commonly be the case that the system of equations whose solution is required is

$$\bar{\boldsymbol{T}}_p'\boldsymbol{h} = \boldsymbol{l}$$

If $\boldsymbol{l}$ belongs to the column space of $\boldsymbol{X}'$, then this is equivalent to setting $h_i = 0$ whenever $t_{ii} = 0$. The matter is considered further in the next section.

3.8. MATRIX INVERSION

For present purposes it is only necessary to consider:

1. The inverse of a matrix of the form $\boldsymbol{S}_p = \boldsymbol{X}'\boldsymbol{X}$.
2. The inverse of an upper triangular matrix $\boldsymbol{T}_p$ which itself is the Cholesky decomposition of a matrix $\boldsymbol{S}_p$, that is, $\boldsymbol{S}_p = \boldsymbol{T}_p'\boldsymbol{T}_p$.

As $\boldsymbol{S}_p = \boldsymbol{T}_p'\boldsymbol{T}_p$ implies that $\boldsymbol{S}_p^{-1} = \boldsymbol{T}_p^{-1}\boldsymbol{T}_p'^{-1}$ (provided that the inverse exists), the chief problem is that of inverting an upper triangular matrix $\boldsymbol{T}_p$. Section 2.3 gives all the necessary details of a simple iterative scheme of calculation, described by eq. (3.2) in Chapter 2.

$$t^{kk} = t_{kk}^{-1}, \quad \boldsymbol{t}^k = -\boldsymbol{T}^{k-1}\boldsymbol{t}_k t^{kk} \tag{8.1}$$

It is assumed that $\boldsymbol{T}^{k-1} = \boldsymbol{T}_{k-1}^{-1}$ has already been formed. The vector $\boldsymbol{t}^k$ consists of elements as far as row $k - 1$ in the next column of $\boldsymbol{T}_p^{-1}$, and t^{kk} is the diagonal element in that column.

If $\boldsymbol{X}$ and so $\boldsymbol{T}_p$ is of full rank then

$$\operatorname{var}(\boldsymbol{b}) = \boldsymbol{T}_p^{-1}\boldsymbol{T}_p'^{-1}\sigma^2 \tag{8.2}$$

Otherwise, it is necessary to achieve the effect of working with

$$\begin{bmatrix} \boldsymbol{X}^{*\prime}\boldsymbol{X}^{*} & \boldsymbol{X}^{*\prime}\boldsymbol{y} \\ \cdot & \boldsymbol{y}'\boldsymbol{y} \end{bmatrix}$$

where $\boldsymbol{X}^*$ has been obtained from $\boldsymbol{X}$ by deleting column i whenever $t_{ii} = 0$. The Cholesky decomposition $\boldsymbol{T}^*$ of this matrix is obtained by deleting row i and column i ($i \leq p$ only) from $\boldsymbol{T}$ whenever $t_{ii} = 0$. The normal equations reduce to

$$\boldsymbol{T}_p^*\boldsymbol{b}^* = \boldsymbol{t}_y^* \tag{8.3}$$

The variances and covariances of elements of $\boldsymbol{b}^*$, that is, the nonzero variances and covariances of elements of $\boldsymbol{b}$, may be obtained by replacing $\boldsymbol{b}$ by $\boldsymbol{b}^*$, and $\boldsymbol{T}_p^{-1}$ by $\boldsymbol{T}_p^{*-1}$, in eq. (8.2). In order to get variances and covariances appearing in the row and column positions appropriate to elements of $\boldsymbol{b}$, the matrix $\boldsymbol{T}_p^-$ may be obtained from $\boldsymbol{T}_p^{*-1}$ by inserting a row and column of zeros wherever a row and column was earlier deleted. The only change to eq. (8.2) is then that $\boldsymbol{T}_p^{-1}$ is replaced by $\boldsymbol{T}_p^-$. Equation (8.1) is easily modified to yield $\boldsymbol{T}_p^-$ directly. The rule is that $t^{kk} = 0$ whenever $t_{kk} = 0$; otherwise, one proceeds as before.

An alternative but less useful expression for $\operatorname{var}(\boldsymbol{b})$ may be derived directly from $\boldsymbol{b} = \overline{\boldsymbol{T}}_p^{-1}\boldsymbol{t}_y$. It is

$$\operatorname{var}(\boldsymbol{b}) = \left(\overline{\boldsymbol{T}}_p'\overline{\boldsymbol{T}}_p\right)^{-1}\boldsymbol{X}'\boldsymbol{X}\left(\overline{\boldsymbol{T}}_p'\overline{\boldsymbol{T}}_p\right)^{-1}\sigma^2$$

Example

Consider again the example of Section 3.5, where

$$\mathbf{T}_4 = \begin{bmatrix} 6 & 2 & 5 & 1 \\ 0 & 4 & -2 & 2 \\ 0 & 0 & 0 & 0 \\ 0 & 0 & 0 & 3 \end{bmatrix}$$

Then, proceeding as before,

$$\mathbf{T}_4^- = \begin{bmatrix} \frac{1}{6} & -\frac{1}{12} & 0 & 0 \\ 0 & \frac{1}{4} & 0 & -\frac{1}{6} \\ 0 & 0 & 0 & 0 \\ 0 & 0 & 0 & \frac{1}{3} \end{bmatrix}$$

The result of setting $t^{33} = 0$ has been to set to zero all elements in row and column number 3.

The variance-covariance matrix for elements of $\mathbf{b} = \bar{\mathbf{T}}_4^{-1}\mathbf{t}_y$ is then

$$\mathbf{T}_4^-\mathbf{T}_4'^-\sigma^2$$

$$= \begin{bmatrix} \left(\frac{1}{6}\right)^2 + \left(\frac{1}{12}\right)^2 + 0 + 0, & -\frac{1}{12} \times \frac{1}{4} + 0 + 0, & 0 + 0, & 0 \\ \cdot & \left(\frac{1}{4}\right)^2 + 0 + \left(\frac{1}{6}\right)^2, & 0 + 0, & -\frac{1}{6} \times \frac{1}{3} \\ \cdot & \cdot & 0 + 0, & 0 \\ \cdot & \cdot & \cdot & \left(\frac{1}{3}\right)^2 \end{bmatrix} \sigma^2$$

With the result set out in this way, it is easy to read off, for example, $\mathbf{T}_3^-\mathbf{T}_3'^-$. The variance of the predicted value $\hat{y} = \boldsymbol{\ell}'\mathbf{b}$ is the sum of squares of elements of $\mathbf{T}_p'^-\boldsymbol{\ell}\sigma$. With fixed σ^2 the successive terms in this sum of squares show how var($\hat{y}$) changes when prediction is based first on x_1 alone, then on x_1 and x_2, and so on. Refer back to Section 2.5.

Further Comments on Dealing with Linear Dependencies

In some applications, notably in the conventional formulation of analysis of variance models, linear dependencies will be anticipated. In that case it will be natural to work with T_p^{*-1} rather than with T_p^-. The point of the preceding discussion, and that of Section 3.3, is that unanticipated linear dependencies can be handled by making quite minor modifications to the scheme of calculation. Rearrangment of storage locations is not necessary.

However a regression equation for $y = x_5$ on earlier variates, derived from data for which $x_3 = x_1 - \frac{1}{2}x_2 + \frac{1}{2}$, cannot be generalized to apply to data for which this relation does hold.

Generalized Matrix Inverses

Observe that

$$T_pT_p^-T_p = T_p, \quad T_p^-T_pT_p^- = T_p^-$$

In the terminology of Rao (1973, p. 26), T_p^- is a *reflexive g-inverse*. The first of these conditions ($T_pT_p^-T_p = T_p$) is all that is necessary to make T_p^- a generalized inverse of T_p. It is well to point out that $T_p^-T_p'^-$ is not a generalized inverse of $T_p'T_p$.

The two forms of generalized inverse of $S = X'X = T_p'T_p$ that arise naturally in the context of calculations described this chapter are:

1. $S^{\sim} = \bar{T}_p^{-1}\bar{T}_p'^{-1}$, where $\bar{T}_p$ is obtained from T_p by replacing any zero diagonal element by a one. This is not a reflexive g-inverse.
2. The Moore-Penrose inverse S^-, introduced in exercises 14, 16, 17 and 19 at the end of this chapter. This is a reflexive g-inverse.

Suppose var(y) $= \sigma^2 I$. A reflexive g-inverse S^- of $S = X'X$ has the pleasant property, not shared by other generalized inverses, that if $b = S^-X'y$, then var(b) $= S^-\sigma^2$. Exercises 13 through 19 at the end of this chapter are intended as a very brief introduction to the theory of generalized matrix inverses.

A great deal of information on generalized matrix inverses is gathered together in Campbell and Meyer (1979).

3.9. EXAMPLE—REGRESSION WHEN X IS NOT FULL RANK

The matrix **S** of Section 3.3 is the CSSP matrix for the data now presented:

$$\begin{array}{ccccc} x_1 & x_2 & x_3 & x_4 & y \end{array}$$
$$\begin{bmatrix} -1 & 0 & -\frac{1}{2} & 1 & 0 \\ 3 & 0 & 3\frac{1}{2} & 1 & 0 \\ 2 & -2 & 3\frac{1}{2} & -2 & -2 \\ -2 & -1 & -1 & 1 & 1 \\ -1 & 1 & -1 & -1 & -1 \\ 3 & 3 & 2 & 1 & 3 \\ 2 & 2 & 1\frac{1}{2} & 2 & 4 \\ -2 & -1 & -1 & -1 & -2 \\ 2 & 1 & 2 & 1 & 3 \end{bmatrix}$$

The CSSP matrix is

$$\mathbf{S} = \begin{bmatrix} 36 & 12 & 30 & 6 & 18 \\ \cdot & 20 & 2 & 10 & 22 \\ \cdot & \cdot & 29 & 1 & 7 \\ \cdot & \cdot & \cdot & 14 & 20 \\ \cdot & \cdot & \cdot & \cdot & 40 \end{bmatrix}$$

As shown in Section 3.3, the Cholesky decomposition of **S** is

$$
\begin{matrix}
& \begin{matrix} x_1 & x_2 & x_3 & x_4 & y \\ \downarrow & \downarrow & \downarrow & \downarrow & \downarrow \end{matrix} \\
\mathbf{T} = & \begin{bmatrix}
6 & 2 & 5 & 1 & 3 \\
0 & 4 & -2 & 2 & 4 \\
0 & 0 & 0 & 0 & 0 \\
0 & 0 & 0 & 3 & 3 \\
0 & 0 & 0 & 0 & \sqrt{6}
\end{bmatrix}
\end{matrix}
$$

Solving in turn for b_4, b_3, b_2, b_1,

$$3b_4 = 3, \quad \text{i.e. } b_4 = 1$$

$$b_3 = 0 \text{ (arbitrarily chosen)}$$

$$4b_2 - 2b_3 + 2b_4 = 4, \quad \text{i.e. } b_2 = \tfrac{1}{2}$$

$$6b_1 + 2b_2 + 5b_3 + b_4 = 3, \quad \text{i.e. } b_1 = \tfrac{1}{6}$$

Thus *with values for all variates measured from the mean*, the regression equation is

$$\hat{y} = \tfrac{1}{6}x_1 + \tfrac{1}{2}x_2 + x_4 \tag{9.1}$$

To obtain information on the linear relation between column 3 and columns 1 and 2 we pick out the leading 2 by 3 submatrix of T:

$$
\begin{bmatrix}
6 & 2 & \bigm| & 5 \\
0 & 4 & \bigm| & -2
\end{bmatrix}
$$

Thus

$$4b_{3.2} = -2, \quad \text{i.e. } b_{3.2} = -\tfrac{1}{2}$$

$$6b_{3.1} + 2b_{3.2} = 5, \quad \text{i.e. } b_{3.1} = 1$$

It follows that $x_3 = x_1 - \tfrac{1}{2}x_2$. Adding $\lambda(x_1 - \tfrac{1}{2}x_2 - x_3)$ to eq. (9.1) for $\hat{y}$, we have

$$\hat{y} = \left(\tfrac{1}{6} + \lambda\right)x_1 + \left(\tfrac{1}{2} - \tfrac{1}{2}\lambda\right)x_2 - \lambda x_3 + x_4 \tag{9.2}$$

Breakdown of the Total Sum of Squares

The following sequential analysis of the variance table is available:

Due to x_1	*(after fitting mean)*	$3^2 = 9$
Due to x_2	*(after fitting mean,* x_1*)*	$4^2 = 16$
Due to x_3	$(\ldots\ mean, x_1, x_2)$	$= 0$
Due to x_4	$[\ldots\ mean, x_1, x_2, (x_3)]$	$3^2 = 9$
Residual		$(\sqrt{6})^2 = 6 \quad (\hat{\sigma}^2 = \frac{6}{5} = 1.2)$

Information on Variances and Covariances

Proceeding as in Section 3.8,

$$\mathbf{T}_4^- = \begin{bmatrix} \frac{1}{6} & -\frac{1}{12} & 0 & 0 \\ 0 & \frac{1}{4} & 0 & -\frac{1}{6} \\ 0 & 0 & 0 & 0 \\ 0 & 0 & 0 & \frac{1}{3} \end{bmatrix}$$

The variance-covariance matrix may thus be written

$$\sigma^2 \mathbf{T}_4^- \mathbf{T}_4'^- = \sigma^2 \begin{bmatrix} \frac{5}{144} & -\frac{1}{48} & 0 & 0 \\ \cdot & \frac{13}{144} & 0 & -\frac{1}{18} \\ \cdot & \cdot & 0 & 0 \\ \cdot & \cdot & \cdot & \frac{1}{9} \end{bmatrix}$$

A Minimum Length Solution for b

There are various alternatives to setting $b_3 = 0$. An approach that may be attractive in some contexts is to choose the vector of regression coefficients in a manner that makes $\hat{y}$ relatively insensitive to accidental changes in the value of any individual variable. Minimization of the sum of squares of elements of the vector of coefficients in eq. (9.2), that is, minimization of

$$\left(\tfrac{1}{6} + \lambda\right)^2 + \left(\tfrac{1}{2} - \tfrac{1}{2}\lambda\right)^2 + \lambda^2$$

leads to $\lambda = 1/27$. Then

$$\hat{y} = \tfrac{11}{54}x_1 + \tfrac{13}{27}x_2 - \tfrac{1}{27}x_3 + x_4$$

Exercise 2 at the end of this chapter shows how to extend the method to cases where there is more than one independent linear relation between the columns of X. See also Exercise 16. Let $\mathbf{S}^-$ be the Moore-Penrose inverse of $\mathbf{S}$. Then the minimum length solution may be written $\mathbf{b} = \mathbf{S}^-\mathbf{X}'\mathbf{y}$.

An alternative and more complicated way of obtaining this minimum length solution is described in Marquardt (1970). In the preceding example his *generalized inverse* solution would be obtained by forming the first three principal components of the CSSP matrix (i.e., those that correspond to nonzero eigenvalues) and regressing upon them. If one column is near (but not equal) to a linear combination of other columns, then the result obtained using the method of the present section will differ slightly from Marquardt's generalized inverse solution.

*3.10. NUMERICAL ERRORS—THE ROLE OF THE SSP OR CSSP MATRIX

With the normal equation methods on which discussion has so far centered, the major numerical errors will almost certainly be those introduced in the formation of $\mathbf{X}_q'\mathbf{X}_q$ from $\mathbf{X}_q$. Errors arising in the subsequent upper triangular reduction, or in the calculation from it of the regression coefficients, can by comparison be ignored. See the references given in Section 3.13 for the relevant error analyses and for a further discussion of these points.

It was pointed out in Section 1.7 that use of the SSP matrix as the basis for calculating $\boldsymbol{T}$ is likely to be unsatisfactory if any of the variables has values that are large relative to the standard deviation. A partial answer is to use the updating formulae, as discussed in that section, to form the CSSP matrix $\mathbf{S}$. Elements of the matrix $\mathbf{T}$, obtained from $\boldsymbol{T}$ by deleting row and column 0, may then be formed as the Cholesky decomposition of $\mathbf{S}$. An attempt will be made to indicate how satisfactory this approach is likely to be.

Precision of the CSSP Matrix

Suppose that the elements of the CSSP matrix are calculated using the updating eqs. (7.4) in Chapter 1. Then the jth diagonal element is calculated in a manner that is pretty well equivalent to using

$$\mathbf{s}_{jj} = \sum_{k=1}^{n} z_{kj}^2$$

where

$$z_{kj} = \sqrt{\frac{k-1}{k}}\left(x_{kj} - \bar{x}_j^{(k-1)}\right) = c\sqrt{\left(x_{kj} - \bar{x}_j^{(k-1)}\right)\left(x_{kj} - \bar{x}_j^{(k)}\right)} \tag{10.1}$$

where $c = \operatorname{sgn}(x_{kj} - \bar{x}_j^{(k-1)})$
The off-diagonal element $\mathbf{s}_{ij}$ is formed as a sum of products

$$\sum z_{ki} z_{kj}$$

Now consider the relative error (RE) in the correlation

$$r_{ij} = \frac{\mathbf{s}_{ij}}{\sqrt{\mathbf{s}_{ii}\mathbf{s}_{jj}}}$$

One has $\text{RE}(r_{ij}) \leq \text{RE}(\mathbf{s}_{ij}) + \frac{1}{2}[\text{RE}(\mathbf{s}_{ii}) + \text{RE}(\mathbf{s}_{jj})]$. Thus the relative errors in $\mathbf{s}_{ii}$ and in $\mathbf{s}_{jj}$ set a lower bound to the relative error in r_{ij}. Results given in Ling (1974) indicate the magnitude of this relative error, with arithmetic handled in single precision on an IBM360 computer, when 1000 data points are taken. On the IBM360 the maximum relative error from truncation to single precision of the result of any arithmetic operation is $\varepsilon \simeq 10^{-6}$; see Section 3.4. For each method of calculation 100 samples of 1000 points were taken. Results from a variant of the updating eq. (10.1) were compared with the "exact" result obtained using double precision and calculating first $\bar{x}$ and then $\Sigma(x - \bar{x})^2$. For comparison results using the centering formula, $\mathbf{s}_{xx} = \Sigma x^2 - n^{-1}(\Sigma x)^2$ are given also. The following are a selection of Ling's results:

Distribution	Average Relative Error	
	Updating Algorithm	Centering Formula
Normal: $\mu = 1, \sigma = 1$	0.0001 (S.D. $\simeq$ 0.00001)	0.00007 (0.00001)
$\mu = 10, \sigma = 1$	0.0001 (0.000025)	0.01 (0.001)
$\mu = 1000, \sigma = 1$	0.007 (0.005)	37 (3)
Systematic: Values 900,000 to 900,100	0.2 (0.001)	518 (7)

Clearly the updating algorithm does not completely obviate the problems involved when the correction formula is used. These problems are a particular concern on machines with a precision equivalent to only six or seven decimal digits. Even with the use of the updating algorithm, one cannot be sure of even three or four decimal digit precision in the elements of the CSSP matrix that results. Either means should first be calculated and then sums of squares and products about the means, or the user should be required to supply working means which are close to the true means and the *centering* formulae used to make a final adjustment.

Points to note in connection with Ling's calculations are:

1. With $\varepsilon < 10^{-6}$, and assuming that the arithmetic is otherwise performed in a similar manner, the likely relative errors are obtained by multiplying the preceding values by $10^6\varepsilon$. Thus to achieve three or four decimal digit precision in the CSSP matrix in the circumstances considered by Ling, ε should be less than about 10^{-9} (or $h = -\log_{10} \varepsilon$ greater than about 10).
2. On a machine that rounds rather than truncates, results are likely to be better than those quoted (see Kuki and Cody, 1973). The reason is that when a sum of positive quantities is formed, each truncation error decreases the value of the result, whereas a rounding error is just as likely to increase as to decrease the result.
3. Loss of precision from the use of the *centering* formula should be compared with that predicted by eq. (7.1) of Chapter 1.

3.11. WHAT PLACE IS THERE FOR NORMAL EQUATION METHODS?

For a large class of problems, especially in such areas as biology, sociology, and economics, the form of least squares equation which is fitted is unlikely to be more than a rough approximation to the underlying relationship. There is no precise theory that determines how variates will, perhaps after transformation, enter into a linear relation. For such applications a CSSP matrix calculated to three or four decimal digit precision, or a correlation matrix calculated to three or four decimal place precision, is likely to be all that the data warrants. If this can be assured, normal equation methods are acceptable, at least as a first recourse. However, it is essential, whatever algorithms are used, to check for the presence of near-linear relations among predictor variates. Such relations lead to both numerical instability (imprecisely formed regression coefficients) and statistical uncertainty (large standard errors). This point is taken up in Section 3.12. Where a near-linear relation is detected, it may be that one or more explanatory variate is redundant and should be omitted. Alternatively, a reparameterization may be called for. Refer back to the comments made in Section 3.4.

Computational Costs Should Be Kept in Perspective

There will be few problems where the additional computational costs from using the generally more precise algorithms of the next chapter are large, relative to other overheads. There is thus a strong argument for using one or other of these orthogonalization algorithms as the standard recourse. In some applications it is convenient to do some initial processing as the data matrix is entered row by row. As an alternative to updating the CSSP matrix, the planar rotation scheme, discussed in Sections 4.1 and 4.2, may be used to update the Cholesky upper triangle matrix $\boldsymbol{T}$ as each new row of data is entered. The fact that the planar rotation algorithm is a little slower will not matter if the user working at a terminal nevertheless gets an almost instantaneous response.

Where calculations are to be repeated a large number of times, so that it is important to use the fastest available algorithm, normal equation methods clearly do have a place. An occasional check of the type discussed in Mullet and Murray (1971), or comparison with results obtained using a more precise algorithm, is then desirable. Mullet and Murray propose entering the problem in another form; Exercise 5 at the end of this chapter gives the theoretical basis for their approach. The two forms of solution are then compared. If the first calculation gives $b_1 = 4.99$ and the second $b_1 = 4.97$, it will be clear that at most two of the digits are numerically meaningful. Simpler devices of this type are:

1. Regress residuals upon the columns of $\boldsymbol{X}$. The numbers obtained will indicate how the coefficients may be perturbed as a result of numerical errors.
2. Regress predicted values upon columns of $\boldsymbol{X}$. This is the approach used by OMNITAB. (Hogben et al., 1971, pp. 148–149.)

See also Wampler (1970, 1980).

*3.12. SENSITIVITY ANALYSIS

Consider now how the calculated regression coefficients are affected by small changes in the elements of $\mathbf{S}_p = \mathbf{X}'\mathbf{X}$ and of $\mathbf{s}_y = \mathbf{X}'\mathbf{y}$. For small changes in $\mathbf{s}_y$

$$\delta\mathbf{b} = \mathbf{S}_p^{-1}\delta\mathbf{s}_y \tag{12.1}$$

For small changes in the elements of $\mathbf{S}_p$

$$\delta \mathbf{b} \simeq -\mathbf{S}_p^{-1}(\delta \mathbf{S}_p)\mathbf{b} \tag{12.2}$$

See Wilkinson (1974).

In either case a *condition number* for $\mathbf{S}_p$ may be used to provide an overall measure of the sensitivity of $\mathbf{b}$ to changes in the elements of $\mathbf{s}_y$ or in those of $\mathbf{S}_p$. Condition numbers will be discussed in this section. Here note that, as any off-diagonal element s^{ij} of $\mathbf{S}_p^{-1}$ has the bound

$$|s^{ij}| \leq (s^{ii}s^{jj})^{1/2}$$

it is the diagonal elements of $\mathbf{S}_p^{-1}$ that are of primary importance in assessments of sensitivity based on eqs. (12.1) and (12.2). Only if one or more diagonal elements of $\mathbf{S}_p^{-1}$ is large, will small changes in the elements of $\mathbf{s}_y$ or of $\mathbf{S}_p$ lead to large changes in the elements of $\mathbf{b}$. If the kth diagonal element of $\mathbf{S}_p^{-1}$ is large, this means that variate k is near to being a linear combination of other explanatory variates. This is a consequence of the result

$$s^{kk} = \mathbf{s}_{kk}^{-1}\left(1 - R^2_{k(1,\ldots,k-1,\,k+1,\ldots,p)}\right)^{-1} \tag{12.3}$$

(See Section 2.2 and Exercise 7, Chapter 2.)

The sensitivity of elements of $\mathbf{b}$ to small changes in the elements of $\mathbf{S}_p$ leads inevitably to a loss of numerical precision in the calculated elements of $\mathbf{b}$. Berk (1977) discusses the problem from this latter viewpoint, and suggests the use of either $\Sigma_{i=1}^{p} s^{ii}\mathbf{s}_{ii}$, or of the maximum of the $s^{ii}\mathbf{s}_{ii}$, in monitoring possible loss of precision in the calculated regression coefficients. The advantage of these statistics over the closely related *spectral condition number* discussed following eq. (12.5) is that (1) eigenvalue calculations are not required, and (2) the scaling of the columns of $\mathbf{X}$ is immaterial. Furthermore these statistics provide bounds for the *spectral condition number* κ_2 of the correlation matrix derived from $\mathbf{X}'\mathbf{X}$:

$$\max_{1 \leq i \leq p} \{s^{ii}\mathbf{s}_{ii}\} \leq \kappa_2 \leq \sum_{i=1}^{p} s^{ii}\mathbf{s}_{ii}$$

Condition Numbers

The discussion will consider *condition numbers* for $\mathbf{S}_p = \mathbf{X}'\mathbf{X}$. Condition numbers for $S_p = \mathbf{X}'\mathbf{X}$, if it is preferred to base the discussion around that matrix, are defined in just the same way. A *condition number* $\kappa(\mathbf{S}_p)$ relates the precision of elements of **b**, measured in some manner that is thought appropriate, to the precision of elements of $\mathbf{S}_p$. A rough guide is that, if the condition number is 10^h and elements of $\mathbf{S}_p$ have t-digit precision, then the elements of **b** will have approximately $(t - h)$-digit precision. Consider first the *condition number* discussed in Dongarra et al. (1979), which aims to relate the maximum relative error of elements of **b** to the maximum relative error of elements of **S**. Thus define $\nu(A)$ to be the maximum of the absolute values of the elements of A, and $\nu(z)$ similarly to be the maximum of absolute values of the elements of z. Dongarra et al. then provide an algorithm for calculating $\kappa_d(\mathbf{S}_p)$ such that

$$\frac{\nu(\delta\mathbf{b})}{\nu(\mathbf{b})} < g_p \kappa_d(\mathbf{S}_p) \frac{\nu(\delta\mathbf{S}_p)}{\nu(\mathbf{S}_p)} \tag{12.4}$$

where $g_p \simeq 1.0$. A similar expression is available that relates $\nu(\delta\mathbf{b})/\nu(\mathbf{b})$ to $\nu(\delta\mathbf{s}_y)/\nu(\mathbf{s}_y)$.

Like all condition numbers based on matrix and vector norms, κ_d will be meaningless unless columns of **X** have been scaled so that all elements of **S** have similar absolute errors. If it happens that numerical error is the main source of error, then κ_d is appropriately calculated from the correlation matrix.

Numerical analysts have generally preferred to start with some *vector norm* $\|z\|$ and to use the *matrix norm* defined by

$$\|A\| = \sup_{|z|=1} \|Az\|$$

Commonly used *vector norms* are the *Euclidean vector norm* $\|z\|_2 = (z'z)^{1/2}$ and the *maximum norm* $\|z\|_\infty = \nu(z)$ defined here. Note that, in general, $\|A\|_\infty > \nu(A)$. Having settled on one or another *matrix norm*, the *condition number* is then defined as

$$\kappa(A) = \|A\|\,\|A^{-1}\|$$

Now take norms of both sides of eq. (12.2), and use the fact that $\|Az\| \le \|A\| \cdot \|z\|$ to give

$$\frac{\|\delta\mathbf{b}\|}{\|\mathbf{b}\|} \le \kappa(\mathbf{S}_p) \frac{\|\delta\mathbf{S}_p\|}{\|\mathbf{S}_p\|} \tag{12.5}$$

If $\|z\| = \|z\|_2$ is the *Euclidean norm*, then $\kappa_2(\mathbf{S}_p)$ is the ratio of the maximum to the minimum eigenvalue of $\mathbf{S}_p$. In this case κ_2 is known as the *spectral condition number*.

Statistical Uncertainty in Variate Values

Davies and Hutton (1975) give an analysis of the manner in which statistical uncertainty in variate values can be expected to affect regression coefficients. Their analysis is relevant when the regression relationship is to be estimated for variate values measured without error.

Example

Consider a dataset that has been obtained from that of Section 3.9 by omitting x_4 and adding $0.01x_4$ to elements of x_3:

x_1	x_2	x_3^*	y
−1	0	−0.49	0
3	0	3.51	0
2	−2	3.48	−2
−2	−1	−0.99	1
−1	1	−1.01	−1
3	3	2.01	3
2	2	1.52	4
−2	−1	−1.01	−2
2	1	2.01	3

Note that $x_3^* \simeq x_1 - \frac{1}{2}x_2 + \frac{1}{2}$. The least squares regression equation of y on x_1, x_2 and x_3^* is

$$\hat{y} = -49.9 - 99.8x_1 + 50.5x_2 + 100.0x_3^*$$

with sequential ANOVA table:

Due to x_1	9.0	
Due to x_2	16.0	
Due to x_3^*	9.0	
Residual	6.0	$(s^2 = \frac{6}{5} = 1.2)$
Total SS about mean	40.0	

Except that x_4 is replaced by x_3^*, this table is identical to that of Section 3.9, for reasons that should be clear upon reflection.

The variance-covariance matrix for the regression coefficients is

$$\mathbf{S}_3^{-1}\sigma^2 = \begin{array}{c} \begin{matrix} b_1 & \qquad b_2 & \qquad b_3 \end{matrix} \\ \begin{bmatrix} 1111(0.0347) & -550(-0.0208) & -1111 \\ \cdot & 272(0.0625) & 550 \\ \cdot & \cdot & 1111 \end{bmatrix} \end{array} \sigma^2$$

The values in brackets are those that obtain if x_3^* is omitted. Use of x_3^* in addition to x_1 and x_2 has a spectacular effect on variances and covariances.

As noted earlier, the large diagonal elements for $\mathbf{S}_3^{-1}$ mean also that changes in the CSSP matrix consequent upon quite small changes in variable values will give large changes in the regression coefficients. If, for example, the 1st, 3rd, ..., 9th values of x_3^* are decreased by 0.01, and even numbered values are increased by 0.01, the regression coefficients become

$$b_1 = -40.7, \quad b_2 = 21.0, \quad b_3 = 40.4$$

*3.13. ANALYSIS OF NUMERICAL ERROR—FURTHER POINTS

Wilkinson's Results

The main result that justifies the opening comments of Section 3.10 is that of Wilkinson (1966). Suppose that $\boldsymbol{T}$ is the *calculated* Cholesky decomposition of $\boldsymbol{S}$ ($= \mathbf{X}_q'\mathbf{X}_q$) and that in fact

$$\boldsymbol{S} + \boldsymbol{E} = \boldsymbol{T}'\boldsymbol{T}$$

In other words, $\boldsymbol{T}$ is regarded as the Cholesky upper triangle decomposition of a perturbed version of $\boldsymbol{S}$. Wilkinson's results give fair assurance that in all ordinary circumstances the elements of $\boldsymbol{E}$ will be no more than a small multiple of the error that may have arisen because elements of $\boldsymbol{E}$ are represented to a finite number of digits. In addition to the Wilkinson reference see Stewart (1973, pp. 153–158). Errors arising in the solution of the upper triangular system of equations $\boldsymbol{T}_p\boldsymbol{b} = \boldsymbol{t}_y$ are likely to be of less importance still; again see Wilkinson (1966, 1974). Errors that appear in the initial formation of $\boldsymbol{S}$ are likely to dwarf in importance errors at either of these later stages.

Suppose that diagonal elements of the SSP matrix $\boldsymbol{S}$ are several hundred or thousand times larger than those of the CSSP matrix $\mathbf{S}$. ($\mathbf{X}_q$ is assumed to have an initial column of ones.) Then a matrix $\boldsymbol{T}$ which is formed from $\boldsymbol{S}$ will give an accurate representation of $\boldsymbol{S}$, but (by deleting its initial row and column) a very much less accurate representation of $\mathbf{S}$. If however $\mathbf{T}$ is formed directly as the Cholesky decomposition of $\mathbf{S}$ then Wilkinson's result implies that elements of $\mathbf{E} = \mathbf{S} - \mathbf{T}'\mathbf{T}$ will be small relative to those of $\mathbf{S}$.

The Virtues of Positive Definite Matrices

Error bounds similar to those given by Wilkinson are available whichever of the standard direct methods is used in solving the normal equations. The results depend on the fact that $\mathbf{X}_q'\mathbf{X}_q$ is positive definite rather than on the details of the method used to reduce it to upper triangular form. A further consequence of working with positive definite matrices is that pivoting is unnecessary, provided that the upper triangle matrix T is used as it stands. (If, for example, the technique to be discussed in Sections 4.2 and 4.3 is subsequently used to give the matrix T^* that results when a column of X other than the last is omitted, precision may be unsatisfactory.)

Effect of Scaling

It is sometimes argued that precision can be improved by suitable scaling. The use of the correlation matrix rather than the CSSP matrix may be advocated for this reason. In fact the only important effect of such scaling would seem to be that it may reduce the chances of underflow or overflow. See Dahlquist and Bjorck (1974, sec. 5.5.5).

Double Precision Accumulation of Inner Products

The CCDA requires the formation of quantities

$$s_{kj}^{(k-1)} = s_{kj} - \sum_{i=0}^{k-1} t_{ik}t_{ij}$$

Precision will be improved if the products $t_{ik}t_{ij}$ are formed in double precision (even though t_{ik} and t_{ij} are in each case stored as single precision values) and the sum $s_{kj} - \sum_{i=0}^{k-1} t_{ik}t_{ij}$ is formed in a double precision working variable. The result is rounded to single precision when $t_{kj} = t_{kk}^{-1}s_{kj}^{(k-1)}$ has been formed. The SCDA does not allow the use of double precision accumulation of inner products in this way.

Double precision accumulation of inner products may similarly be used to an advantage in the solution of upper and lower triangular systems of equations [eqs. (7.1) and (7.2)], in forming T_p^{-1} [eq. (3.2) in Chapter 2], and in forming $T_p^{-1}T_p'^{-1}$.

See Chambers (1977, sec. 4.c) for a discussion of the implementation of double precision accumulation of inner products.

*3.14. THEORY—PROJECTION MATRICES

The simplest way to form a *projection matrix* is

$$\boldsymbol{P} = \boldsymbol{X}(\boldsymbol{X}'\boldsymbol{X})^{-1}\boldsymbol{X}'$$

For any vector $\boldsymbol{y}$, the vector $\hat{\boldsymbol{y}} = \boldsymbol{P}\boldsymbol{y}$ is the least squares projection of $\boldsymbol{y}$ upon the space spanned by the columns of $\boldsymbol{X}$. For if $\boldsymbol{b}$ is the solution of the least squares normal equations $\boldsymbol{X}'\boldsymbol{X}\boldsymbol{b} = \boldsymbol{X}'\boldsymbol{y}$, then

$$\begin{aligned}\hat{\boldsymbol{y}} &= \boldsymbol{X}\boldsymbol{b} = \boldsymbol{X}(\boldsymbol{X}'\boldsymbol{X})^{-1}\boldsymbol{X}'\boldsymbol{y} \\ &= \boldsymbol{P}\boldsymbol{y}\end{aligned}$$

The matrix $\boldsymbol{P}$ is also known as the *hat matrix* (because it puts the hat on $\boldsymbol{y}$.)

The preceding method assumes that $\boldsymbol{X}$ is of full rank. If $\boldsymbol{X}$ is not of full rank then, as in the discussion following eq. (7.1), we take

$$\boldsymbol{b} = \overline{\boldsymbol{T}}_p^{-1}\overline{\boldsymbol{T}}_p'^{-1}\boldsymbol{X}'\boldsymbol{y}$$

$$\hat{\boldsymbol{y}} = \boldsymbol{X}\boldsymbol{b} = \boldsymbol{X}\left(\overline{\boldsymbol{T}}_p'\overline{\boldsymbol{T}}_p\right)^{-1}\boldsymbol{X}'\boldsymbol{y} \tag{14.1}$$

In this case the appropriate projection matrix is

$$\boldsymbol{P} = \boldsymbol{X}\left(\overline{\boldsymbol{T}}_p'\overline{\boldsymbol{T}}_p\right)^{-1}\boldsymbol{X}' \tag{14.2}$$

The second part of the argument of Section 3.3, designed to establish the existence of the Cholesky decomposition when $\boldsymbol{X}$ is not full rank, left various gaps. These can now be filled. Suppose that $\boldsymbol{T}_{k-1}$ and $\boldsymbol{t}_{j(k)}$ $(j \geq k)$ have been formed such that

$$\overline{\boldsymbol{T}}_{k-1}'\boldsymbol{T}_{k-1} = \boldsymbol{S}_{k-1} \tag{14.3}$$

$$\overline{\boldsymbol{T}}_{k-1}'\boldsymbol{t}_{j(k-1)} = \boldsymbol{s}_{j(k-1)} \quad (j \geq k) \tag{14.4}$$

as in eqs. (3.1′) and (3.2′). Recall that $\boldsymbol{T}_{k-1}$ has all elements in row ℓ equal to zero, whenever $t_{\ell\ell} = 0$ $(\ell < k)$, and that $\overline{T}_{k-1}$ is obtained from $\boldsymbol{T}_{k-1}$ by replacing any zero diagonal element by 1.0. Let $Z_{k-1} = \bar{T}_{k-1} - T_{k-1}$, that is, Z_{k-1} has 1.0 on the diagonal wherever T_{k-1} has a zero and otherwise has all its elements equal to zero. Observe that

$$\bar{T}'_{k-1}\bar{T}_{k-1} = T'_{k-1}T_{k-1} + Z_{k-1}$$

Now define

$$P_{k-1} = X_{k-1}\left(\bar{T}'_{k-1}\bar{T}_{k-1}\right)^{-1}X'_{k-1}$$

It will be proved that

1. $P_{k-1}h = h$ if and only if h is a linear combination of the columns of X_{k-1}.
2. $P^2_{k-1} = P_{k-1}$.

First observe that $P_{k-1}h = h$ implies $h = X_{k-1}c$ with $c = (\bar{T}'_{k-1}\bar{T}_{k-1})^{-1}X'_{k-1}h$. (See the hint to Exercise 8, Chapter 1.) Conversely, suppose $h = X_{k-1}c$. Without loss of generality it will be supposed that, whenever a column x_ℓ of X_{k-1} is a linear combination of earlier columns so that the corresponding diagonal element of Z_{k-1} is nonzero, then the element c_ℓ of c is zero. For if $c_\ell \neq 0$, the term $x_\ell c_\ell$ may be rewritten as a linear combination of earlier columns of X. Then

$$\begin{aligned} P_{k-1}h &= X_{k-1}\left(\bar{T}'_{k-1}\bar{T}_{k-1}\right)^{-1}X'_{k-1}X_{k-1}c \\ &= X_{k-1}\left(\bar{T}'_{k-1}\bar{T}_{k-1}\right)^{-1}\left(\bar{T}'_{k-1}\bar{T}_{k-1} - Z_{k-1}\right)c \\ &= X_{k-1}c \end{aligned}$$

as $Z_{k-1}c = 0$; this follows because the nonzero diagonal elements of Z_{k-1} are matched with zero elements of c. □

To prove **2**, observe that each column of X_{k-1} is trivially a linear combination of columns of X_{k-1}; thus $P_{k-1}X_{k-1} = X_{k-1}$. Hence

$$\begin{aligned} P^2_{k-1} &= P_{k-1}X_{k-1}\left(\bar{T}'_{k-1}\bar{T}_{k-1}\right)^{-1}X'_{k-1} \\ &= X_{k-1}\left(\bar{T}'_{k-1}\bar{T}_{k-1}\right)^{-1}X'_{k-1} \\ &= P_{k-1} \end{aligned}$$

It follows immediately that $(I - P_{k-1})^2 = I - P_{k-1}$. □

We saw in Section 3.3 that the diagonal element t_{kk} must satisfy

$$t_{kk}^2 = \boldsymbol{x}_k'(\boldsymbol{I} - \boldsymbol{P}_{k-1})\boldsymbol{x}_k$$

$$= \boldsymbol{x}_k'(\boldsymbol{I} - \boldsymbol{P}_{k-1})^2\boldsymbol{x}_k$$

which is the sum of squares of elements of $(\boldsymbol{I} - \boldsymbol{P}_{k-1})\boldsymbol{x}_k$ and is zero if and only if $\boldsymbol{x}_k = \boldsymbol{P}_{k-1}\boldsymbol{x}_k$, that is, if and only if $\boldsymbol{x}_k$ is a linear combination of earlier columns of $\boldsymbol{X}$. Having determined t_{kk}, elements t_{kj} for $j > k$ are determined by eq. (3.4), Section 3.3. If $t_{kk} = 0$, then the convention followed is that $t_{kj} = 0$ $(j > k)$. Thus the argument by induction in Section 3.3 goes through when $\boldsymbol{S}_p$ is singular as well.

Projection Matrices and the Vector of Residuals

The vector of predicted values when $\boldsymbol{y}$ is regressed upon columns of $\boldsymbol{X}$ is, from eq. (14.1), $\hat{\boldsymbol{y}} = \boldsymbol{P}\boldsymbol{y}$. Hence the vector of residuals may be written

$$\boldsymbol{e} = (\boldsymbol{I} - \boldsymbol{P})\boldsymbol{y}$$

Assuming that the variance-covariance matrix for $\boldsymbol{y}$ is $\sigma^2\boldsymbol{I}$, it follows that

$$\operatorname{var}(\boldsymbol{e}) = (\boldsymbol{I} - \boldsymbol{P})^2\sigma^2$$

$$= (\boldsymbol{I} - \boldsymbol{P})\sigma^2$$

Because $\hat{\boldsymbol{y}} = \boldsymbol{P}\boldsymbol{y}$, the name hat matrix has been given to $\boldsymbol{P}$. The ith diagonal element of $\boldsymbol{P}$ may be interpreted as the leverage of the ith data point; see Section 4.11. Hoaglin and Welch (1978) discuss and demonstrate the interpretation of off-diagonal as well as diagonal elements of $\boldsymbol{P}$.

3.15. FURTHER READING AND REFERENCES

A good reference for numerical matrix algebra is Stewart (1973). An excellent brief account is that in Chapter 5 of Dahlquist and Bjorck (1974). See also Forsythe, Malcolm, and Moler (1977) for a discussion of floating point computation, with some reference to specific machines. Chambers (1977) has a broad-ranging discussion, at a technical level, of all aspects of linear least squares computations. The best source of FORTRAN subroutines is Dongarra el al. (1979).

3.16. EXERCISES

1. Assume a linear relation $\boldsymbol{Xc} = \mathbf{0}$ between the columns of $\boldsymbol{X}$. The vector of regression coefficients has the form $\boldsymbol{b} + \lambda\boldsymbol{c}$, where λ is arbitrary. Show that the length of this vector, that is, the square root of $(\boldsymbol{b} + \lambda\boldsymbol{c})'(\boldsymbol{b} + \lambda\boldsymbol{c})$, is minimized for $\lambda = -\boldsymbol{b}'\boldsymbol{c}/\boldsymbol{c}'\boldsymbol{c}$.

2. (Continuation of Exercise 1) Suppose more generally that there are linear relations $\boldsymbol{XC} = \mathbf{0}$, given by the columns of $\boldsymbol{C}$, between the columns of $\boldsymbol{X}$. The vector of regression coefficients has the form $\boldsymbol{b} + \boldsymbol{C\lambda}$, where $\boldsymbol{\lambda}$ is arbitrary. Show that the choice of $\boldsymbol{\lambda}$ which minimizes the length of this vector is given by solving $\boldsymbol{C}'\boldsymbol{C\lambda} = -\boldsymbol{C}'\boldsymbol{b}$.

3. Consider the data set

Observation Number	x_1	x_2	x_3	x_4	y
1	0	−1	0	0.1	4
2	1	1	−4	−3.9	1
3	5	3	−2	−2.1	6
4	4	1	2	2.0	8
5	3	2	−3	−3.1	3
6	3	0	3	3.0	4
7	4	0	5	4.9	7

The Cholesky decomposition of the CSSP matrix is as follows:

$$\mathbf{T} = \begin{bmatrix} 4.342 & 2.040 & 2.566 & 2.378 & 4.079 \\ 0 & 2.588 & -7.764 & -7.776 & -2.939 \\ 0 & 0 & 0 & 0 & 0 \\ 0 & 0 & 0 & 0.115 & 0.870 \\ 0 & 0 & 0 & 0 & 3.065 \end{bmatrix}$$

Determine

(i) The linear relation between x_3 and x_1 and x_2.

(ii) The least squares regression of x_4 upon x_1 and x_2.

(iii) The least squares regression of y upon x_1 and x_2.

Assuming that x_4 should be taken as a linear combination of x_1 and x_2, give the general form of the vector of regression coefficients in the regression of y upon x_1, x_2, x_3, and x_4, and obtain the minimum length solution.

4. Give the general form of the vector of standardized regression coefficients in the regression of y upon x_1, x_2, x_3 and x_4 in Exercise 3. (See

Section 1.10 for the definition of the *standardized coefficients*.) Determine the minimum length solution for the standardized coefficients.

5. Suppose that in the usual least squares model $\boldsymbol{y} = \boldsymbol{X\beta} + \boldsymbol{\varepsilon}$ with $\text{var}(\boldsymbol{\varepsilon}) = \sigma^2 \boldsymbol{I}_n$, $\boldsymbol{X}$ is replaced by $\boldsymbol{X}^* = \boldsymbol{XC}^{-1}$ and $\boldsymbol{\beta}$ is replaced by $\boldsymbol{\beta}^* = \boldsymbol{C\beta}$. Thus $\boldsymbol{y} = \boldsymbol{X}^*\boldsymbol{\beta}^* + \boldsymbol{\varepsilon}$. Prove that, if $\boldsymbol{b}^*$ is the least squares estimate of $\boldsymbol{\beta}^*$, then

$$\text{var}(\boldsymbol{b}^*) = \boldsymbol{C}(\boldsymbol{X}'\boldsymbol{X})^{-1}\boldsymbol{C}'\sigma^2$$

and that the choice $\boldsymbol{C} = \boldsymbol{T}$, where $\boldsymbol{T}$ is upper triangular such that $\boldsymbol{T}'\boldsymbol{T} = \boldsymbol{X}'\boldsymbol{X}$, makes this a diagonal matrix.

6. Suppose that $\text{var}(y_i) = cw_i^{-1}$. Let $\boldsymbol{W} = \text{diag}[w_i]$. Let $\hat{\boldsymbol{y}}$ be the vector of fitted values in the regression of $\boldsymbol{y}$ upon columns of $\boldsymbol{X}$ that uses $\boldsymbol{W}$ as a matrix of weights. Prove that, assuming $\boldsymbol{X}$ has rank $p + 1$, the expected value of the residual sum of squares is

$$\text{E}\left[(\hat{\boldsymbol{y}} - \boldsymbol{y})'(\hat{\boldsymbol{y}} - \boldsymbol{y})\right] = (n - p - 1)c$$

7. Suppose that $\boldsymbol{b}$ is the vector of regression coefficients in the least squares regression of $\boldsymbol{y}$ upon columns of $\boldsymbol{X} = [\boldsymbol{x}_0, \boldsymbol{x}_1, \ldots, \boldsymbol{x}_p]$. Show that the vector of regression coefficients, when $\boldsymbol{y} + \alpha\boldsymbol{x}_i$ is regressed upon columns of $\boldsymbol{X}$, is obtained from $\boldsymbol{b}$ by replacing b_i by $b_i + \alpha$. Show that the vector of residuals and the residual sum of squares are unchanged (Mullet and Murray, 1971).

8. Prove that if $\text{E}(\boldsymbol{y}) = \boldsymbol{\theta}$, $\text{var}(\boldsymbol{y}) = \sigma^2 I_n$, then

$$\text{E}(\boldsymbol{y}'\boldsymbol{A}\boldsymbol{y}) = \boldsymbol{\theta}'\boldsymbol{A}\boldsymbol{\theta} + \sigma^2 \text{ trace}(\boldsymbol{A})$$

(Hint: write $\boldsymbol{y}'\boldsymbol{A}\boldsymbol{y} = \Sigma\Sigma a_{ij} y_i y_j$.)

9. Prove that, if $\boldsymbol{b}$ is the least squares estimate of $\boldsymbol{\beta}$ in the model $\boldsymbol{y} = \boldsymbol{X\beta} + \boldsymbol{\varepsilon}$ with $\text{var}(\boldsymbol{\varepsilon}) = \sigma^2 \boldsymbol{I}_n$, then

$$\text{E}[\boldsymbol{b}'\boldsymbol{b}] = \boldsymbol{\beta}'\boldsymbol{\beta} + \sigma^2 \text{ trace}(\boldsymbol{X}'\boldsymbol{X})^{-1}$$

10. The position of a point in a plane is determined as the intersection of several lines furnished by observation. Owing to errors of observation, the lines are not exactly concurrent. The required point is therefore taken as the one which minimizes the sum $\Sigma_{i=1}^n d_i^2$ of squared distances from the point to the respective lines. Given lines

$a_i x + b_i y + c_i = 0$ $(i = 1, 2, \ldots, n)$ show that the required point (x, y) satisfies

$$\sum_{i=1}^{n} \frac{a_i(a_i x + b_i y + c_i)}{a_i^2 + b_i^2} = 0$$

$$\sum_{i=1}^{n} \frac{b_i(a_i x + b_i y + c_i)}{a_i^2 + b_i^2} = 0$$

(Whittaker and Robinson, 1944, pp. 248–249.) Show how to modify the foregoing equations if the weighted sum of squares $\Sigma_{i=1}^{n} w_i d_i^2$ is to be minimized.

11. A circle is to be fitted to the set of points (x_i, y_i) $(i = 1, \ldots, n)$, by minimizing

$$\sum_{i=1}^{n} \left[(x_i - a)^2 + (y_i - b)^2 - r^2 \right]^2$$

$$= \sum_{i=1}^{n} \left(x_i^2 + y_i^2 - 2ax_i - 2by_i - c \right)^2$$

where $c = r^2 - a^2 - b^2$. Show that a and b may be determined as coefficients in the regression of $z = \frac{1}{2}(x^2 + y^2)$ upon x and y, and that the estimate of r^2 is

$$\hat{r}^2 = \frac{1}{n} \left(\sum_{i=1}^{n} (x_i - a)^2 + \sum_{i=1}^{n} (y_i - b)^2 \right)$$

Show that this formula is equivalent to minimizing the sum of squares of the *powers* of the points with respect to the circle. [Suppose that a line through the point $P_i(x_i, y_i)$ intersects the circle in points A and B. Then the *power* of the point P_i with respect to the circle is $AP_i \cdot BP_i$, independent of the particular line taken.] (K. R. Pledger and R. H. Fletcher contributed to this exercise.)

***12.** Give conditions under which the choice of a, b, c, f, g, h to minimize

$$\sum_{i=1}^{n} \left[ax_i^2 + by_i^2 + 2hx_i y_i + 2gx_i + 2fy_i + c \right]^2$$

will yield an ellipse.

Exercises 13 through 16 give a brief account of some aspects of the theory of generalized matrix inverses.

*13. Let A be any real matrix; show that it may be written $A = GH$, where G is of full column rank and H is of full row rank. Show that one then may define

$$A^{-} = H'(HH')^{-1}(G'G)^{-1}G$$

and that A^{-} then satisfies $AA^{-}A = A$. [Let G consist of a subset of the columns of A which form a basis for the column space of A. Then, for a suitable choice of H, it follows that $A = GH$. But $A = GH$ also expresses the rows of A as linear combinations of the rows of H. Hence $\text{rank}(A) = \text{rank}(G) \le \text{rank}(H)$.]

*14. Any matrix A^{-} such that $AA^{-}A = A$ is known as a *generalized inverse* of A. Show that the matrix A^{-} defined in Exercise 13 also satisfies:

(i) $A^{-}AA^{-} = A^{-}$.

(ii) $(AA^{-})' = AA^{-}$; $(A^{-}A)' = A^{-}A$.

The (unique) generalized inverse that satisfies conditions (i) and (ii) is known as the *Moore-Penrose inverse*.

*15. Show that, if the system of equations $Ab = c$ is consistent (i.e., has at least one solution b^*), and A^{-} is a generalized inverse of A, then $b = A^{-}c$ is also a solution. Show that the general form of solution is

$$b = A^{-}c + (I - A^{-}A)h$$

where h is an arbitrary real vector.

*16. Show that if A^{-} is the Moore-Penrose inverse of A; then $b = A^{-}c$ makes $b'b$ a minimum among all vectors b such that $Ab = c$.

*17. Calculate the Moore-Penrose inverse of the matrix $\mathbf{S}_4$ of Section 3.8.

*18. Let T be upper triangular such that $T'T = X'X$, and let $\bar{T}$ be obtained from T by replacing any zero diagonal element by 1.0, and let $Z = \bar{T} - T$. Let C, with full column rank, be such that its columns form a complete set of linear relations $XC = 0$. Let $F = C(C'C)^{-1}C'$. Prove that

(i) $\bar{T}'^{-1}Z = Z$.

(ii) C may be formed by taking those columns of $\bar{T}^{-1}Z$ which are not identically zero.

(iii) $X(\bar{T}'\bar{T})^{-1}Z = \mathbf{0}$.

(iv) $(I - F)(\bar{T}'\bar{T})^{-1}Z = \mathbf{0}$.

***19.** Using the notation and results of exercises 2, 14, and 18 show that

$$(I - F)(\bar{T}'\bar{T})^{-1}(I - F)$$

is the Moore-Penrose inverse of $X'X$.

20. Algorithms are available for solving both lower and upper triangular systems of equations. Given X and the upper triangular matrix T such that $T'T = X'X$ show how to use these two algorithms to form $X(X'X)^{-1}$. [This allows simultaneous calculation of all sets of partial residuals, as explained in Section 3.6.]

21. Use the following subroutines to put together a BASIC program for inverting a positive definite symmetric matrix:

(i) For the Cholesky decomposition (lines 130–180 in Fig. 10.8; line 210 and lines 3450–3720 in Fig. 10.7).

(ii) For inversion of an upper triangular matrix (lines 4750–4880 in Fig. 10.8).

(iii) For postmultiplication of an upper triangular matrix by its transpose (lines 8800–8980 in Fig. 10.8 with S2 set equal to one).

What happens if the matrix is not positive definite? [Note that the matrix T formed by the Cholesky decomposition overwrites the upper triangle of the input matrix. Diagonal elements of T^{-1} are calculated only when required in lines 8880 and 8890. Other elements of T^{-1} are stored in the lower triangle of the input matrix. Lines 8800–8980 calculate and print out the matrix elements but do not store them. Lines 9500–9550 are required for the printing.]

CHAPTER 4

Orthogonal Reduction to Upper Triangular Form

Several methods are available that form the upper triangular matrix $\boldsymbol{T}$ such that $\boldsymbol{T}'\boldsymbol{T} = \mathbf{X}_q'\mathbf{X}_q$ directly by applying orthogonal transformations to the columns of $\mathbf{X}_q$. These methods are easily adapted to handle the addition or deletion of rows of data. This provides a convenient context in which to discuss the effect of adding or deleting data points and to introduce the notion of leverage.

4.1. ORTHOGONAL REDUCTION METHODS—GENERAL

Orthogonal reduction methods were briefly discussed in Section 3.2. Given $\mathbf{X}_q = [\boldsymbol{X}, \boldsymbol{y}]$, suppose that a sequence of orthogonal rotations, together equivalent to premultiplication by an orthogonal matrix $\boldsymbol{Q}$, is applied to the columns of $\mathbf{X}_q$ to give

$$\boldsymbol{Q}\mathbf{X}_q = \begin{bmatrix} \boldsymbol{T} \\ \mathbf{0} \end{bmatrix} \tag{1.1}$$

where $\boldsymbol{T}$ is upper triangular. Then, as $\mathbf{X}_q'\boldsymbol{Q}'\boldsymbol{Q}\mathbf{X}_q = \mathbf{X}_q'\mathbf{X}_q = \boldsymbol{T}'\boldsymbol{T}$, the matrix $\boldsymbol{T}$ is the Cholesky decomposition of $\mathbf{X}_q'\mathbf{X}_q$. The matrix $\boldsymbol{T}$ can then be made the basis of calculations for determining the least squares regression of $\boldsymbol{y}$ upon columns of $\boldsymbol{X}$, just as in Chapters 1 through 3.

The Use of Planar (Jacobi or Givens Rotations)

Planar rotations provide a straightforward means for achieving the desired orthogonal reduction. As a simple special case consider

$$[X, y] = \left[\begin{array}{cc|c} a & g & y_1 \\ b & h & y_2 \end{array}\right]$$

Let $d = (a^2 + b^2)^{1/2}$, $c = ad^{-1}$, $s = bd^{-1}$, and

$$Q = \begin{bmatrix} c & s \\ -s & c \end{bmatrix}$$

Observe that $c = \cos\theta$, $s = \sin\theta$, with $\theta = \arccos(ad^{-1})$. Then

$$QX = \begin{bmatrix} d & d^{-1}(ag + bh) \\ 0 & d^{-1}(-bg + ah) \end{bmatrix}, \quad Qy = \begin{bmatrix} d^{-1}(ay_1 + by_2) \\ d^{-1}(-by_1 + ay_2) \end{bmatrix}$$

The use of such two-dimensional or planar rotations (*Jacobi* or *Givens* rotations) to reduce arbitrary real matrices to upper triangular form will be discussed in Section 4.3. In general, a number of such rotations are required. Each rotation operates on two rows only of the matrix and reduces to zero just one below diagonal element.

At this stage the reader may wish to look at the numerical example which is given in Section 4.3.

More General Types of Orthogonal Rotation

It is natural to consider how planar rotations can be generalized to give computationally convenient forms of orthogonal rotation in an arbitrary number of dimensions. This leads to the *Householder* algorithm, and to the *modified Gram-Schmidt* (MGS) algorithm which in the present account will be regarded as a variant of Householder's. In either case an individual rotation may be used to reduce to zero all below diagonal elements in one column of the matrix.

Whatever algorithm is used, the complete reduction requires a sequence of orthogonal transformations represented by matrices $Q_0, Q_1, \ldots, Q_m$, say. The matrix of the finally achieved reduction is then

$$Q = Q_m \cdots Q_1 Q_0$$

which is orthogonal because it is a product of orthogonal matrices.

A Minor Change in the Basic Scheme

Suppose $\boldsymbol{Q}$ is such that

$$\boldsymbol{QX} = \begin{bmatrix} \boldsymbol{T}_p \\ \boldsymbol{0} \end{bmatrix} \tag{1.2}$$

that is, below diagonal elements of $\boldsymbol{X}$ are reduced to zero. Applying the same sequence of orthogonal transformations to $\boldsymbol{y}$, let

$$\boldsymbol{Qy} = \begin{bmatrix} \boldsymbol{t}_y \\ \boldsymbol{z} \end{bmatrix}$$

where $\boldsymbol{t}_y$ has the same number of rows as $\boldsymbol{T}_p$. Then as the further transformations that are applied to the matrix

$$\begin{bmatrix} \boldsymbol{T}_p & \boldsymbol{t}_y \\ \boldsymbol{0} & \boldsymbol{z} \end{bmatrix}$$

to reduce it to upper triangular form affect only the elements of $\boldsymbol{z}$; the submatrix $[\boldsymbol{T}_p \quad \boldsymbol{t}_y]$ is identical with the matrix obtained by deleting the final row from the matrix $\boldsymbol{T}$ of eq. (1.1). Furthermore the final diagonal element of $\boldsymbol{T}$ is $t_{yy} = (\boldsymbol{z}'\boldsymbol{z})^{1/2}$.

There is available a direct proof that the vector $\boldsymbol{b}$ of regression coefficients is the solution of the equation $\boldsymbol{T}_p\boldsymbol{b} = \boldsymbol{t}_y$. Let $\boldsymbol{e} = \boldsymbol{y} - \boldsymbol{Xb}$; the vector $\boldsymbol{b}$ of regression coefficients is chosen to minimize

$$\boldsymbol{e}'\boldsymbol{e} = (\boldsymbol{Qe})'\boldsymbol{Qe} \tag{1.3}$$

This is the sum of squares of elements of

$$\begin{aligned} \boldsymbol{Qe} &= \boldsymbol{Qy} - \boldsymbol{QXb} \\ &= \begin{bmatrix} \boldsymbol{t}_y - \boldsymbol{T}_p\boldsymbol{b} \\ \boldsymbol{z} - \boldsymbol{0} \end{bmatrix} \end{aligned} \tag{1.4}$$

It follows that the sum of squares (1.3) is greater than or equal to $\boldsymbol{z}'\boldsymbol{z}$, with equality if and only if $\boldsymbol{T}_p\boldsymbol{b} = \boldsymbol{t}_y$. Note, incidentally, that whereas $\boldsymbol{t}_y$ is unique, $\boldsymbol{z}$ is not and depends on the manner in which $\boldsymbol{X}$ is reduced to upper triangular form.

The Order of Columns of X

The usual descriptions of the Householder and modified Gram-Schmidt (MGS) algorithms assume that complete columns are modified at each step. This means that the decision on which of the remaining columns of X will be taken to form column k of T can be delayed till earlier columns of T have all been formed. Common criteria for ordering the columns of X are:

1. Maximize the amount by which the residual sum of squares is reduced at each stage, as in forward stepwise regression.
2. Choose at each stage the column of X which maximizes $s_{kk}^{-1}t_{kk}^2$ or a related quantity.

Use of the second criterion ensures that any very small values of t_{kk} come at the end. A benefit of this is that omission of the corresponding variates from subsequent calculations, as may be desirable, is then straightforward. The algorithms are numerically stable whether or not columns are ordered in some such manner.

Section 4.4 shows how to modify T, once it has been formed, in a manner that corresponds to any desired reordering of the columns of X.

4.2. RESIDUALS

From eq. (1.4) the vector e of residuals is such that

$$Qe = \begin{bmatrix} 0 \\ z \end{bmatrix}, \quad e = Q'\begin{bmatrix} 0 \\ z \end{bmatrix} \tag{2.1}$$

An arbitrary vector in the column space of X may be written Xc; applying the same orthogonal rotation Q

$$QXc = \begin{bmatrix} T_p c \\ 0 \end{bmatrix}$$

Thus e lies in an $(n - p - 1)$-dimensional subspace of n-space, orthogonal to the space spanned by the columns of X.

It is pertinent to ask how the elements z_i of z relate to the usual residuals e_i. The MGS algorithm needs to be considered separately; it applies orthogonal rotations to a matrix obtained from $\mathbf{X}_q$ by adding as many initial rows of zeros as X has columns, and the elements of z are the ordinary residuals. In other cases the details of the relationship depend on the matrix $\boldsymbol{Q}$ which has been used to reduce $\boldsymbol{X}$ to upper triangular form. However, under the usual model assumptions as in Section 1.13, and for algorithms other than MGS, the z_i are independently and identically distributed with common variance σ^2. This contrasts with

$$\operatorname{var}(\boldsymbol{e}) = (\boldsymbol{I} - \boldsymbol{P})\sigma^2 \tag{2.2}$$

where $\boldsymbol{P} = \boldsymbol{X}(\boldsymbol{X}'\boldsymbol{X})^{-1}\boldsymbol{X}'$.

Calculation of the Elements of e

Complication is kept to a minimum if the elements of $[\boldsymbol{X}, \boldsymbol{y}]$ are stored, allowing residuals to be calculated as $\boldsymbol{e} = \boldsymbol{y} - \boldsymbol{X}\boldsymbol{b}$. An alternative approach is to overwrite $\boldsymbol{X}$ with information from which the matrix $\boldsymbol{Q}$ can be reconstructed. Let $\boldsymbol{Q}' = [\boldsymbol{Q}'_{(1)}, \boldsymbol{Q}'_{(2)}]$, where $\boldsymbol{Q}_{(2)}$ has the same number of rows as $\boldsymbol{z}$ has elements. Then, by eq. (2.1),

$$\boldsymbol{e} = \boldsymbol{Q}'\begin{bmatrix} \boldsymbol{0} \\ \boldsymbol{z} \end{bmatrix} = \boldsymbol{Q}'_{(2)}\boldsymbol{z}$$

[See Chambers (1977).]

Plots of Residuals

In order to check on normality assumptions, it is common to compare the distribution of the e_i, perhaps standardized to take account of variation in the diagonal elements of the matrix $\boldsymbol{I} - \boldsymbol{P}$ in eq. (2.2), with that for a theoretical normal distribution. The normal probability plot is a favorite device for this purpose. Other types of check rely on looking for a pattern in the plot of residuals against one or another variate. Correlations between the e_i will rarely matter for either type of plot. If $n - p$ is large relative to p, correlations will be small; whereas if $n - p$ is small, sampling variation is likely to dwarf the effect of correlation.

In deciding whether to use the z_i in some appropriate way in place of the e_i, attention should be centered more on the different perspective that different versions of the z_i may provide, not on the absence of correlation. Thus when planar rotations are used, the squares of the z_i (which are now known as the *Givens* residuals) show how the residual sum of squares changes as data points are added one by one to the model in the chosen order. If, for example, data points appear in order of time, the *Givens* residuals may be useful in detecting changes of model with time.

4.3. THE METHOD OF GIVENS (PLANAR ROTATIONS)

An individual planar rotation, applied to the current version of the matrix $\boldsymbol{X}$, is equivalent to premultiplication by a matrix of the form

$$\begin{bmatrix} 1 & & & & & & & & & \\ & \ddots & & & & & & & & \\ & & 1 & & & & & & & \\ & & & c & & & s & & & \\ & & & & 1 & & & & & \\ & & & & & \ddots & & & & \\ & & & & & & 1 & & & \\ & & & -s & & & c & & & \\ & & & & & & & 1 & & \\ & & & & & & & & \ddots & \\ & & & & & & & & & 1 \end{bmatrix} \begin{matrix} \\ \\ \\ \leftarrow \text{Row } i \\ \\ \\ \\ \leftarrow \text{Row } \ell \\ \\ \\ \\ \end{matrix}$$

where $c = \cos\theta$, $s = \sin\theta$. Except for the s and the $-s$, all off-diagonal elements are zero. The transformation leaves unchanged all rows except row i and row ℓ.

Suppose that earlier transformations have reduced to zero below diagonal elements in rows 1 to ℓ of $\boldsymbol{X}$. Then, in order to reduce to zero below diagonal elements in the $(\ell+1)$th row of $\boldsymbol{X}$, it is rotated successively with the first, second,... rows, thus

$$\begin{bmatrix} c & s \\ -s & c \end{bmatrix} \begin{bmatrix} \times & \times & \times & \cdots & \times \\ \times & \times & \times & \cdots & \times \end{bmatrix} \begin{matrix} \leftarrow \text{Initial row} \\ \leftarrow (\ell+1)\text{th row} \end{matrix}$$

$$= \begin{bmatrix} \times & \times & \times & \cdots & \times \\ 0 & \times & \times & \cdots & \times \end{bmatrix} \begin{matrix} \leftarrow \text{New initial row} \\ \leftarrow \text{New } (\ell+1)\text{th row} \end{matrix}$$

The new $(\ell+1)$th row is then rotated with the second row, in order to reduce its second element to zero:

$$\begin{bmatrix} c' & s' \\ -s' & c' \end{bmatrix} \begin{bmatrix} 0 & \times & \times & \cdots & \times \\ 0 & \times & \times & \cdots & \times \end{bmatrix} \begin{matrix} \leftarrow \text{Second row} \\ \leftarrow \text{Current } (\ell+1)\text{th row} \end{matrix}$$

$$= \begin{bmatrix} 0 & \times & \times & \cdots & \times \\ 0 & 0 & \times & \cdots & \times \end{bmatrix}$$

This process continues until all below diagonal elements in the $(\ell+1)$th row have been reduced to zero.

Example

Consider

$$[X, y] = \begin{bmatrix} 3 & -1 & -3 \\ 0 & 1.5 & 1 \\ 4 & 2 & 1 \\ 5 & 1 & 3 \end{bmatrix}$$

The associated least-squares problem is that of finding $\boldsymbol{b}$ (elements b_1, b_2) to minimize the sum of squares of elements of $\boldsymbol{y} - \boldsymbol{Xb}$. Notice that no constant term has been included in the regression equation. Calculations are shown in Table 4.1. A rotation involving rows 1 and 2 is unnecessary as the first element in row 2 is already zero. For each of rows 3 and 4 the algorithm then requires:

1. A rotation with row 1, to reduce the first element in the row to zero.
2. A rotation with row 2, to reduce the second element in the row to zero. In Table 4.1 row 4 does not require this rotation, because its second element is at this point already zero.

Table 4.1. Use of Planar Rotations to Reduce the First Two Columns of a 4 by 3 Matrix to Upper Triangular Form

$$\begin{matrix} \rightarrow \\ \\ \rightarrow \\ \\ \end{matrix} \begin{bmatrix} \boxed{3} & -1 & -3 \\ 0 & 1.5 & 1 \\ \boxed{4} & 2 & 1 \\ 5 & 1 & 3 \end{bmatrix} \xrightarrow{c = \frac{3}{5},\ s = \frac{4}{5}} \begin{matrix} \\ \rightarrow \\ \rightarrow \\ \\ \end{matrix} \begin{bmatrix} 5 & 1 & -1 \\ 0 & \boxed{1.5} & 1 \\ 0 & \boxed{2} & 3 \\ 5 & 1 & 3 \end{bmatrix}$$

$$\xrightarrow{c = \frac{3}{5},\ s = \frac{4}{5}}$$

$$\begin{matrix} \rightarrow \\ \\ \\ \rightarrow \end{matrix} \begin{bmatrix} \boxed{5} & 1 & -1 \\ 0 & 2.5 & 3 \\ 0 & 0 & 1 \\ \boxed{5} & 1 & 3 \end{bmatrix} \xrightarrow{c = \frac{1}{\sqrt{2}},\ s = \frac{1}{\sqrt{2}}} \begin{bmatrix} 5\sqrt{2} & \sqrt{2} & \sqrt{2} \\ 0 & 2.5 & 3 \\ 0 & 0 & 1 \\ 0 & 0 & 2\sqrt{2} \end{bmatrix}$$

Note: At each step arrows pointing to the right identify the pair of rows involved in the next rotation.

The Givens Residuals

Using the notation of eq. (2.1):

$$z = \begin{bmatrix} 1 \\ 2\sqrt{2} \end{bmatrix}$$

The residual sum of squares is thus $1^2 + (2\sqrt{2})^2 = 9$. Now observe that, if the final row of (X, y) is omitted and one proceeds as before, calculations cease upon obtaining the matrix

$$\begin{bmatrix} 5 & 1 & -1 \\ 0 & 2.5 & 3 \\ 0 & 0 & 1 \end{bmatrix}$$

This would give a residual sum of squares equal to 1 (i.e., the square of the element in row 3, column 3). Omission of the final data point reduces the residual sum of squares by an amount equal to the square of the final element of z.

For the discussion which now follows it will be convenient to assume that columns of $[X, y]$ are numbered from 1 to $p + 1$. In general, planar rotations applied to a matrix $[X, y]$, lead to

$$\begin{bmatrix} T_p & t_y \\ \mathbf{0} & z \end{bmatrix}$$

where

$$z = \begin{bmatrix} z_{p+1,\,p+1} \\ \vdots \\ z_{n,\,p+1} \end{bmatrix}$$

Suppose that calculations proceed as shown, and that rows (data points) $1, 2, \ldots, \ell - 1$ of $[X, y]$, where $\ell > p$, have been rotated into the upper triangular scheme. Subsequent rotations then involve rows $1, 2, \ldots, p$ (which contain an intermediate stage in the formation of T_p and t_y) and in turn the rows $\ell, \ell + 1, \ldots, n$ of $[X, y]$. Figure 4.1 gives a diagrammatic representation, at the point immediately before carrying out calculations involving the ℓth row of $[X, y]$. The residual sum of squares after bringing in the first $\ell - 1$ data points is

$$\sum_{j=p+1}^{\ell-1} z_{j,\,p+1}^2$$

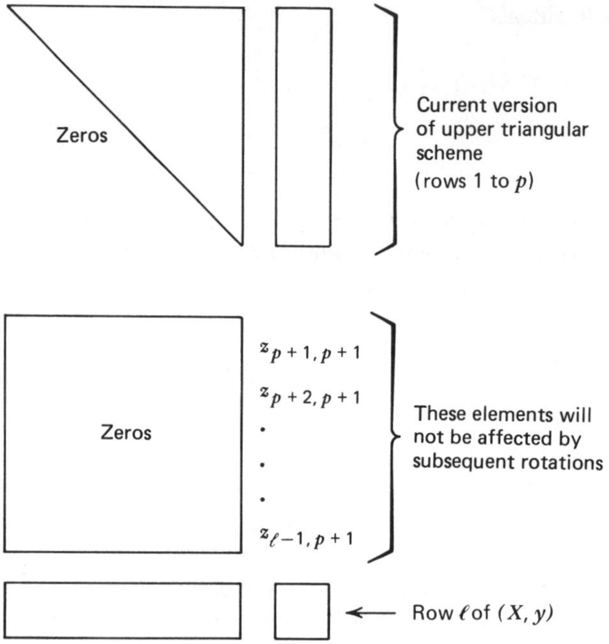

Figure 4.1. The formation of Givens residuals when planar rotations are used to reduce all except the final column of $[\mathbf{X}, \mathbf{y}]$ to upper triangular form.

and the inclusion of the ℓth data point will increase this by an amount $z_{\ell, p+1}^2$. Thus the "Givens" residuals give a data point by data point breakdown of the residual sum of squares. Any sudden change as one proceeds down the rows of $[\boldsymbol{X}, \boldsymbol{y}]$, so that the linear model applicable for the first $\ell - 1$ data points is no longer applicable for the ℓth and subsequent data points, should lead to a sudden increase in the magnitude of the Givens residuals.

Reordering and Deleting Columns of X

When planar rotations are used it is usually best to retain the initial order of columns of $\boldsymbol{X}$ until all rows of $\boldsymbol{X}$ have been processed. If on the basis of the evidence then available it is decided that columns should be taken in some different order, the method of the next section may be used to modify $[\boldsymbol{T}_p, \boldsymbol{t}_y]$ in the manner required. Or in order to fit a regression through the origin, one may want to modify $[\boldsymbol{T}_p, \boldsymbol{t}_y]$ to correspond to the matrix obtained by deleting an initial column of ones from $[\boldsymbol{X}, \boldsymbol{y}]$.

An important case is that in which it becomes apparent that column k of $\boldsymbol{X}$, that is, $\boldsymbol{x}_k$, is close to a linear combination of earlier columns. In its calculations the computer takes, not $\boldsymbol{x}_k$, but a vector $\boldsymbol{x}_k + \boldsymbol{\eta}_k$, where $\boldsymbol{\eta}_k$ is used to represent the effect of rounding error in the subsequent calculations. Because $\boldsymbol{x}_k$ adds little or nothing to the information in earlier columns of $\boldsymbol{X}$, the values in row k of $\boldsymbol{T}$ depend quite critically on $\boldsymbol{\eta}_k$ and are nonsense. [Eq. (5.3) in Chapter 3 makes this clear.] However, if now column k is moved to the final column position by the method of the next section, the vector $\boldsymbol{x}_k + \boldsymbol{\eta}_k$ is transposed, and all is well. This final column can then be ignored in subsequent calculations. Or alternatively all elements in row positions k and later in the new final column may be set to zero, the column placed back in column position k, and a row of zeros interposed in row position k. See the example in Section 4.5. The upper triangle matrix then obtained can be made the basis for subsequent calculations in the manner discussed in Chapter 3.

4.4. THE REARRANGEMENT OF COLUMNS OF X

Suppose that $\boldsymbol{X}^*$ has the same columns as $\boldsymbol{X}$, but taken in a different order, and that the same reordering has been applied to the columns of $\boldsymbol{T}_p$ to give $\tilde{\boldsymbol{T}}_p$. Then

$$\boldsymbol{QX} = \begin{bmatrix} \boldsymbol{T}_p \\ \boldsymbol{0} \end{bmatrix} \quad \text{implies} \quad \boldsymbol{QX}^* = \begin{bmatrix} \tilde{\boldsymbol{T}}_p \\ \boldsymbol{0} \end{bmatrix}$$

Thus consider

$$\boldsymbol{Q}\begin{bmatrix} 1 & -2 & 0 \\ 1 & -1 & 2 \\ 1 & 2 & 5 \\ 1 & 7 & 3 \end{bmatrix} = \begin{bmatrix} 2 & 3 & 5 \\ 0 & 7 & 2 \\ 0 & 0 & 3 \\ 0 & 0 & 0 \end{bmatrix}$$

This may be rewritten, with columns 2 and 3 interchanged,

$$\boldsymbol{Q}\begin{bmatrix} 1 & 0 & -2 \\ 1 & 2 & -1 \\ 1 & 5 & 2 \\ 1 & 3 & 7 \end{bmatrix} = \begin{bmatrix} 2 & 5 & 3 \\ 0 & 2 & 7 \\ 0 & 3 & 0 \\ 0 & 0 & 0 \end{bmatrix}$$

Now suppose orthogonal rotations (represented by $\tilde{\boldsymbol{Q}}$) are applied to columns of $\tilde{\boldsymbol{T}}_p$ to reduce it to an upper triangular matrix $\boldsymbol{T}_p^*$. Then

$$\begin{aligned} \begin{bmatrix} \boldsymbol{T}_p^* \\ 0 \end{bmatrix} &= \tilde{\boldsymbol{Q}}\begin{bmatrix} \tilde{\boldsymbol{T}}_p \\ 0 \end{bmatrix} \\ &= \tilde{\boldsymbol{Q}}\boldsymbol{QX}^* \\ &= \boldsymbol{Q}^*\boldsymbol{X}^* \end{aligned}$$

Thus T_p^* is the matrix that would result from reducing X^* to upper triangular form.

In the example just given, $\tilde{T}_p$ may be reduced to upper triangular form by applying a suitable planar rotation to rows 2 and 3:

$$T_p^* = \begin{bmatrix} 1 & 0 & 0 \\ 0 & \cos\theta & \sin\theta \\ 0 & -\sin\theta & \cos\theta \end{bmatrix} \begin{bmatrix} 2 & 5 & 3 \\ 0 & 2 & 7 \\ 0 & 3 & 0 \end{bmatrix}$$

We require

$$-2\sin\theta + 3\cos\theta = 0$$

that is,

$$\cos\theta = 2/\sqrt{13}, \quad \sin\theta = 3/\sqrt{13}$$

so that

$$T_p^* = \begin{bmatrix} 2 & 5 & 3 \\ 0 & \sqrt{13} & 14/\sqrt{13} \\ 0 & 0 & -21/\sqrt{13} \end{bmatrix}$$

Having determined the upper triangular matrix for one ordering of the columns of X, it is thus straightforward to obtain the upper triangular matrix that corresponds to some different ordering of the columns. Note that if the original orthogonal transformations have also been applied to the vector y to give Qy, the additional transformation may be applied to Qy to give $\tilde{Q}Qy = Q^*y$.

Regression Through the Origin

Calculations can if desired begin by reducing to upper triangular form a matrix $[X, y]$ which has an initial column of ones. The above reordering method is then used to obtain the upper triangular form that corresponds to placing the column of ones in the final column position of X. In that position the column can be excluded from the calculations.

4.5. LINEAR DEPENDENCY—AN EXAMPLE

The example taken is that of Section 3.5, where $x_3 = x_1 - \frac{1}{2}x_2 + \frac{1}{2}$. Calculations were carried out in single precision on an LSI11/23, which lies at the lower end of the Digital Equipment PDP11 range of computers. Section 10.8 gives the BASIC program that was used.

For simplicity the complete matrix $\mathbf{X}_q = [X, y]$ is reduced to upper triangular form, yielding

$$
\begin{array}{cccccc}
\text{constant} & x_1 & x_2 & x_3 & x_4 & y
\end{array}
$$

$$
T = \begin{bmatrix}
3 & 2 & 1 & 3 & 1 & 2 \\
 & 6 & 2 & 5 & 1 & 3 \\
 & & 4 & -2 & 2 & 4 \\
 & & & 3.8542 \times 10^{-7} & 0.29888 & 1.3406 \\
 & & & & 2.9851 & 2.8808 \\
 & & & & & 2.2145
\end{bmatrix}
$$

The five most significant digits of each result are printed.

If the number in the (3, 3) position were meaningful, it would imply that

$$
R^2_{3(1,2)} \simeq \frac{5^2 + 2^2}{5^2 + 2^2 + (0.5614 \times 10^{-6})^2}
$$

$$
\simeq 1 - 1.09 \times 10^{-14}
$$

There need be no doubt that, to within machine precision, x_3 is a linear combination of x_1 and x_2. Using the method of Section 4.4 to move x_3 to the final column position yields

$$
\begin{array}{cccccc}
\text{constant} & x_1 & x_2 & x_4 & y & x_3
\end{array}
$$

$$
T^* = \begin{bmatrix}
3 & 2 & 1 & 1 & 2 & 3 \\
 & 6 & 2 & 1 & 3 & 5 \\
 & & 4 & 2 & 4 & -2 \\
 & & & 3 & 3 & 0 \\
 & & & & 2.4495 & 0 \\
 & & & & & 0
\end{bmatrix}
$$

Note that $2.4495 \simeq \sqrt{6}$.

The final row and column may now be deleted from T^*, in which case x_3 is left completely aside. Or alternatively the column that corresponds to x_3 may be placed back in column position 3, and a row of zeros interposed in row position 3. This leads to a matrix in which columns correspond to variates in their original order, and for which the methods of Chapter 3 may be used. Thus one obtains

$$\bar{T} = \begin{array}{c} \begin{array}{cccccc} \text{constant} & x_1 & x_2 & x_3 & x_4 & y \end{array} \\ \left[\begin{array}{cccccc} 3 & 2 & 1 & 3 & 1 & 2 \\ & 6 & 2 & 5 & 1 & 3 \\ & & 4 & -2 & 2 & 4 \\ & & & 0 & 0 & 0 \\ & & & & 3 & 3 \\ & & & & & 2.4495 \end{array}\right] \end{array}$$

This is in essence the matrix **T** of Section 3.9; here an initial row and column corresponding to the constant term is included.

*4.6. SQUARE-ROOT-FREE VARIANTS OF PLANAR ROTATIONS

Instead of forming T directly, square-root-free variants form $U = MT$, where $M = \text{diag}[\sqrt{m_0}, \sqrt{m_1}, \ldots, \sqrt{m_p}]$. As calculations proceed, the m_i change and are updated along with the elements of U. With this type of algorithm the number of arithmetic operations required can be reduced, when p is large, by a factor of nearly 2 by comparison with the direct use of planar rotations.

The two rows to which the next rotation is to be applied will be taken as

$$\begin{bmatrix} 0,\ldots,0, & u_i,\ldots,u_p \\ 0,\ldots,0, & v_i,\ldots,v_p \end{bmatrix} \begin{array}{l} \leftarrow \text{Scale factor } \sqrt{\mu} \\ \leftarrow \text{Scale factor } \sqrt{\nu} \end{array} \tag{6.1}$$

Here μ is the current value of m_i. (The use of planar rotations would give $\sqrt{\mu}\, u_j$ in place of u_j, and $\sqrt{\nu}\, v_j$ in place of v_j.) The rows just given are then replaced by

$$\begin{bmatrix} 0,\ldots,0, & u_i', & u_{i+1}',\ldots,u_p' \\ 0,\ldots,0, & 0, & v_{i+1}',\ldots,v_p' \end{bmatrix} \begin{array}{l} \leftarrow \text{Scale factor } \sqrt{\mu'} \\ \leftarrow \text{Scale factor } \sqrt{\nu'} \end{array}$$

Following Hammarling (1974), a suitable choice is

$$\mu' = \frac{\mu}{\left(1 + \dfrac{\nu}{\mu} \cdot \dfrac{v_i^2}{u_i^2}\right)}, \quad \nu' = \frac{\nu}{\left(1 + \dfrac{\nu}{\mu} \cdot \dfrac{v_i^2}{u_i^2}\right)}$$

$$u_j' = u_j + \left(\frac{\nu}{\mu} \cdot \frac{v_i}{u_i}\right) v_j, \quad j = i,\ldots,p$$

$$v_j' = v_j - \frac{v_i}{u_i} u_j, \quad j = i,\ldots,p$$

which is one of several possibilities mentioned by Hammarling. Equation (6.1) or its equivalent enables translation to and from results obtained using planar rotations.

*4.7. THE USE OF A VARIANT OF HOUSEHOLDER'S REFLECTIONS

The algorithm here proposed is completely described by eqs. (7.7) and (7.8). The reader who wishes may observe the notation of eqs. (7.1) and (7.2), and then proceed directly to eqs. (7.7) and (7.8) and to the example that follows.

An individual genuine *Householder reflection*, applied to columns of the matrix $\mathbf{X}_q = [X, y]$, is equivalent to premultiplication of $\mathbf{X}_q$ by a matrix

$$Q_w = \frac{I - 2ww'}{\tau}$$

where $\tau = \|w\|^2 = w'w$. Observe that $Q_w' Q_w = I$. The vector $Q_w a$, for any a, is a reflection of a in the plane through the origin which is at right angles to w. See Fig. 4.2. In the variant proposed the reflection may be followed by a change of sign in the first nonzero element of $(I - 2ww'/\tau)a$. The matrix of the transformation is thus

$$Q_w = I_*\left(\frac{I - 2ww'}{\tau}\right) \tag{7.1}$$

where I_* is either the identity matrix or is obtained from the identity by changing the appropriate diagonal element from 1 to -1.

The first Householder reflection is designed to reduce to zero the elements after the first in the initial column of X. The next transformation

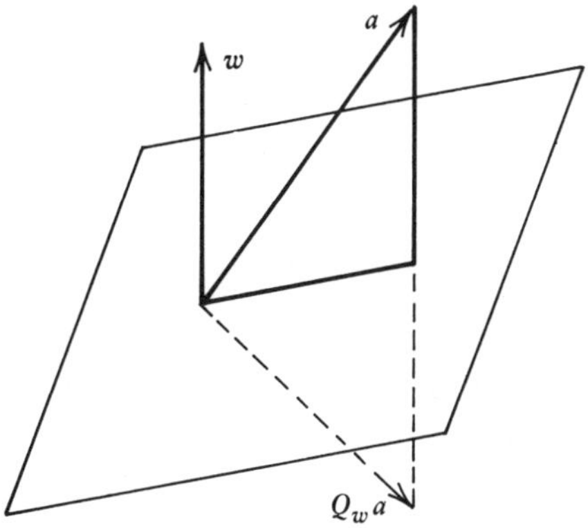

Figure 4.2. A Householder reflection.

reduces to zero the below diagonal elements (elements after the second) in the next column, leaving the initial row of the current version of the matrix unchanged. Following the use of transformations designed to reduce to zero the below diagonal elements in columns $0, 1, \ldots, k-1$, suppose that $\boldsymbol{X}$ is replaced by

$$\boldsymbol{X}^{(k-1)} = \begin{bmatrix} \boldsymbol{T}_{11} & \boldsymbol{T}_{12} \\ 0 & \boldsymbol{A}^{(k-1)} \end{bmatrix} \tag{7.2}$$

The matrix $\boldsymbol{T}_{11}$ is more properly written $\boldsymbol{T}_{11(k-1)}$, and similarly for $\boldsymbol{T}_{12}$. Let

$$\boldsymbol{A}^{(k-1)} = \left[\boldsymbol{a}_k, \boldsymbol{a}_{k+1}, \ldots, \boldsymbol{a}_p\right] \tag{7.3}$$

where $\boldsymbol{a}_j (j \geq k)$ is more properly written $\boldsymbol{a}_j^{(k-1)}$. As the transformation which must now be applied to the columns of $\boldsymbol{X}^{(k-1)}$ leaves $\boldsymbol{T}_{11}$ and $\boldsymbol{T}_{12}$ unchanged, the vector $\boldsymbol{w}$ in eq. (7.1) has elements 0 to $k-1$ all equal to zero. Let $\boldsymbol{w}$ be the vector obtained when the initial zeros of $\boldsymbol{w}$ are deleted. The discussion will be simplified if attention is restricted to the Householder reflection $(\boldsymbol{I} - 2\boldsymbol{w}\boldsymbol{w}'/\tau)$, applied to columns of $\boldsymbol{A}^{(k-1)}$.

The vector $\boldsymbol{w}$ is chosen to ensure that

$$\boldsymbol{I}_*\left(\frac{\boldsymbol{I} - 2\boldsymbol{w}\boldsymbol{w}'}{\tau}\right)\boldsymbol{A}^{(k-1)} = \begin{bmatrix} t_{kk}, \ldots, t_{kp} & \\ 0 & \\ \vdots & \boldsymbol{A}^{(k)} \\ 0 & \end{bmatrix} \tag{7.4}$$

where $\boldsymbol{I}_*$ differs from the identity matrix, if at all, in a manner that will ensure that $t_{kk} \geq 0$. It is readily verified that a suitable choice of $\boldsymbol{w}$ is

$$\boldsymbol{w} = \begin{bmatrix} \alpha_k \\ a_{k+1,k} \\ \vdots \\ a_{n,k} \end{bmatrix} \tag{7.5}$$

where $\alpha_k = a_{kk} + \|\boldsymbol{a}_k\| \operatorname{sgn}(a_{kk})$. Then

$$\tau = \|\boldsymbol{w}\|^2 = 2\|\boldsymbol{a}_k\| \cdot \left(|a_{kk}| + \|\boldsymbol{a}_k\|\right) \tag{7.6}$$

Algebraically, correct results will be obtained if the second term in the expression for α_k is subtracted and τ is changed accordingly. However, this would lead to an algorithm that is numerically unstable.

From eqs. (7.4), (7.5), and (7.6) one can then show, with a little careful algebra, that

$$t_{kk} = \|\boldsymbol{a}_k\|, \quad t_{kj} = \frac{\boldsymbol{a}'_k \boldsymbol{a}_j}{t_{kk}}, \quad (j > k) \tag{7.7}$$

$$a_{ij}^{(k)} = a_{ij}^{(k-1)} - a_{ik}^{(k-1)} \cdot |\alpha_k|^{-1} \cdot \left[a_{kj}^{(k-1)} \cdot \operatorname{sgn}(a_{kk}) + t_{kj} \right], \quad (i, j > k) \tag{7.8}$$

where $|\alpha_k| = |a_{kk}| + t_{kk}$.

Table 4.2. Use of a Variant of Householder Reflections to Reduce a Matrix to Upper Triangular Form

$$\begin{bmatrix} 1 & -2 & 0 & -3 \\ 1 & -1 & 2 & 1 \\ 1 & 2 & 5 & 2 \\ 1 & 7 & 3 & 6 \end{bmatrix}$$

$$t_{00} = \|\boldsymbol{a}_0\| = \sqrt{(1^2 + 1^2 + 1^2 + 1^2)} = 2, \qquad \alpha_0 = 2 + 1 = 3$$

$$t_{01} = 2^{-1}\boldsymbol{a}'_0\boldsymbol{a}_1 = 3, \qquad t_{02} = 2^{-1}\boldsymbol{a}'_0\boldsymbol{a}_2 = 5, \qquad t_{03} = 2^{-1}\boldsymbol{a}'_0\boldsymbol{a}_3 = 3$$

$$|\alpha_0|^{-1}\left[a_{0j}\operatorname{sgn}(a_{00}) + t_{0j}\right] = (*, \underset{j=1}{\tfrac{1}{3}}, \underset{j=2}{\tfrac{5}{3}}, \underset{j=3}{0})$$

$$\begin{bmatrix} 2 & 3 & 5 & 3 \\ 0 & -\frac{4}{3} & \frac{1}{3} & 1 \\ 0 & \frac{5}{3} & \frac{10}{3} & 2 \\ 0 & \frac{20}{3} & \frac{4}{3} & 6 \end{bmatrix}$$

Row 0: $t_{00}, t_{01}, t_{02}, t_{03}$, as above

Row 1: $(-1, 2, 1) - 1 \times (\frac{1}{3}, \frac{5}{3}, 0) = (-\frac{4}{3}, \frac{1}{3}, 1)$

Row 2: $(2, 5, 2) - 1 \times (\frac{1}{3}, \frac{5}{3}, 0) = (\frac{5}{3}, \frac{10}{3}, 2)$

Row 3: $(7, 3, 6) - 1 \times (\frac{1}{3}, \frac{5}{3}, 0) = (\frac{20}{3}, \frac{4}{3}, 6)$

$$t_{11} = \|\boldsymbol{a}_1\| = \sqrt{\left[\tfrac{1}{9}(4^2 + 5^2 + 20^2)\right]} = 7, \qquad |\alpha_1| = 7 + \tfrac{4}{3} = \tfrac{25}{3}$$

$$t_{12} = 7^{-1}\boldsymbol{a}'_1\boldsymbol{a}_2 = 2, \qquad t_{13} = 7^{-1}\boldsymbol{a}'_1\boldsymbol{a}_3 = 6$$

$$|\alpha_1|^{-1}\left[a_{1j}\operatorname{sgn}(a_{11}) + t_{1j}\right] = (*, *, \tfrac{1}{5}, \tfrac{3}{5})$$

$$\begin{bmatrix} 2 & 3 & 5 & 3 \\ 0 & 7 & 2 & 6 \\ 0 & 0 & 3 & 1 \\ 0 & 0 & 0 & 2 \end{bmatrix}$$

Row 1: t_{11}, t_{12}, t_{13}, as above

Row 2: $(\frac{10}{3}, 2) - \frac{5}{3} \times (\frac{1}{5}, \frac{3}{5}) = (3, 1)$

Row 3: $(\frac{4}{3}, 6) - \frac{20}{3} \times (\frac{1}{5}, \frac{3}{5}) = (0, 2)$

As the final element in column 2 (column 3) is already zero, calculations are complete.

Note: Calculations are demonstrated using the data from the example in Section 1.9 and elsewhere.

Table 4.2 demonstrates calculations based on these equations with the data of Section 1.9, for which

$$[X, y] = \begin{bmatrix} 1 & -2 & 0 & -3 \\ 1 & -1 & 2 & 1 \\ 1 & 2 & 5 & 2 \\ 1 & 7 & 3 & 6 \end{bmatrix}$$

*4.8. THE MODIFIED GRAM-SCHMIDT ALGORITHM

In the present development it is natural to view this as a variant of the Householder algorithm. It is left to the reader to work out the connection with the classical Gram-Schmidt algorithm which was used for theoretical purposes in Section 1.11. See Longley (1981) for a comparison of the two algorithms.

Suppose that orthogonal transformations are to be applied to $\mathbf{X}_q = [X, y]$ so that the n by $p + 1$ matrix X is reduced to upper triangular form. Let

$$G = \begin{bmatrix} \mathbf{0} \\ \mathbf{X}_q \end{bmatrix} \begin{matrix} \} \; q \text{ by } q + 1 \text{ matrix of zeros} \\ \} \; n \text{ by } q + 1 \end{matrix}$$

(Note: $q = p + 1$.) Reduction of X to upper triangular form by the use of the modified Gram-Schmidt (MGS) algorithm is equivalent to the reduction of columns 0 to p of G to upper triangular form by the use of the variant of the Householder scheme which was described in the last section.

At the stage where the Householder scheme has completed calculations associated with the formation of row $k - 1$ of T, the matrix G has been reduced to

$$G^{(k-1)} = \begin{bmatrix} T_{11} & T_{12} \\ \mathbf{0} & A^{(k-1)} \end{bmatrix} = \begin{bmatrix} T_{11} & T_{12} \\ \mathbf{0} & \mathbf{0} \\ \mathbf{0} & E^{(k-1)} \end{bmatrix} \begin{matrix} \left.\vphantom{\begin{matrix} a \\ b \end{matrix}}\right\} q \text{ rows} \\ \} \; n \text{ rows} \end{matrix}$$

Write

$$E^{(k-1)} = \left[e_k^{(k-1)}, e_{k+1}^{(k-1)}, \ldots, e_q^{(k-1)} \right]$$

The column $e_j^{(k-1)}$ is the current version of column j of $\mathbf{X}_q$.

For purposes of applying eqs. (7.7) and (7.8) of Section 4.4 the elements $a_{kj}^{(k-1)}$ are the elements of the initial row of $A^{(k-1)}$ and are thus all zero.

These equations then lead to

$$t_{kk} = \left\| e_k^{(k-1)} \right\|, \quad t_{kj} = \frac{e_k^{(k-1)\prime} e_j^{(k-1)}}{t_{kk}} \tag{8.1}$$

$$e_j^{(k)} = e_j^{(k-1)} - e_k^{(k-1)} \cdot \frac{t_{kj}}{t_{kk}}, \quad j \geq k \tag{8.2}$$

Observe that this implies $\boldsymbol{e}_k^{(k)} = \mathbf{0}$.

Table 4.3 demonstrates calculations based on the use of these equations. The matrix which is reduced to upper triangular form is

$$[X, y] = \begin{bmatrix} 1 & -4 & 0 & -6 \\ 1 & -2 & 4 & 2 \\ 1 & 4 & 10 & 4 \\ 1 & 14 & 6 & 12 \end{bmatrix}$$

MGS Described in Terms of Repeated Simple Regressions

In eq. (8.2) set $\boldsymbol{x} = \boldsymbol{e}_k^{(k-1)}$, $\boldsymbol{y} = \boldsymbol{e}_j^{(k-1)}$ $(j > k)$. Then in the regression $\hat{\boldsymbol{y}} = b\boldsymbol{x}$ of $\boldsymbol{y}$ upon $\boldsymbol{x}$, the regression coefficient is

$$b = \frac{\boldsymbol{x}'\boldsymbol{y}}{\boldsymbol{x}'\boldsymbol{x}} = \frac{t_{kj}}{t_{kk}}$$

Hence $\boldsymbol{e}_j^{(k)} = \hat{\boldsymbol{y}} - b\boldsymbol{x}$ is the vector of residuals when $\boldsymbol{e}_j^{(k-1)}$ is regressed upon $\boldsymbol{e}_k^{(k-1)}$.

Now suppose, for the purposes of an argument by induction, that $\boldsymbol{e}_j^{(k-1)}$ $(j = k, k+1, \ldots, p)$ is the vector of residuals from the regression of $\boldsymbol{x}_j$ upon columns of $\boldsymbol{X}_{k-1}$. The result of Section 2.14 then implies that $\boldsymbol{e}_j^{(k)}$ is the vector of residuals from the regression of $\boldsymbol{x}_j$ upon columns of $\boldsymbol{X}_k$.

*4.9. HYBRID METHODS

Suppose that, using Householder, we have formed

$$\begin{bmatrix} \boldsymbol{T}_{11} & \boldsymbol{T}_{12} & \boldsymbol{t}_{1y} \\ \mathbf{0} & \boldsymbol{A}^{(k-1)} & \boldsymbol{a}_y^{(k-1)} \end{bmatrix}$$

Then calculations may be continued from this stage by forming the matrix

Table 4.3. Use of the Modified Gram-Schmidt Algorithm to Reduce a Matrix to Upper Triangular Form

$$\begin{bmatrix} 0 & 0 & 0 & 0 \\ 0 & 0 & 0 & 0 \\ 0 & 0 & 0 & 0 \\ 1 & -4 & 0 & -6 \\ 1 & -2 & 4 & 2 \\ 1 & 4 & 10 & 4 \\ 1 & 14 & 6 & 12 \end{bmatrix}$$

$$t_{00} = \sqrt{(1^2 + 1^2 + 1^2 + 1^2)} = 2$$

$$t_{01} = (-4 - 2 + 4 + 14)/t_{00} = 6; \qquad t_{01}/t_{00} = 3$$

$$t_{02} = (0 + 4 + 10 + 6)/t_{00} = 10; \qquad t_{02}/t_{00} = 5$$

$$t_{03} = (-6 + 2 + 4 + 12)/t_{00} = 6; \qquad t_{03}/t_{00} = 3$$

$$\begin{bmatrix} -4 \\ -2 \\ 4 \\ 14 \end{bmatrix} - \begin{bmatrix} 1 \\ 1 \\ 1 \\ 1 \end{bmatrix} \frac{t_{01}}{t_{00}} = \begin{bmatrix} -7 \\ -5 \\ 1 \\ 11 \end{bmatrix} \quad \begin{bmatrix} 0 \\ 4 \\ 10 \\ 6 \end{bmatrix} - \begin{bmatrix} 1 \\ 1 \\ 1 \\ 1 \end{bmatrix} \frac{t_{02}}{t_{00}} = \begin{bmatrix} -5 \\ -1 \\ 5 \\ 1 \end{bmatrix} \quad \begin{bmatrix} -6 \\ 2 \\ 4 \\ 12 \end{bmatrix} - \begin{bmatrix} 1 \\ 1 \\ 1 \\ 1 \end{bmatrix} \frac{t_{03}}{t_{00}} = \begin{bmatrix} -9 \\ -1 \\ 1 \\ 9 \end{bmatrix}$$

$$\downarrow$$

$$\begin{bmatrix} 2 & 6 & 10 & 6 \\ 0 & 0 & 0 & 0 \\ 0 & 0 & 0 & 0 \\ 0 & -7 & -5 & -9 \\ 0 & -5 & -1 & -1 \\ 0 & 1 & 5 & 1 \\ 0 & 11 & 1 & 9 \end{bmatrix}$$

$$t_{11} = \sqrt{(7^2 + 5^2 + 1^2 + 11^2)} = 14$$

$$t_{12} = (7 \times 5 + 5 \times 1 + 1 \times 5 + 11 \times 1)/t_{11} = 4; \qquad t_{12}/t_{11} = \tfrac{2}{7}$$

$$t_{13} = (7 \times 9 + 5 \times 1 + 1 \times 1 + 11 \times 9)/t_{11} = 12; \qquad t_{13}/t_{11} = \tfrac{6}{7}$$

$$\begin{bmatrix} -5 \\ -1 \\ 5 \\ 1 \end{bmatrix} - \begin{bmatrix} -7 \\ -5 \\ 1 \\ 11 \end{bmatrix} \frac{t_{12}}{t_{11}} = \begin{bmatrix} -3 \\ \frac{3}{7} \\ 4\frac{5}{7} \\ -2\frac{1}{7} \end{bmatrix} \quad \begin{bmatrix} -9 \\ -1 \\ 1 \\ 9 \end{bmatrix} - \begin{bmatrix} -7 \\ -5 \\ 1 \\ 11 \end{bmatrix} \frac{t_{13}}{t_{11}} = \begin{bmatrix} -3 \\ 3\frac{2}{7} \\ \frac{1}{7} \\ -\frac{3}{7} \end{bmatrix}$$

$$\downarrow$$

$$\begin{bmatrix} 2 & 6 & 10 & 6 \\ 0 & 14 & 4 & 12 \\ 0 & 0 & 0 & 0 \\ 0 & 0 & -3 & -3 \\ 0 & 0 & \frac{3}{7} & 3\frac{2}{7} \\ 0 & 0 & 4\frac{5}{7} & \frac{1}{7} \\ 0 & 0 & -2\frac{1}{7} & -\frac{3}{7} \end{bmatrix}$$

$$t_{22} = \sqrt{\left(3^2 + \tfrac{3}{7}^2 + 4\tfrac{5}{7}^2 + 2\tfrac{1}{7}^2\right)} = 6$$

$$t_{23} = \left(3 \times 3 + \tfrac{3}{7} \times 3\tfrac{2}{7} + 4\tfrac{5}{7} \times \tfrac{1}{7} + 2\tfrac{1}{7} \times \tfrac{3}{7}\right)/t_{22} = 2; \qquad t_{23}/t_{22} = \tfrac{1}{3}$$

$$\downarrow$$

$$\begin{bmatrix} -3 \\ 3\frac{2}{7} \\ \frac{1}{7} \\ -\frac{3}{7} \end{bmatrix} - \begin{bmatrix} -3 \\ \frac{3}{7} \\ 4\frac{5}{7} \\ -2\frac{1}{7} \end{bmatrix} \frac{t_{23}}{t_{22}} = \begin{bmatrix} -2 \\ 3\frac{1}{7} \\ -1\frac{3}{7} \\ \frac{2}{7} \end{bmatrix}$$

$$\begin{bmatrix} 2 & 6 & 10 & 6 \\ 0 & 14 & 4 & 12 \\ 0 & 0 & 6 & 2 \\ 0 & 0 & 0 & -2 \\ 0 & 0 & 0 & 3\frac{1}{7} \\ 0 & 0 & 0 & -1\frac{3}{7} \\ 0 & 0 & 0 & \frac{2}{7} \end{bmatrix}$$

(Last four rows:) Residuals from the regression of y on columns of $\boldsymbol{X}$

Note: The matrix used differs slightly from that in Table 4.2. Values in the final three columns have all been doubled in order to simplify the arithmetic.

of partial sums of squares and products

$$\begin{bmatrix} \boldsymbol{S}_{22.1} & \boldsymbol{s}_{y2.1} \\ \cdot & s_{yy.1} \end{bmatrix} = \begin{bmatrix} \boldsymbol{A}^{(k-1)\prime}\boldsymbol{A}^{(k-1)} & \boldsymbol{A}^{(k-1)\prime}\boldsymbol{a}_y^{(k-1)} \\ \cdot & \boldsymbol{a}_y^{(k-1)\prime}\boldsymbol{a}_y^{(k-1)} \end{bmatrix}$$

Calculations can then be completed by forming the Cholesky decomposition of this:

$$\begin{bmatrix} \boldsymbol{T}_{22} & \boldsymbol{t}_{2y} \\ \boldsymbol{0}' & t_{yy} \end{bmatrix}$$

A similar changeover is possible when the modified Gram-Schmidt (MGS) algorithm is used; only $\boldsymbol{E}^{(k-1)}$ replaces $\boldsymbol{A}^{(k-1)}$, and $\boldsymbol{e}_y^{(k-1)}$ replaces $\boldsymbol{a}_y^{(k-1)}$.

More generally, let

$$\boldsymbol{X} = \left[\boldsymbol{X}_{(1)}, \boldsymbol{X}_{(2)}\right], \quad \boldsymbol{b} = \begin{bmatrix} \boldsymbol{b}_{(1)} \\ \boldsymbol{b}_{(2)} \end{bmatrix}$$

as in Section 2.14.

The result of that section then shows that the vector $\boldsymbol{b}_{(2)}$ may be obtained in the following manner:

1. Each column of $[\boldsymbol{X}_{(2)}, \boldsymbol{y}]$ is regressed upon columns of $\boldsymbol{X}_{(1)}$, giving the matrix of residuals $[\boldsymbol{X}_{(2)} - \boldsymbol{X}_{(1)}\boldsymbol{B}, \boldsymbol{y} - \boldsymbol{X}_{(1)}\boldsymbol{b}_{(1)}^0]$.
2. The vector $\boldsymbol{b}_{(2)}$ of regression coefficients may then be obtained by regressing $\boldsymbol{y} - \boldsymbol{X}_{(1)}\boldsymbol{b}_{(1)}^0$ upon columns of $\boldsymbol{X}_{(2)} - \boldsymbol{X}_{(1)}\boldsymbol{B}$.
3. The residuals from this regression are the residuals that would have been obtained if $\boldsymbol{y}$ had been regressed upon columns of $\boldsymbol{X}$.
4. A knowledge of $\boldsymbol{B}$ and of $\boldsymbol{b}_{(1)}^0$ allows $\boldsymbol{b}_{(1)}$ to be determined; in fact, $\boldsymbol{b}_{(1)} = \boldsymbol{b}_{(1)}^0 - \boldsymbol{B}\boldsymbol{b}_{(2)}$.
5. $\operatorname{Var}(\boldsymbol{b}_{(2)}) = \boldsymbol{T}_{22}^{-1}\boldsymbol{T}_{22}^{\prime -1}\sigma^2$, where $\boldsymbol{T}_{22}$ may be obtained either by applying orthogonal reductions to $\boldsymbol{X}_{(2)} - \boldsymbol{X}_{(1)}\boldsymbol{B}$, or as the Cholesky decomposition of $\boldsymbol{S}_{22.1} = (\boldsymbol{X}_{(2)} - \boldsymbol{X}_{(1)}\boldsymbol{B})'(\boldsymbol{X}_{(2)} - \boldsymbol{X}_{(1)}\boldsymbol{B})$. (The notation is that of Section 3.5.)
6. In Section 2.13, eq. (13.10) it was shown that $\boldsymbol{b}_{(2)}$ is independent of $\boldsymbol{b}_{(1)}^0$. Hence the variance-covariance information for $\boldsymbol{b}_{(1)}$ may be

obtained from that for $\boldsymbol{b}^0_{(1)}$ by using

$$\operatorname{var}(\boldsymbol{b}_{(1)}) = \operatorname{var}(\boldsymbol{b}^0_{(1)}) + \operatorname{var}(\boldsymbol{B}\boldsymbol{b}_{(2)})$$

$$= \operatorname{var}(\boldsymbol{b}^0_{(1)}) + \boldsymbol{B}\boldsymbol{T}_{22}^{-1}\boldsymbol{T}_{22}'^{-1}\boldsymbol{B}'\sigma^2$$

Furthermore

$$\operatorname{cov}(\boldsymbol{b}_{(1)}, \boldsymbol{b}_{(2)}) = \operatorname{cov}(-\boldsymbol{B}\boldsymbol{b}_{(2)}, \boldsymbol{b}_{(2)})$$

$$= -\boldsymbol{B}\operatorname{var}(\boldsymbol{b}_{(2)})$$

The two sets of Regressions 1 and 2 may be carried out by quite different methods. It is in this way possible to take advantage of special methods for handling calculations for the orthogonal or balanced part of an experimental design [represented perhaps by columns of $\boldsymbol{X}_{(1)}$], while using general regression type calculations to handle other terms in the model.

4.10. THE ADDITION AND DELETION OF ROWS OF THE DATA MATRIX

The Addition of Rows

Suppose that

$$\boldsymbol{Q}\mathbf{X}_q = \begin{bmatrix} \boldsymbol{T} \\ \mathbf{0} \end{bmatrix} \tag{10.1}$$

and that a further row $\boldsymbol{l}'$ is to be added to $\mathbf{X}_q$. From the point of view of the discussion of Section 4.2 the matrix to be reduced to upper triangular form is

$$\begin{bmatrix} \mathbf{X}_q \\ \boldsymbol{l}' \end{bmatrix}$$

The rows of $\mathbf{X}_q$ have already been reduced to upper triangular form, so what is now required is to perform the calculations that incorporate $\boldsymbol{l}'$. This is readily achieved by the use of a sequence of Givens rotations so that

$$\begin{bmatrix} \boldsymbol{T} \\ \mathbf{0} \\ \boldsymbol{l}' \end{bmatrix}, \quad \text{or equivalently} \quad \begin{bmatrix} \boldsymbol{T} \\ \boldsymbol{l}' \end{bmatrix}$$

is reduced to upper triangular form.

If a number of further rows, those of $\mathbf{X}_+$, say, are to be added, this may be done one row at a time. Alternatively, Householder of Gram-Schmidt type rotations may be used to reduce all at once the rows of the matrix

$$\begin{bmatrix} \boldsymbol{T} \\ \boldsymbol{X}_+ \end{bmatrix}$$

to upper triangular form.

The Deletion of Rows

Suppose that

$$\mathbf{X}_q \overset{Q}{\rightarrow} \begin{bmatrix} \boldsymbol{T} \\ \mathbf{0} \end{bmatrix} \overset{\tilde{Q}}{\rightarrow} \begin{bmatrix} \tilde{\boldsymbol{T}} \\ \mathbf{0} \\ \boldsymbol{l}' \end{bmatrix} \tag{10.2}$$

where $\boldsymbol{l}'$ is a row of $\mathbf{X}_q$. Thus

$$\tilde{Q}Q\mathbf{X}_q = \begin{bmatrix} \tilde{\boldsymbol{T}} \\ \mathbf{0} \\ \boldsymbol{l}' \end{bmatrix}$$

that is, the total effect of the transformations represented by Q and $\tilde{Q}$ is to reduce to upper triangular form the matrix that results when the row $\boldsymbol{l}'$ is deleted from $\mathbf{X}_q$.

The rows of zeros which $\tilde{Q}$ leaves unchanged can be omitted from consideration. Suppose that the matrix of the orthogonal transformation is then $\overline{Q}$, so that

$$\overline{Q}\begin{bmatrix} \boldsymbol{T} \\ \mathbf{0}' \end{bmatrix} = \begin{bmatrix} \tilde{\boldsymbol{T}} \\ \boldsymbol{l}' \end{bmatrix} \tag{10.3}$$

To achieve the effect of deleting from $\mathbf{X}_q$ the row $\boldsymbol{l}'$ it is necessary only to find a sequence of orthogonal transformations, represented by $\tilde{Q}$ in eq. (10.2) and by $\overline{Q}$ in eq. (10.3), that leaves below diagonal zeros in the matrix $\boldsymbol{T}$ undisturbed and replaces the final row of zeros by the row vector $\boldsymbol{l}'$.

Let $\boldsymbol{v}' = (\boldsymbol{v}_q', v_{q+1})$ be the final row of $\overline{Q}$. Variates are as always numbered $1, 2, \ldots, p$, $q = p + 1$. Then, equating the final rows on the two sides of eq. (10.3),

$$\boldsymbol{v}_q'\boldsymbol{T} = \boldsymbol{l}' \tag{10.4}$$

This condition is sufficient to ensure that the row of zeros is replaced by $\boldsymbol{l}'$. The vector $\boldsymbol{v}_q$ may be obtained by solving the lower triangular system of equations $\boldsymbol{T}'\boldsymbol{v}_q = \boldsymbol{l}$ by forward substitution. Then $\boldsymbol{v}'\boldsymbol{v} = 1$ implies that $v_{q+1} = (1 - \boldsymbol{v}_q'\boldsymbol{v}_q)^{1/2}$. As the rows of $\overline{\boldsymbol{Q}}$ are mutually orthogonal, the condition that the final row is $\boldsymbol{v}'$ is equivalent to

$$\overline{\boldsymbol{Q}}\boldsymbol{v} = \begin{bmatrix} 0 \\ \vdots \\ 0 \\ 1 \end{bmatrix} \tag{10.5}$$

[To verify that eq. (10.5) implies that the final row of $\overline{\boldsymbol{Q}}$ is $\boldsymbol{v}'$, premultiply both sides by $\overline{\boldsymbol{Q}}'$.]

Finally observe that any sequence of planar rotations that rotates a row whose elements start as zeros successively with rows $q, q-1, \ldots, 0$ of $\boldsymbol{T}$ will leave below diagonal zeros of $\boldsymbol{T}$ undisturbed. The requirements of eq. (10.5) will be met if the first such rotation is designed to replace the last but one element of $\boldsymbol{v}$ by zero, the second to set to zero the last but two element, and so on.

Checks for Numerical Instability When a Row is Deleted

The major concern is with cases where $\boldsymbol{v}_q'\boldsymbol{v}_q$ is close to 1.0. Suppose that ε is defined as in Sections 3.4 and 3.9; it may alternatively be taken as the smallest number such that $1 + \varepsilon$ is calculated as greater than 1. Even if elements of $\boldsymbol{v}_q$ were accurate to machine precision, the calculated value of v_{q+1}^2 would still have a maximum absolute error of the order of $\boldsymbol{v}_q'\boldsymbol{v}_q\varepsilon \simeq \varepsilon$. Hence it is necessary to check that v_{q+1}^2 is substantially larger than ε.

For a detailed error analysis, see Stewart (1979). Stewart's arguments suggest that v_{q+1}^{-1} provides a good indication of the factor by which the relative error in elements of $\boldsymbol{T}$ may, on average, be increased as a result of the calculations. The deletion of a row may, if v_{q+1}^{-1} is large, lead to a substantial increase in the relative error of some or all elements of $\boldsymbol{T}$.

If, however, one's interest is in $\boldsymbol{T}$ as a representation of $\mathbf{X}_q'\mathbf{X}_q$, the prospects are much more encouraging. The relative error in elements of $\mathbf{X}_q'\mathbf{X}_q$ will be substantially increased only if deletion of the row leads to a substantial decrease in the largest eigenvalue of $\mathbf{X}_q'\mathbf{X}_q$. The effect on precision will be no worse than if the SSP matrix $\mathbf{X}_q'\mathbf{X}_q$ is modified in a manner that corresponds to deletion of a row, and the new Cholesky upper triangle matrix is formed from this.

Example

$$[\boldsymbol{X}, \boldsymbol{y}] = \begin{bmatrix} 1 & -2 & 0 & -3 \\ 1 & -1 & 2 & 1 \\ 1 & 2 & 5 & 2 \\ 1 & 7 & 3 & 6 \\ 1 & 4 & 2 & 9 \end{bmatrix}$$

A fifth row has been added to the data matrix of Section 1.9. The upper triangle matrix $\boldsymbol{T}$ thus obtained is as follows; underneath is printed the result of solving for $\boldsymbol{v}_q$ in $\boldsymbol{T}'\boldsymbol{v}_q = \boldsymbol{l}$, where $\boldsymbol{l}' = [1, 4, 2, 9]$. Also given is $v_{q+1} = (1 - \boldsymbol{v}_q'\boldsymbol{v}_q)^{1/2}$.

$$\boldsymbol{T} = \begin{bmatrix} 2.2361 & 4.4721 & 5.3666 & 6.7082 \\ & 7.3485 & 1.7961 & 7.7567 \\ & & 3.1734 & -0.5427 \\ & & & 5.0536 \end{bmatrix}$$

$$\boldsymbol{v}_q = \begin{bmatrix} 0.4472 \\ 0.2722 \\ -0.2788 \\ 0.7397 \end{bmatrix}$$

$$v_{q+1} = 0.3188$$

Observe that 0.4472 is the result of solving $2.2361 v_1 = 1$ for v_1, and so on.

Planar rotations are now applied to the 5 by 5 matrix

$$\begin{bmatrix} \boldsymbol{T} & \boldsymbol{v}_q \\ \boldsymbol{0}' & v_{q+1} \end{bmatrix}$$

The first rotation

$$\begin{bmatrix} c & -s \\ s & c \end{bmatrix}$$

has $s = 0.7397/d$, $c = 0.3188/d$, $d = (0.7397^2 + 0.3188^2)^{1/2}$, and rotates the final row with the final row but one. This rotation replaces 5.0536 by $t_{33} = 2$, replaces v_q by 0, and replaces v_{q+1} by $d = 0.8055$. The second rotation rotates the final row with the final row but two, and so on. The total effect (after four rotations) is to retrieve the upper triangular matrix

$$\tilde{\boldsymbol{T}} = \begin{bmatrix} 2 & 3 & 5 & 3 \\ & 7 & 2 & 6 \\ & & 3 & 1 \\ & & & 2 \end{bmatrix}$$

*4.11. THE CONTRIBUTION FROM INDIVIDUAL ROWS

The Contribution to the Residual Sum of Squares

The results that follow are quoted, in most cases without proof, from Gentleman and Wilk (1975). See also Belsey et al. (1980).

If just the ith row $[\boldsymbol{l}_i', y_i]$ of $[\boldsymbol{X}, \boldsymbol{y}]$ is omitted, then it may be shown that the residual sum of squares decreases by an amount $(1-h_i)^{-1}e_i^2$, where $h_i = \boldsymbol{l}_i'(\mathbf{X}'\mathbf{X})^{-1}\boldsymbol{l}_i$, and e_i is the ith residual. As $\text{var}(e_i) = (1-h_i)\sigma^2$, the decrease is proportional to the ith standardized residual. Outliers are points for which e_i, or equivalently the change in the residual sum of squares upon deletion of the ith row, are unexpectedly large in magnitude.

As an extension of this type of analysis, the decreases in the residual sum of squares from all $\binom{n}{k}$ possibilities for omitting data points k at a time may be examined, where perhaps $k = 2$ or 3. If any of the $\binom{n}{k}$ decreases is unexpectedly large, this will call for scrutiny. The hope is that this type of analysis will detect cases where two or more outliers have conspired to conceal their existence. If $\boldsymbol{e}_*$ is the vector of residuals that corresponds to the k omitted values, and $\boldsymbol{P}_*$ is defined as in eq. (11.1) below, it may be shown that the residual sum of squares decreases by an amount $\boldsymbol{e}_*'(\boldsymbol{I}-\boldsymbol{P}_*)^{-1}\boldsymbol{e}_*$ when the data points in question are omitted. Gentleman and Wilk (1975) suggest, among other possibilities, taking, say, 50 or 100 of the largest values of this statistic and plotting them against the 50 or 100 largest values of the same statistic for data that have been simulated to follow the model $\boldsymbol{y} = \boldsymbol{X}\boldsymbol{\beta} + \boldsymbol{\varepsilon}$. Gentleman and Wilk's results will now be reported in more detail.

For convenience, initially, it is assumed that the first k rows are to be omitted. In the full fitted model $\boldsymbol{y} = \boldsymbol{X}\boldsymbol{b} + \boldsymbol{e}$, let

$$\boldsymbol{e} = \begin{bmatrix} \boldsymbol{e}_* \\ \boldsymbol{e}_0 \end{bmatrix}, \quad \boldsymbol{X} = \begin{bmatrix} \boldsymbol{X}_* \\ \boldsymbol{X}_0 \end{bmatrix}, \quad \boldsymbol{y} = \begin{bmatrix} \boldsymbol{y}_* \\ \boldsymbol{y}_0 \end{bmatrix}$$

Let $\tilde{\boldsymbol{e}}_0$ be the vector of residuals in the reduced fitted model $\boldsymbol{y}_0 = \boldsymbol{X}_0\boldsymbol{b}_0 + \tilde{\boldsymbol{e}}_0$, and let $\boldsymbol{\delta}_* = \boldsymbol{y}_* - \boldsymbol{X}_*\boldsymbol{b}_0$ be the vector of residuals this model gives for the rows that have been deleted. Then, setting

$$\boldsymbol{P}_* = \boldsymbol{X}_*(\boldsymbol{X}'\boldsymbol{X})^{-1}\boldsymbol{X}_*' \tag{11.1}$$

it may be shown that

$$(\boldsymbol{I}_k - \boldsymbol{P}_*)\boldsymbol{\delta}_* = \boldsymbol{e}_* \tag{11.2}$$

and that omission of the rows of $[\boldsymbol{X}_*, \boldsymbol{y}_*]$ reduces the residual sum of squares by an amount

$$\begin{aligned} \boldsymbol{e}'\boldsymbol{e} - \tilde{\boldsymbol{e}}_0'\tilde{\boldsymbol{e}}_0 &= \boldsymbol{\delta}_*'(\boldsymbol{I}_k - \boldsymbol{P}_*)\boldsymbol{\delta}_* \\ &= \boldsymbol{e}_*'(\boldsymbol{I}_k - \boldsymbol{P}_*)^{-1}\boldsymbol{e}_* \end{aligned} \tag{11.3}$$

In general, $\boldsymbol{I}_k - \boldsymbol{P}_*$ is obtained by picking out the requisite k rows and columns from $\boldsymbol{I} - \boldsymbol{P}$, where $\boldsymbol{P} = \boldsymbol{X}(\boldsymbol{X}'\boldsymbol{X})^{-1}\boldsymbol{X}'$.

The matrix $\boldsymbol{P}$ may be calculated as $\boldsymbol{X}\boldsymbol{T}_p^{-1}(\boldsymbol{X}\boldsymbol{T}_p^{-1})'$. If $\boldsymbol{T}_p$ has one or more zero diagonal elements, then $\boldsymbol{T}_p^{-1}$ should be replaced by $\bar{\boldsymbol{T}}_p^{-1}$, where $\bar{\boldsymbol{T}}_p$ is obtained from $\boldsymbol{T}_p$ by replacing any diagonal zero by one. As in Section 3.14, $\boldsymbol{P}$ is defined by the requirement that $\boldsymbol{Ph} = \boldsymbol{h}$ if and only if $\boldsymbol{h}$ is a linear combination of the columns of $\boldsymbol{X}$. For calculation of $\boldsymbol{I} - \boldsymbol{P}$ the method given in Section 4.12 is the one that is to be preferred on numerical grounds.

Leverage Points

Omission of the ith data point causes the fitted value associated with the omitted point to change by an amount

$$\begin{aligned} \delta_i - e_i &= \frac{e_i}{1 - h_i} - e_i \quad \text{[by eq. (11.2)]} \\ &= \frac{h_i}{1 - h_i} e_i \end{aligned} \tag{11.4}$$

The relative increase in the ith residual is $h_i(1 - h_i)^{-1}$, which will be large if h_i is close to one. Large values for h_i are thus associated with points that noticeably lever the fitted plane in their direction. Such points are known as *leverage* points, and h_i is the *leverage* associated with the ith point.

Now let h_i be the leverage value when the calculation uses the centered data:

$$\mathsf{h}_i = \ell_i'(\mathbf{X}'\mathbf{X})^{-1}\ell_i$$

where ℓ_i consists of all elements except the first from $\boldsymbol{l}_i - \bar{\boldsymbol{l}}$. Then noting that h_i is the sum of squares of elements of $\mathbf{T}'^{-1}\ell_i$, and the similar expression for h_i, it may be shown that

$$\mathsf{h}_i = h_i - \frac{1}{n} \tag{11.5}$$

[See Exercise 8(iii), Chapter 2, with $T_{12} = \sqrt{n}(\bar{x}_1, \bar{x}_2, \ldots, \bar{x}_p)$.] Thus $h_i = h_i - n^{-1}$ may be interpreted as a squared metric measuring *distance* from l_i' to $\bar{l}'$. For a given Euclidean distance $(\ell_i'\ell_i)^{1/2}$, h_i is largest when ℓ_i is proportional to the eigenvector of $\mathbf{X}'\mathbf{X}$ which corresponds to its smallest eigenvalue. (See Exercise 8, Chapter 6.)

Belsey et al. (1980) include an extensive discussion on the use of the leverage statistics h_i. As a rough rule they suggest that particular attention should be paid to data points whose leverage h_i is greater than $2\tilde{p}/n$, where $\tilde{p} = p + 1$ is the rank of X. (The average leverage is $\tilde{p}/n$.)

A statistic suggested by Cook (refer to Cook and Weisberg, 1980) has been widely used as a measure of the influence of the ith data point. It takes the squared Euclidean distance that the vector $\hat{y}$ of fitted values moves when the ith data point is omitted, and divides this by $\tilde{p}s^2$. It may be calculated as:

$$\frac{e_i^2 h_i}{\tilde{p}s^2(1 - h_i)^2}$$

The Givens Residuals

Also readily available is information on the effect of omitting rows from the regression one by one in a predefined order. This provides a method, alternative to that of Sections 4.2 and 4.3, for calculating the Givens residuals; however, there it is assumed that the last row is omitted first. Let V be an orthogonal matrix such that $V(I - P) = U$ is upper triangular, and let U_* be the leading k by k submatrix of U. Then

$$U_*'U_* = I_k - P_* \tag{11.6}$$

and from eq. (11.2)

$$\delta_*'(I_k - P_*)\delta_* = (U_*\delta_*)'U_*\delta_*$$

$$= (U_*'^{-1}e_*)'U_*'^{-1}e_*$$

where if U_* is singular, U_*^{-1} should be replaced by $\bar{U}_*^{-1}$. But this is just the sum of squares of the first k elements of $\bar{U}'^{-1}e$; that is, the squares of the successive elements of $\bar{U}'^{-1}e$ give the amounts by which the residual sum of squares changes as rows are successively deleted.

*4.12. THE MATRIX $\boldsymbol{Q}$ OF THE ORTHOGONAL REDUCTION

Section 4.10 discussed a use for the matrix $\boldsymbol{Q}$ of the orthogonal reduction. In many applications it is convenient to use below diagonal storage locations, which in the mathematical description are filled with zeros, for keeping information used to reconstruct the sequence of transformations.

With the MGS algorithm the elements in column k of the current version of $\boldsymbol{X}$ at the stage where earlier columns have been filled with zeros determine the next transformation. All that is needed is to leave these [the elements of $\boldsymbol{e}_k^{(k-1)}$] in place. When Householder's reflections are used, the elements of $\boldsymbol{a}_k^{(k-1)}$ need to be kept. Elements subsequent to the (k, k) element can be left in place. The (k, k) position is required for t_{kk}, so that either $a_{kk}^{(k-1)}$ or α_k must be stored separately.

A convenient approach to calculation when Givens rotations are used is to start with an initial version of the matrix $\boldsymbol{T}$ with all elements zero. The c $(= \cos\theta)$ of Section 4.1 is then calculated as t/d, where t is the current version of the appropriate diagonal element of $\boldsymbol{T}$. It follows that $c \geq 0$ at each rotation. Hence it is only necessary to store the value of s for each rotation; there is one such value for each element of $\boldsymbol{X}$.

Submatrices of $\boldsymbol{Q}$

With

$$\boldsymbol{QX} = \begin{bmatrix} \boldsymbol{T}_p \\ \boldsymbol{0} \end{bmatrix}$$

let

$$\boldsymbol{Q} = \begin{bmatrix} \boldsymbol{Q}_{(1)} \\ \boldsymbol{Q}_{(2)} \end{bmatrix}$$

where $\boldsymbol{Q}_{(1)}$ has the same number of rows as $\boldsymbol{T}_p$. Then, premultiplying by $\boldsymbol{Q}'$,

$$\boldsymbol{X} = \boldsymbol{Q}'_{(1)}\boldsymbol{T}_p \tag{12.1}$$

If $\boldsymbol{T}_p$ is of full rank this implies that

$$\boldsymbol{XT}_p^{-1} = \boldsymbol{Q}'_{(1)} \tag{12.2}$$

If $\boldsymbol{T}_p$ is not full rank, most applications will require that any column of $\boldsymbol{Q}'_{(1)}$

that corresponds to $t_{ii} = 0$ should be transferred to $Q'_{(2)}$. Recall that $X\bar{T}_p^{-1}$ has a column of zeros whenever $t_{ii} = 0$. Note finally that

$$\begin{aligned} I - P &= I - XT_p^{-1}\left(XT_p^{-1}\right)' \\ &= I - Q'_{(1)}Q_{(1)} \\ &= Q'_{(2)}Q_{(2)} \end{aligned}$$

Thus the matrix U_* in eq. (11.6) may be formed by the orthogonal reduction of $Q_{(2)}$ to upper triangular form.

*4.13. ERROR ANALYSIS

Suppose that orthogonal reduction of the n by $p + 1$ matrix X to upper triangular form yields the matrix

$$\begin{bmatrix} T_p \\ 0 \end{bmatrix}$$

Then, if Q is the matrix of the orthogonal reduction, set

$$E = Q'\begin{bmatrix} T_p \\ 0 \end{bmatrix} - X$$

that is

$$Q(X + E) = \begin{bmatrix} T_p \\ 0 \end{bmatrix}$$

Chambers (1977) gives a relatively simple form of bound for E which is available for Householder and for MGS, provided inner products are accumulated in double precision (Section 3.13). It is

$$\|E\| \leq cp\varepsilon\|X\| \tag{13.1}$$

where $\|\cdot\|$ is the Euclidean norm (i.e., $\|\cdot\|^2$ is the sum of squares of the matrix elements); c is a number close to 1; and as in Section 3.4, ε is the relative precision of the arithmetic. Planar rotations, or the use of Householder or MGS without double precision accumulation of inner products, lead to somewhat poorer error bounds. The right-hand side should be multiplied by n. For details of the error analysis, see Wilkinson (1965) and Lawson and Hanson (1974).

By contrast to eq. (13.1) all that normal equation methods can achieve is to ensure that rounding errors in the calculations are small by comparison with elements of $X'X$ or of $\mathbf{X}'\mathbf{X}$.

4.14. FURTHER READING AND REFERENCES

In addition to the references given in Section 3.15, see Golub and Styan (1973) and Lawson and Hanson (1974). The implementation of the Householder algorithm described in Section 4.7 is, except for the change which ensures $t_{kk} \geq 0$, that described in Nai-Kuan Tsao (1975). Gill et al. (1974) discuss algorithms for updating T when rows are deleted from $[X, y]$ which make use of knowledge of the orthogonal transformation used to obtain T.

For a detailed discussion of regression diagnostics, see Belsey et al. (1980) and Atkinson (1982). Dempster and Gasko-Green (1981) give a unified theoretical framework for discussing a number of alternative statistics which have been suggested as measures of discrepancy from a linear model.

4.15. EXERCISES

1. Define $z_{ik} = \sqrt{(k-1)/k}\,(x_{ik} - \bar{x}_i^{(k-1)})$ $(i = 1,\ldots,q)$, where $\bar{x}_i^{(k-1)}$ is the mean of the first $k-1$ values of the ith variate. Let Z be the matrix whose elements are z_{ik} $(i = 1,\ldots,q;\ k = 1,\ldots,n)$. Show that orthogonal reduction of the matrix Z to upper triangular form yields the Cholesky decomposition of the CSSP matrix for the q variates $x_1, x_2,\ldots,x_n$. (Cf., eq. (10.1) of Section 3.10.)

2. Suppose that orthogonal reduction of $[X, Y]$ to upper triangular form yields

$$T = \begin{bmatrix} T_{11} & T_{12} \\ \mathbf{0} & T_{22} \end{bmatrix}$$

where the partitioning corresponds to the partitioning of columns of $[X, Y]$. Show that if $[X, Y]$ is replaced by $[XC, Y]$, where C is a square non-singular matrix, the same matrix T_{22} is obtained.

3. (Continuation of Exercise 2)
Show that the orthogonal reduction to upper triangular form of

$$\begin{bmatrix} \boldsymbol{T}_{12} \\ \boldsymbol{T}_{22} \end{bmatrix}$$

yields a matrix $\boldsymbol{U}$ such that $\boldsymbol{U}'\boldsymbol{U} = \boldsymbol{Y}'\boldsymbol{Y}$.

4. Consider a variation of the planar rotation scheme where all below diagonal elements in the first column are reduced to zero before reducing to zero any below diagonal element in the second column. What advantages or disadvantages does such an algorithm have

(i) By comparison with the version of the algorithm described in Section 4.3;

(ii) By comparison with the use of Householder's algorithm?

***5.** Let

$$\boldsymbol{X} = \begin{bmatrix} 1 & 0 & 0 & 2 & 1 \\ 1 & 0 & 0 & 1 & 3 \\ 1 & 0 & 0 & 3 & 2 \\ 1 & 0 & 0 & 6 & 6 \\ 0 & 1 & 0 & 4 & 3 \\ 0 & 1 & 0 & 7 & 3 \\ 0 & 1 & 0 & 2 & 2 \\ 0 & 1 & 0 & 9 & 2 \\ 0 & 0 & 1 & 3 & 0 \\ 0 & 0 & 1 & 3 & 6 \\ 0 & 0 & 1 & 4 & 1 \\ 0 & 0 & 1 & 6 & 1 \end{bmatrix}$$

Form the upper triangle matrix $\boldsymbol{T}$ such that $\boldsymbol{T}'\boldsymbol{T} = \mathbf{X}'\mathbf{X}$:

(i) As the Cholesky decomposition of $\mathbf{X}'\mathbf{X}$.

(ii) By applying the MGS algorithm to $\mathbf{X}$.

***6.** Let

$$\boldsymbol{X} = \begin{bmatrix} \boldsymbol{X}_* \\ \boldsymbol{X}_0 \end{bmatrix},$$

let

$$\boldsymbol{y} = \begin{bmatrix} \boldsymbol{y}_* \\ \boldsymbol{y}_0 \end{bmatrix},$$

and suppose that observations $\boldsymbol{y}_*$ are missing. Show that the following are equivalent:

(i) Choose $\boldsymbol{b}$ to minimize $(y_0 - \boldsymbol{X}_0\boldsymbol{b})'(y_0 - \boldsymbol{X}_0\boldsymbol{b})$, and take $\hat{y}_* = \boldsymbol{X}_*\boldsymbol{b}$.

(ii) Fit the model $\boldsymbol{y} = \boldsymbol{X}\boldsymbol{b} + \boldsymbol{d} + \boldsymbol{e}$, where $\boldsymbol{d}' = [\boldsymbol{d}'_*, \boldsymbol{0}']$, elements of $\boldsymbol{y}_*$ are chosen arbitrarily, and elements of $\boldsymbol{d}'_*$ are to be estimated. Show how, in this case, the fitted values $\hat{y}_*$ may be obtained. Show that the vector $\boldsymbol{b}$ satisfies $\boldsymbol{X}'\boldsymbol{X}\boldsymbol{b} = \boldsymbol{X}'\boldsymbol{y}$, where $\boldsymbol{y}' = [\hat{y}'_*, y'_0]$.

7. Let $\boldsymbol{K} = \text{diag}\left(\sqrt{k_1}, \sqrt{k_2}, \ldots, \sqrt{k_p}\right)$ be a diagonal matrix. Show how the matrix that results from orthogonal reduction to upper triangular form of

$$\begin{bmatrix} \mathbf{X} & \mathbf{y} \\ \boldsymbol{K} & \mathbf{0} \end{bmatrix}$$

may be used to solve the ridge regression problem

$$(\mathbf{X}'\mathbf{X} + \boldsymbol{K}^2)\boldsymbol{b} = \mathbf{X}'\mathbf{y}$$

Assuming $\text{var}(\boldsymbol{y}) = \sigma^2 \boldsymbol{I}_n$, show how to find $\text{var}(\boldsymbol{b})$. (Ridge regression may be a suitable recourse when explanatory variables are strongly intercorrelated. In its simplest form $k_1 = k_2 = \cdots = k_p = k$. The aim in choosing k is to ensure that $\boldsymbol{b}'\boldsymbol{b}$ is not too much larger than $\boldsymbol{\beta}'\boldsymbol{\beta}$, while not introducing an unacceptable bias into $\boldsymbol{b}$.)

CHAPTER 5

Models for Data from Designed Experiments

The discussion of models for designed experiments is inevitably selective and incomplete. The intention is to make the reader aware of the special character of these models and to indicate approaches to handling the computations. The large amount of control exercised in the gathering of the data, allowing the use of randomization and the design of symmetries into the pattern of combinations of factor levels, makes possible clear-cut interpretation of the results. Dummy variates or their equivalent are used to indicate whether or not a particular factor is present at a particular level. These appear in addition to any quantitative variates, which in this context are likely to be called covariates. The elements of the model matrix X will usually be mostly zeros and ones, with zeros predominating. It is desirable to use a form of analysis that reflects and takes advantage of the sparseness and pattern in the model matrix.

Symmetries present in the design usually mean that $X'X$ has a small number of distinct eigenvalues. This and the sparseness of X make the method of conjugate gradients a potentially attractive method. Details of its use in the context of conventional experimental designs, and allowing for more than one error stratum, have still to be worked out. Other efficient general methods for analyzing experimental designs are discussed only briefly. Much work remains to be done in implementing and investigating competing methods before clear preferences can be established.

The methods and models discussed are used more widely than has so far been indicated, for observational data cast in a suitable form as well as for data from designed experiments. It is then difficult to reach firm conclusions. Just as in the case of general regression methods apparent associations may arise from the overriding influence of variates that were not observed.

5.1. MODELS FOR QUALITATIVE EFFECTS

It will help to base discussion on a simple example. Six tomato plants are allocated at random, two to each of three treatments. Weights in grams of the root systems at the end of the experiment were:

Treated with concentrated nutrient	1.5, 2.1
Treated with thrice the concentrated nutrient	0.7, 0.9
Water only	1.3, 1.9.

Among the alternative ways of writing a model that assumes each result is the sum of a treatment effect and an *error term* specific to that particular plant are:

$$y_{ij} = \mu_i + \varepsilon_{ij} \quad (i = 1, 2, 3;\ j = 1, 2) \tag{1.1}$$

and

$$y_{ij} = \mu + \alpha_i + \varepsilon_{ij} \quad (i = 1, 2, 3;\ j = 1, 2) \tag{1.2}$$

Setting $\mu_i = \mu + \alpha_i$ in eq. (1.2) reduces it to eq. (1.1). Conversely, μ_i in eq. (1.1) may be written $\mu + \alpha_i$ with μ chosen arbitrarily. Equation (1.1) is often called the *cell means* model.

The fitted model which corresponds in form to eq. (1.2) is

$$y_{ij} = m + a_i + e_{ij} \quad (i = 1, 2, 3;\ j = 1, 2) \tag{1.3}$$

It is cumbersome to give a form of solution for the fitted parameters that includes an arbitrary constant. Instead it is usual to impose an *identifiability* restriction on the a_i, so that they are uniquely determined. Commonly used alternative forms of condition are (with i running from 1 to I)

$$\sum_{i=1}^{I} a_i = 0$$

or $a_I = 0$ or $a_1 = 0$. It is convenient though unnecessary to apply the same restriction, however chosen, to the α_i as well.

Consider now the alternative forms of least squares solution which are available. In addition to the form of solution appropriate to the model formulation of eq. (1.1), two forms of solution will be given (one with $\sum_{i=1}^{I} a_i = 0$, the other with $a_I = 0$) which are allied to the model formulation of eq. (1.2). A further form of solution is that allied to the model

$$y_{ij} = \beta_0 + \beta_1\phi_1(i) + \beta_2\phi_2(i) + \varepsilon_{ij} \tag{1.4}$$

where $\phi_1(i)$ and $\phi_2(i)$ are orthogonal polynomials of degree 1 and 2, respectively, in the treatment number i. This is often a useful way to represent the treatment levels if there is an ordering in the treatment levels to which the treatment numbers have been made to correspond. The natural ordering for the tomato plant would be:

$i = 1$ for water only.
$i = 2$ for concentrated nutrient.
$i = 3$ for $3 \times$ concentrated nutrient.

(More generally, one may define orthogonal polynomial functions of any relevant quantity which is different for different treatments.)

Alternative Forms of Parameter Estimates—Tomato Plant Example

(i) *The Cell Means Model*

The cell means model is the model of eq. (1.1) that is,

$$y_{ij} = \mu_i + \varepsilon_{ij} \quad (i = 1, 2, 3; j = 1, 2)$$

In matrix notation

$$\begin{bmatrix} y_{11} \\ y_{12} \\ y_{21} \\ y_{22} \\ y_{31} \\ y_{32} \end{bmatrix} = \begin{bmatrix} 1 & 0 & 0 \\ 1 & 0 & 0 \\ 0 & 1 & 0 \\ 0 & 1 & 0 \\ 0 & 0 & 1 \\ 0 & 0 & 1 \end{bmatrix} \begin{bmatrix} \mu_1 \\ \mu_2 \\ \mu_3 \end{bmatrix} + \begin{bmatrix} \varepsilon_{11} \\ \varepsilon_{12} \\ \varepsilon_{21} \\ \varepsilon_{22} \\ \varepsilon_{31} \\ \varepsilon_{32} \end{bmatrix}$$

or $\boldsymbol{y} = \boldsymbol{X\mu} + \boldsymbol{\varepsilon}$. The least squares estimate of μ_i will be written as m_i.

Writing

$$\sum_{i=1}^{3}\sum_{j=1}^{2} e_{ij}^2 = \sum_{i=1}^{3}\sum_{j=1}^{2} (y_{ij} - \bar{y}_{i.} + \bar{y}_{i.} - m_i)^2$$

$$= \sum_{i=1}^{3}\left[\sum_{j=1}^{2} (y_{ij} - \bar{y}_{i.})^2 + 2(\bar{y}_{i.} - m_i)^2\right] \tag{1.5}$$

in the standard manner, it then follows that $m_i = \bar{y}_{i.}$ is the least squares estimate of m_i. Thus

$$m_1 = \tfrac{1}{2}(1.5 + 2.1) = 1.8$$

$$m_2 = \tfrac{1}{2}(0.7 + 0.9) = 0.8$$

$$m_3 = \tfrac{1}{2}(1.3 + 1.9) = 1.6$$

(ii) ***Usual ANOVA Formulation—The Sigma Restrictions***

The usual way the ANOVA model is formulated in statistical texts is

$$y_{ij} = \mu + \alpha_i + \varepsilon_{ij} \quad (i = 1, 2, 3; j = 1, 2) \tag{1.6}$$

where $\Sigma_{i=1}^{3}\alpha_i = 0$. The *sigma* restriction $\Sigma_{i=1}^{3}\alpha_i = 0$ is a particular choice of *identifiability* restriction. The α_i are sometimes known as *effects*. In matrix notation $\boldsymbol{y} = \boldsymbol{X\mu} + \boldsymbol{\varepsilon}$, with

$$\boldsymbol{X} = \begin{bmatrix} 1 & 1 & 0 & 0 \\ 1 & 1 & 0 & 0 \\ 1 & 0 & 1 & 0 \\ 1 & 0 & 1 & 0 \\ 1 & 0 & 0 & 1 \\ 1 & 0 & 0 & 1 \end{bmatrix}, \quad \boldsymbol{\mu} = \begin{bmatrix} \mu \\ \alpha_1 \\ \alpha_2 \\ \alpha_3 \end{bmatrix}$$

where $\Sigma_{i=1}^{3}\alpha_i = 0$. Here the least squares estimates may be obtained by setting $m_i = m + a_i$ in eq. (1.5), and replacing $\Sigma_{i=1}^{3} 2(\bar{y}_i - m_i)^2$ by

$$\sum_{i=1}^{3} 2(\bar{y}_{i.} - \bar{y}_{..} - a_i + \bar{y}_{..} - m)^2 = \sum_{i=1}^{3} 2(\bar{y}_{i.} - \bar{y}_{..} - a_i)^2 + 6(\bar{y}_{..} - m)^2 \tag{1.7}$$

Again there is orthogonality; the crossproduct term vanishes. The least squares estimates are $m = \bar{y}_{..}$ and $a_i = \bar{y}_{i.} - \bar{y}_{..}$:

$$m = \bar{y}_{..} = 1.4$$

$$a_1 = \bar{y}_{1.} - \bar{y}_{..} = 1.8 - 1.4 = 0.4$$

$$a_2 = 0.8 - 1.4 = -0.6$$

$$a_3 = 1.6 - 1.4 = 0.2$$

Fitted values (which are the same for all model formulations) provide a link between the different model formulations. Parameter estimates in an orthogonal model formulation which satisfies the sigma restrictions are readily expressed as differences between fitted values at successive stages of the fit when model terms are fitted sequentially. Suppose that at the point where the α's are included [e.g., in eq. (1.6) after fitting only the constant term] the vector of fitted values changes from $\hat{\mathbf{y}}^{(0)}$ to $\hat{\mathbf{y}}^{(\alpha)}$. Then because of orthogonality, any element of $\hat{\mathbf{y}}^{(\alpha)} - \hat{\mathbf{y}}^{(0)}$ is the estimate of the parameter α_i appropriate to that element. This is true whatever model terms may already be included. The fitted values required for such a calculation may be obtained using any model formulation, perhaps the model formulation that follows. Standard errors are readily calculated using the fact that $\hat{\mathbf{y}}^{(\alpha)} - \hat{\mathbf{y}}^{(0)}$ is independent of $\hat{\mathbf{y}}^{(0)}$. See Section 2.13. This discussion is continued in Section 5.4.

(iii) *Differences from Final Factor Level as Parameters*

Consider the following model:

$$y_{ij} = \mu^* + \alpha_i^* + e_{ij} \quad (i = 1, 2, 3; j = 1, 2)$$

where now $\alpha_3^* = 0$. In matrix notation

$$\mathbf{y} = \begin{bmatrix} 1 & 1 & 0 \\ 1 & 1 & 0 \\ 1 & 0 & 1 \\ 1 & 0 & 1 \\ 1 & 0 & 0 \\ 1 & 0 & 0 \end{bmatrix} \begin{bmatrix} \mu^* \\ \alpha_1^* \\ \alpha_2^* \end{bmatrix} + \boldsymbol{\varepsilon} \tag{1.8}$$

Direct use of an equation such as eq. (1.5) or (1.7) is not now possible. However, it is straightforward to work from the estimates for the m_i obtained earlier and to use the fact that $m^* = m_3$. Thus

$$m^* = m_3 = 1.6 \quad (m_i \text{ are the parameters in } (\mathbf{i}))$$

$$a_1^* = m_1 - m^* = 1.8 - 1.6 = 0.2$$

$$a_2^* = m_2 - m^* = 0.8 - 1.6 = -0.8$$

Because the a_i^*'s estimate differences from the final factor level, this formulation is suitable for use in analyses where the final factor level is a control or standard which is the basis for comparison. The GLM procedure in SAS (SAS Institute, 1979) sets parameters associated with final factor levels to zero, just as described. It is easy to see how one might alternatively set parameters associated with initial factor levels to zero. This is the formulation used in the statistical package/language GLIM (Baker and Nelder, 1978) and in generalized linear model analysis in GENSTAT (Alvey et al., 1982). (Some further details on these packages are given in Chapter 10.)

(iv) *Parameterization Using Orthogonal Contrasts*

Writing the previous model formulation given by eq. (1.8) as $y = X\mu + \varepsilon$, the matrix X is obtained by taking each row twice in the basic pattern represented by the matrix

$$F = \begin{bmatrix} 1 & 1 & 0 \\ 1 & 0 & 1 \\ 1 & 0 & 0 \end{bmatrix}$$

The model may equally well be formulated

$$\begin{aligned} y &= XCC^{-1}\mu + \varepsilon \\ &= X_*\beta + \varepsilon \end{aligned}$$

where C is any nonsingular matrix. The matrix X_* may be obtained by taking each row twice in the matrix $F_* = FC$. Because F is nonsingular, F_* may be taken as any nonsingular 3 by 3 matrix; then $C = F^{-1}F_*$.

For example, F_* might be taken so that columns after the first correspond to contrasts which are of interest and are orthogonal to each other; for example,

$$F_* = \begin{bmatrix} 1 & -1 & 1 \\ 1 & 0 & -2 \\ 1 & 1 & 1 \end{bmatrix}$$

Orthogonality, so that the pairwise scalar products of columns are zero, means that X'_*X_* is a diagonal matrix; this greatly simplifies the calculations. Determination of the matrix C in this case is left as an exercise for the reader.

An important type of orthogonal contrast is that associated with orthogonal polynomials. The contrasts used here may be interpreted as linear and quadratic orthogonal polynomial functions, respectively, of the equally spaced factor levels $i = 1, 2, 3$. The model may be written

$$y_{ij} = \beta_0^* + \beta_1^*\phi_1(i) + \beta_2^*\phi_2(i) + \varepsilon_{ij} \tag{1.9}$$

where $\phi_1(i) = i - 2$ and $\phi_2(i) = 3(i-2)^2 - 2$.

Seber (1977) has a useful brief discussion of orthogonal polynomials, including details of recurrence relations that may be used for generating them. See also Exercise 7 at the end of this chapter. Such recurrence relations are available for unequally as well as for equally spaced factor levels.

Interactions

Consider a complete factorial design, with a factor A at three levels and a factor B at two levels. Then, using a model formulation of type given by eq. (1.8), one has

$$y_{ij} = \mu^* + \alpha_i^* + \beta_j^* + \gamma_{ij}^* + \varepsilon_{ij} \tag{1.10}$$

where $\alpha_3^* = 0$, $\beta_2^* = 0$, $\gamma_{12}^* = \gamma_{22}^* = \gamma_{32}^* = \gamma_{31}^* = 0$. Then

$$\begin{array}{cc} & \begin{matrix} \mu^* & \alpha_1^* & \alpha_2^* & \beta_1^* & \gamma_{11}^* & \gamma_{21}^* \end{matrix} \\ \boldsymbol{X} = & \begin{bmatrix} 1 & 1 & 0 & 1 & 1 & 0 \\ 1 & 1 & 0 & 0 & 0 & 0 \\ 1 & 0 & 1 & 1 & 0 & 1 \\ 1 & 0 & 1 & 0 & 0 & 0 \\ 1 & 0 & 0 & 1 & 0 & 0 \\ 1 & 0 & 0 & 0 & 0 & 0 \end{bmatrix} \end{array} \tag{1.11}$$

Notice that $\gamma_{11}^* = 1$ only if $\alpha_1^* = 1$ and $\beta_1^* = 1$, and is zero otherwise. Similarly, $\gamma_{21}^* = 1$ only if $\alpha_2^* = 1$ and $\beta_1^* = 1$. Thus the columns corresponding to interactions may be obtained by multiplying together element by element the corresponding columns for main effects. This gives a mechanical scheme for generating the columns associated with interactions. However the method fails if, with the model formulated as in eq. (1.8), one or more rows that contains a one in an interaction position is missing. Omission of such a row, effectively replacing it by a row of zeros, reduces the space spanned by the columns of $\boldsymbol{X}$ to a subspace of lower dimension. One solution is to restore the original column space by appending to $\boldsymbol{X}$ a suitably chosen column that corresponds to an interaction involving a final level of one of the factors. For further discussion of this point see Searle and Henderson (1983).

If orthogonal polynomials are used, one will want the interactions to represent linear (in A) by linear (in B) and quadratic by linear effects, respectively. Hence again it is appropriate to multiply element by element corresponding main effects columns to obtain interactions. One has, writing

l for linear and q for quadratic,

$$
X_* = \begin{array}{c} \begin{array}{cccccc} A(l) & A(q) & B(l) & A(l)\times B(l) & A(q)\times B(l) \end{array} \\ \begin{bmatrix} 1 & -1 & 1 & -1 & 1 & -1 \\ 1 & 0 & -2 & -1 & 0 & 2 \\ 1 & 1 & 1 & -1 & -1 & -1 \\ 1 & -1 & 1 & 1 & -1 & 1 \\ 1 & 0 & -2 & 1 & 0 & -2 \\ 1 & 1 & 1 & 1 & 1 & 1 \end{bmatrix} \end{array}
$$

These arguments are readily extended to handle the generation of higher-order interactions in more complex cases.

*5.2. METHODS FOR INCORPORATING IDENTIFIABILITY RESTRICTIONS

Suppose that, using a model formulation of the type given by eq. (1.6) in Section 5.1, $e'e$ is to be minimized in

$$y = Xm + e \qquad (2.1)$$

where $Hm = 0$. In the particular example given

$$H = [0,1,1,1], \quad m' = [m, a_1, a_2, a_3]$$

An identifiability restriction has the form of a linear constraint; however, it does not restrict the set of possible fitted values. Stated formally

$$\{ f \mid f = Xm \} = \{ f \mid f = Xm, \quad \text{where } Hm = 0 \} \qquad (2.2)$$

Making Identifiability Restrictions Automatic

Suppose that the model is formulated so that the fitted value is

$$\hat{y} = \tilde{m}_0 + \tilde{m}_1(x_1 - d_1) + \tilde{m}_2(x_2 - d_2) + \cdots + \tilde{m}_p(x_p - d_p) \qquad (2.3)$$

where useful choices of d_i $(i = 1,\ldots,p)$ will become evident shortly. Suppose that terms 1 to k correspond to a single factor in the model. Thus suppose $x_i = 1$ $(1 \le i \le k - 1)$ if the ith level of this first factor is present, and zero otherwise. It is convenient to take x_k as identically zero, to set $\tilde{m}_k = 0$, and to omit the corresponding column from the model matrix.

Then the additive effects that correspond to the k parameter levels are

$$a_1 = \tilde{m}_1 - \sum_{j=1}^{k} \tilde{m}_j d_j$$
$$a_2 = \tilde{m}_2 - \sum_{j=1}^{k} \tilde{m}_j d_j$$
$$\cdots\cdots\cdots\cdots\cdots$$
$$a_k = \tilde{m}_k - \sum_{j=1}^{k} \tilde{m}_j d_j$$

(In order to obtain a_1, take $x_1 = 1$, $x_2 = x_3 = \cdots = x_k = 0$, etc.) Then $\sum_{j=1}^{k} a_j = 0$ if one takes $d_j = k^{-1}$ for $j = 1, 2, \ldots, k$. Similarly for any other factor or interaction.

Now write the fitted model as $\boldsymbol{y} = \tilde{\boldsymbol{X}}\tilde{\boldsymbol{m}} + \boldsymbol{e}$, and suppose that $\boldsymbol{T}$ is the Cholesky decomposition of $\tilde{\boldsymbol{X}}'\tilde{\boldsymbol{X}}$, and that $\boldsymbol{a} = \boldsymbol{L}'\tilde{\boldsymbol{m}}$. It is a straightforward matter to write down the elements of $\boldsymbol{L}$. Then $\text{var}[\boldsymbol{a}] = \boldsymbol{G}'\boldsymbol{G}$, where $\boldsymbol{G}$ is obtained by solving $\boldsymbol{T}'\boldsymbol{G} = \boldsymbol{L}$.

Omission of Superfluous Parameters

The method now to be discussed will in fact handle any linear constraints; for this see Exercise 8 at the end of the chapter. It relies on writing

$$\boldsymbol{H}\boldsymbol{m} = \boldsymbol{H}_1\boldsymbol{m}_{(1)} + \boldsymbol{H}_2\boldsymbol{m}_{(2)} \tag{2.4}$$

where $\boldsymbol{H}_1$ consists of a linearly independent set of columns of $\boldsymbol{H}$ which form a basis for the column space of $\boldsymbol{H}$. It is convenient to insist that $\boldsymbol{H}_1$ be either upper or lower triangular; here $\boldsymbol{H}_1$ will be chosen to be upper triangular. If $\boldsymbol{H}$ is not already in a form that makes such a choice possible, a preliminary use of Gaussian elimination will be necessary, with any rows of zeros (corresponding to redundant conditions) being discarded. The matrix $\boldsymbol{H}_1$ then consists of those columns of $\boldsymbol{H}$ that contain the first nonzero element in one of the rows. Note that, in order to recover $\boldsymbol{H}$, columns of $\boldsymbol{H}_1$ must be interleaved with columns of $\boldsymbol{H}_2$; a similar comment applies to $\boldsymbol{m}$.

Because $\boldsymbol{H}\boldsymbol{m} = \boldsymbol{0}$

$$\boldsymbol{m}_{(1)} = -\boldsymbol{H}_1^{-1}\boldsymbol{H}_2\boldsymbol{m}_{(2)}$$

$$= -\boldsymbol{G}\boldsymbol{m}_{(2)} \tag{2.5}$$

where $\boldsymbol{G}$ is the solution to $\boldsymbol{H}_1\boldsymbol{G} = \boldsymbol{H}_2$. The matrix $\boldsymbol{G}$ gives the identifiability conditions in a form that is convenient for subsequent calculations. Then, apportioning columns of $\boldsymbol{X}$ between $\boldsymbol{X}_1$ and $\boldsymbol{X}_2$ in the obvious way,

$$\begin{aligned} \boldsymbol{y} - \boldsymbol{X}\boldsymbol{m} &= \boldsymbol{y} - \boldsymbol{X}_1\boldsymbol{m}_{(1)} - \boldsymbol{X}_2\boldsymbol{m}_{(2)} \\ &= \boldsymbol{y} - \tilde{\boldsymbol{X}}_2\boldsymbol{m}_{(2)} \end{aligned} \tag{2.6}$$

where

$$\tilde{\boldsymbol{X}}_2 = \boldsymbol{X}_2 - \boldsymbol{X}_1\boldsymbol{G} \tag{2.7}$$

The problem has now been reduced to that of minimizing the sum of squares of elements of $\boldsymbol{y} - \tilde{\boldsymbol{X}}_2\boldsymbol{m}_{(2)}$. Having obtained the least squares estimate for $\boldsymbol{m}_{(2)}$, and associated variance-covariance information, calculations are completed by the use of eq. (2.5). Note that once $\tilde{\boldsymbol{X}}_2$ has been formed, it is unnecessary to retain $\boldsymbol{X}_1$.

A Method That Gives All Elements of *m* in One Fell Swoop

An approach more commonly canvased in textbooks relies on minimizing the sum of squares

$$(\boldsymbol{y} - \boldsymbol{X}\boldsymbol{m})'(\boldsymbol{y} - \boldsymbol{X}\boldsymbol{m}) + (\boldsymbol{H}\boldsymbol{m})'\boldsymbol{H}\boldsymbol{m} \tag{2.8}$$

Setting $\boldsymbol{H}\boldsymbol{m} = \boldsymbol{0}$ in order to minimize the second term does not restrict the possible values of the first term; this is a consequence of eq. (2.2). The size of the least squares problem is now determined by the number of elements in $\boldsymbol{m}$ rather than by the number of elements in $\boldsymbol{m}_{(2)}$. If a number of factors are at two or three levels, this greatly increases the amount of storage and the amount of computation required. An additional complication is that the upper triangular system of equations which is solved to determine $\boldsymbol{m}$ does not yield variance-covariance information in any very direct manner.

The approach based on eq. (2.8) will be pursued in the exercises.

5.3. ESTIMABLE FUNCTIONS AND RELATED TOPICS

According to the theory of estimable functions, the particular choice of identifiability restrictions is a matter of computational convenience. Interest centers on those linear combinations of parameter estimates that have the same values independently of the parameterization. A linear function $\boldsymbol{l}'\boldsymbol{\mu}$ of the parameters is defined to be estimable if for some $\boldsymbol{c}$ it is identically equal to $\boldsymbol{c}'E[y] = \boldsymbol{c}'\boldsymbol{X}\boldsymbol{\mu}$, where $\boldsymbol{y}$ is the vector of observations. Thus $\boldsymbol{l}' = \boldsymbol{c}'\boldsymbol{X}$.

Now let $\boldsymbol{Q}$ be an orthogonal matrix such that $\boldsymbol{QX}$ is upper triangular. Then $\boldsymbol{l}' = \boldsymbol{c}'\boldsymbol{Q}'\boldsymbol{QX} = \boldsymbol{h}'\boldsymbol{T}_p$, where $\boldsymbol{T}_p$ is the Cholesky decomposition of $\boldsymbol{X}'\boldsymbol{X}$. Clearly any choice of $\boldsymbol{h}$ can be made to yield an estimable function. Thus take the ith element of $\boldsymbol{h}$ equal to 1 and all other elements zero; in this case $\boldsymbol{l}$ is the vector whose elements are those of the ith row of $\boldsymbol{T}_p$. The rows of $\boldsymbol{T}_p$ with nonzero diagonal elements are linearly independent, and hence provide a linearly independent set of estimable functions.

Example

Consider the two-way additive model

$$y_{ij} = \mu + \alpha_i + \beta_j + \varepsilon_{ij} \quad (i = 1,2,3; j = 1,2) \tag{3.1}$$

with model matrix

$$\boldsymbol{X} = \begin{bmatrix} 1 & 1 & 0 & 0 & 1 & 0 \\ 1 & 1 & 0 & 0 & 0 & 1 \\ 1 & 0 & 1 & 0 & 1 & 0 \\ 1 & 0 & 1 & 0 & 0 & 1 \\ 1 & 0 & 0 & 1 & 1 & 0 \\ 1 & 0 & 0 & 1 & 0 & 1 \end{bmatrix}$$

The Cholesky decomposition of $\boldsymbol{X}'\boldsymbol{X}$ is

$$\boldsymbol{T}_5 = \begin{bmatrix} \sqrt{6} & \sqrt{\frac{2}{3}} & \sqrt{\frac{2}{3}} & \sqrt{\frac{2}{3}} & \sqrt{\frac{3}{2}} & \sqrt{\frac{3}{2}} \\ 0 & \sqrt{\frac{4}{3}} & -\sqrt{\frac{1}{3}} & -\sqrt{\frac{1}{3}} & 0 & 0 \\ 0 & 0 & 1 & -1 & 0 & 0 \\ 0 & 0 & 0 & 0 & 0 & 0 \\ 0 & 0 & 0 & 0 & \sqrt{\frac{3}{2}} & -\sqrt{\frac{3}{2}} \\ 0 & 0 & 0 & 0 & 0 & 0 \end{bmatrix}$$

Each row of this matrix (except those entirely filled with zeros) gives the coefficients for one of a linearly independent set of estimable functions. Each coefficient is multiplied by the parameter corresponding to its own column. Each estimable function can then be simplified, as seems appropriate, by dividing through by a suitable constant. This leads to the following linearly independent set of estimable functions (divisors used are, in order; $\sqrt{6}, \sqrt{\frac{1}{3}}, 1, \sqrt{\frac{3}{2}}$):

$$\begin{array}{l} m + \frac{1}{3}(\alpha_1 + \alpha_2 + \alpha_3) + \frac{1}{2}(\beta_1 + \beta_2) \\ \quad 2\alpha_1 - \alpha_2 - \alpha_3 \\ \qquad \alpha_2 - \alpha_3 \\ \qquad\qquad\qquad \beta_1 - \beta_2 \end{array}$$

The *R*-Notation

The *R*-notation has been developed to describe succinctly sums of squares that enter into an ANOVA table. Thus $R(\beta|\mu, \alpha)$ is the sum of squares for β, after fitting μ and α. More simply $R(\alpha|\mu)$ is the sum of squares for α, after fitting μ. In the two-way additive model considered earlier, $R(\alpha|\mu)$ tests

$$2\alpha_1 - \alpha_2 - \alpha_3 = \alpha_2 - \alpha_3 = 0, \quad \text{i.e. } \alpha_1 = \alpha_2 = \alpha_3$$

The sum of squares $R(\beta|\mu, \alpha)$ tests $\beta_1 - \beta_2 = 0$. Because the design is balanced, $R(\beta|\mu) = R(\beta|\mu, \alpha)$ and $R(\alpha|\mu) = R(\alpha|\mu, \beta)$.

Now consider a 2 by 2 additive model in which the cell numbers are $n_{11} = 7$, $n_{12} = 1$, $n_{21} = 3$, and $n_{22} = 5$. The model matrix $\boldsymbol{X}$, with cell numbers (which are taken as weights) given alongside each row, is then

$$\boldsymbol{X} = \begin{bmatrix} 1 & 1 & 0 & 1 & 0 \\ 1 & 1 & 0 & 0 & 1 \\ 1 & 0 & 1 & 1 & 0 \\ 1 & 0 & 1 & 0 & 1 \end{bmatrix} \quad \begin{array}{l} n_{11} = 7 \\ n_{12} = 1 \\ n_{21} = 3 \\ n_{22} = 5 \end{array}$$

Set $\boldsymbol{W} = \text{diag}[w_i]$. The Cholesky decomposition of $\boldsymbol{X}'\boldsymbol{W}\boldsymbol{X}$ is

$$\boldsymbol{T}_4 = \begin{bmatrix} 4 & 2 & 2 & 2\frac{1}{2} & 1\frac{1}{2} \\ 0 & 2 & -2 & 1 & -1 \\ 0 & 0 & 0 & 0 & 0 \\ 0 & 0 & 0 & \frac{1}{2}\sqrt{11} & -\frac{1}{2}\sqrt{11} \\ 0 & 0 & 0 & 0 & 0 \end{bmatrix}$$

Thus $R(\alpha|\mu)$ tests $2(\alpha_1 - \alpha_2) + \beta_1 - \beta_2 = 0$, and $R(\beta|\mu, \alpha)$ tests $\beta_1 - \beta_2 = 0$. The hypothesis that $R(\alpha|\mu)$ tests will rarely be of interest. It is left to the reader (by interchanging the order of α and β before calculating $\boldsymbol{T}_p$) to determine the estimable function to which $R(\beta|\mu)$ corresponds in the model under discussion.

If now the interaction term γ is included in a two-way unbalanced model, the only R-statistic of the type so far considered that is likely to correspond to a hypothesis of genuine interest is $R(\gamma|\mu, \alpha, \beta)$. The safest approach is to base inference directly on the cell means. If some cells are empty, it becomes even more desirable to work directly with the cell means.

Another form of R-statistic which may be of interest for the two-way unbalanced design with all cells filled is designated $R^*(\dot{\alpha}|\dot{\mu}, \dot{\beta}, \dot{\gamma})_\Sigma$ in Searle et al. (1981). This statistic is the decrease in the residual sum of squares from setting the α_i to zero and estimating the remaining parameters while retaining the usual anova sigma restrictions:

$$\sum_{i=1}^{I} \dot{\alpha}_i = 0$$

$$\sum_{j=1}^{J} \dot{\beta}_j = 0$$

$$\sum_{i=1}^{I} \dot{\gamma}_{ij} = 0 \quad (\text{all } j)$$

$$\sum_{j=1}^{J} \dot{\gamma}_{ij} = 0 \quad (\text{all } i)$$

These restrictions are now in fact constraints. The equations refer to the constrained parameters, with dots placed over them. It can be shown that the same fitted values are obtained if, starting with the unrestricted parameters, one applies the constraints

$$\alpha_1 + \bar{\gamma}_{1.} = \alpha_2 + \bar{\gamma}_{2.} = \cdots = \alpha_I + \bar{\gamma}_{I.}$$

Testable Hypotheses

The hypothesis $H: \boldsymbol{K}'\boldsymbol{\mu} = \boldsymbol{d}$ is said to be testable if $\boldsymbol{K}' = \tilde{\boldsymbol{G}}'\boldsymbol{X}$ for some matrix $\tilde{\boldsymbol{G}}$. Equivalently $\boldsymbol{K}' = \boldsymbol{G}'\boldsymbol{T}_p$ for some matrix $\boldsymbol{G}$. Estimability can be checked by verifying that $\boldsymbol{T}_p'\boldsymbol{G} = \boldsymbol{K}$ can be solved for $\boldsymbol{G}$. (If there is no solution, then at some stage $t_{ii}g_{ij}$, with $t_{ii} = 0$, will be set equal to a nonzero right-hand side.) Then the sum of squares associated with the hypothesis tested is $\boldsymbol{u}'\boldsymbol{u}$, where if $\boldsymbol{m}$ is the estimate of $\boldsymbol{\mu}$, then $\boldsymbol{u}$ is obtained by solving $\boldsymbol{G}'\boldsymbol{u} = \boldsymbol{K}'\boldsymbol{m} - \boldsymbol{d}$.

5.4. SPECIAL METHODS FOR ORTHOGONAL MODELS

An orthogonal model has the characteristic that the sum of squares to be minimized may be written as a sum of positive terms, such that each term is a function of at most one parameter. Equations (1.5) and (1.7) illustrated this property.

As noted in eq. (1.7), parameter estimates that satisfy the sigma restrictions can be readily calculated for an orthogonal design even when a model formulation of another type, such as eq. (1.8), has been used in the calculations. Mathematical details will now be given.

Using an obvious generalization of the notation of Section 2.13, let

$$\boldsymbol{X} = \left[\boldsymbol{X}_\mu, \ldots, \boldsymbol{X}_\gamma, \ldots\right] \tag{4.1}$$

where columns of $\boldsymbol{X}_\gamma$ correspond to just one term (= factor or interaction) in the model. Let $\boldsymbol{l}' = [\boldsymbol{l}'_\mu, \ldots, \boldsymbol{l}'_\gamma, \ldots]$ be an actual or possible row of $\boldsymbol{X}$. Let

$$\boldsymbol{d}' = \left(\boldsymbol{T}_p'^{-1}\boldsymbol{l}\right)' = \left[\boldsymbol{d}'_\mu, \ldots, \boldsymbol{d}'_\gamma, \ldots\right] \tag{4.2}$$

$$\boldsymbol{t}'_y = \left[\boldsymbol{t}'_{\mu y}, \ldots, \boldsymbol{t}'_{\gamma y}, \ldots\right] \tag{4.3}$$

where these vectors are partitioned in the same manner as $\boldsymbol{X}$. By a slight generalization of the argument of Section 2.13, it then follows that the change to a predicted value when columns of $\boldsymbol{X}_\gamma$ are fitted in addition to earlier columns is

$$\Delta \hat{y}_\gamma = \boldsymbol{d}'_\gamma \boldsymbol{t}_\gamma \tag{4.4}$$

Assume now that the elements of $\boldsymbol{y}$ are independently distributed with common variance σ^2. As this implies that the elements of $\boldsymbol{t}_y$ are independently distributed with the same variance σ^2 [see Chapter 2, eq. (13.3)], it follows that

$$\operatorname{var}\left(\Delta \hat{y}_\gamma\right) = \boldsymbol{d}'_\gamma \boldsymbol{d}_\gamma \sigma^2 \tag{4.5}$$

independent of the predicted value obtained by fitting earlier columns of $\boldsymbol{X}$.

In order to see how this result may be used, consider computations based on the 3 by 2 complete factorial design

$$y = X\mu^* + \varepsilon$$

represented by eq. (1.10). Here X is partitioned

$$X = [X_\mu, X_\alpha, X_\beta, X_\gamma]$$

where X_μ has one column, X_α has two columns, X_β has one column, and X_γ has two columns. The vectors d and t_y are partitioned similarly. Estimates (and variances) for μ, α_1, β_1, and γ_{11} in the orthogonal version where $\Sigma\alpha_i = 0$, and so on, may be obtained by taking

$$l' = [1; 1, 0; 1; 1, 0]$$

Semicolons separate elements corresponding to different terms. Then

$$m = d_\mu' t_{\mu y}, \quad a_1 = d_\alpha' t_{\alpha y}, \quad b_1 = d_\beta' t_{\beta y}, \quad g_{11} = d_\gamma' t_{\gamma y}$$

With $l' = [1; 1, 0; 0; 0, 0]$, the elements μ, α_1, β_2, and γ_{12} can be estimated in this way, and so on.

Efficient Algorithms for Analyzing Orthogonal Designs

The preceding discussion has been included mainly because of the possibilities it suggests for using a general purpose least squares regression program to fit orthogonal models. If the least squares fitting is handled using an algorithm that takes advantage of the sparseness and balance of X, the resulting algorithm will handle problems of moderate size with acceptable efficiency.

For the special case of balanced orthogonal designs, in which each combination of factor levels occurs an equal number of times, a better approach is that discussed in Kennedy and Gentle (1980), based on ideas of Yates, Hemmerle, and others. Or one of the more general algorithms discussed in the next section may be used.

5.5. MORE GENERAL MODELS

The examples in Cochran and Cox (1957) have been widely used as patterns in the development of computer programs for the analysis of particular models. At the other extreme a number of computer programs cast the analysis in general linear model form and proceed as in general linear least squares regression. In between are methods designed to handle a restricted but wide class of models for designed experiments. Examples are Hemmerle's (1982) algorithm (ACM Algorithm 591), G. N. Wilkinson's algorithm (discussed in a little more detail in the next paragraph), and the algorithm of Kuiper and Corsten which will be described here. Section 5.8 will describe a method (conjugate gradients) that will handle any least squares regression problem but is particularly efficient with models for designed experiments. Before going on to discuss the method of Kuiper and Corsten, brief comments will be made on the analysis of variance facilities in some individual widely used packages. Other aspects of several of these packages are discussed briefly in Chapter 10.

Computer Packages for the Analysis of Designed Experiments

Most of the standard packages now use a notation for describing models similar to that in Wilkinson and Rogers (1973). SAS (SAS Institute, 1979) relies for the most part on general least squares methods for carrying out the analysis. The user is expected to set down the required form of ANOVA table. A special feature of the SAS GLM (general linear model) analysis is its presentation (upon request) of a linearly independent set of estimable functions associated with each sum of squares in the analysis of variance table, along the lines discussed in Section 5.3. The BMDP program BMDP-4V (Dixon et al., 1981), GLIM (Baker and Nelder, 1978), and the FIT command in GENSTAT (Nelder et al., 1981, and Alvey et al., 1982) are similarly designed to give a general linear model form of analysis. In addition BMDP offers the user a choice of more specialized programs, whereas the ANOVA command in GENSTAT, PSTAT (see Heiberger, 1982) and the forthcoming GLIM4/PRISM AOV module (see Gilchrist, ed., 1982) use a single analytic approach to handle a wide class of designs which includes most of those in Cochran and Cox (1957).

GENSTAT (and GLIM4/PRISM) use an algorithm due to G. N. Wilkinson. For a discussion of this algorithm, and further references, see Payne and Wilkinson (1977) and McIntosh (1982). The GENSTAT implementation caters only for designs where all contrasts involving two given treatments are estimated with equal precision. Following the ideas of Nelder (1965), the algorithm handles the analysis of a special class of models (with *orthogonal block structure*) in which there are several error strata. Any block structure (e.g., plots within blocks, subplots within plots) with an equal number of subunits within each unit is of this type. The fixed effects (the *treatment* structure) and the random effects (the *block* structure) are specified using a separate Wilkinson and Rogers type of model formula. Output is by default given in a form similar to that of the Cochran and Cox examples, with additional information available upon request. The theory associated with models with orthogonal block structure is presented in an accessible form in Speed's article on "General Balance" in volume 3 of the *Encyclopedia of Statistical Sciences* (Kotz, Johnson, and Read, 1982).

The general linear model approach to the analysis of designed experiments has received attention earlier in this chapter. A discussion of Wilkinson's algorithm is outside the scope of this book. Instead an approach will be outlined very briefly that is based on work by Kuiper and Corsten, described in Corsten (1958). The method appears capable of extension to handle any design with orthogonal block structure and is well suited to providing a GENSTAT type of output.

The Method of Kuiper and Corsten

Consider the two-way additive model, which in its fitted form appears as

$$y_{ij} = m + a_i + b_j + e_{ij} \quad (i = 1,\ldots,I; j = 1,\ldots,J) \tag{5.1}$$

The following sequence of calculations may be used:

1. Set $m = \bar{y}$, and subtract from each observation, leaving $y_{ij} - \bar{y}$ as residuals from the overall mean.
2. Set $a_i = \bar{y}_{i.} - \bar{y}$, obtained by taking the average of the residuals $y_{ij} - \bar{y}$ at the previous step for each value of i. Subtract from the residuals at the previous step leaving $y_{ij} - \bar{y}_{i.}$.
3. Calculate $b_j = \bar{y}_{.j} - \bar{y}$ by taking the average of the residuals $y_{ij} - \bar{y}_{i.}$ at the previous step for each value of j. Subtract from the residuals at the previous step, leaving $y_{ij} - \bar{y}_{i.} - \bar{y}_{.j} + \bar{y}$. Because the design is orthogonal, this is the end of the calculations. This can be checked by repeating the Steps 1 through 3 on the residuals at Step 3, giving estimates of increments Δm, Δa_i, and Δb_j which are zero.

If the design is not orthogonal then repetition of the sequence of Steps 1 through 3 on the residuals $y_{ij} - \bar{y}_{i.} - \bar{y}_{.j} + \bar{y}$ from the first cycle of calculations will give nonzero increments Δm, Δa_i, and Δb_j to be added to the current values of the parameters, together with a new set of residuals at Step 3. The sum of squares of residuals is reduced at each step of each cycle of iterations, unless estimates have already converged to their least squares values. It will be demonstrated that this is true in general.

Suppose the model is parameterized so that corresponding to a given level of a factor there is a column $\boldsymbol{x}_i$ of $\boldsymbol{X}$ that contains a one whenever the factor is present at that level and otherwise has elements equal to zero. The same is the case for interactions. Suppose that the vector of residuals is currently $\boldsymbol{e}^{()}$, and that the vector of current values of parameter estimates is $\boldsymbol{m}^{()}$. Then unless iterates have converged, $\boldsymbol{X}'\boldsymbol{X}\boldsymbol{m}^{()} - \boldsymbol{X}'\boldsymbol{y} = \boldsymbol{X}'\boldsymbol{e}^{()} \neq \boldsymbol{0}$. Thus there is at least one column $\boldsymbol{x}_i$ of $\boldsymbol{X}$ such that $\boldsymbol{x}_i'\boldsymbol{e}^{()} \neq 0$ for some i. Consider the set of k_i elements which correspond to ones in $\boldsymbol{x}_i$. Calculations proceed, let us assume, by replacing $\boldsymbol{e}^{()}$ by $\boldsymbol{e}^{(*)} = \boldsymbol{e}^{()} - [\boldsymbol{x}_i'\boldsymbol{e}^{()}/k_i]\boldsymbol{x}_i$, so that $m_i^{()}$ is replaced by $m_i^{(*)} = m_i^{()} + \boldsymbol{x}_i'\boldsymbol{e}^{()}/k_i$. Then a little algebra will show that

$$\boldsymbol{e}^{(*)\prime}\boldsymbol{e}^{(*)} = \boldsymbol{e}^{()\prime}\boldsymbol{e}^{()} - k_i^{-1}\left(\boldsymbol{x}_i'\boldsymbol{e}^{()}\right)^2 \tag{5.2}$$

Thus the sum of squares of residuals must reduce at each step unless the normal equations are already satisfied.

For the method as described, convergence may be very slow. D. N. Hunt and C. M. Triggs, in work thus far unpublished, suggest speeding up convergence by the use of a form of Aitken extrapolation (see Section 7.3). For the calculation of variances and covariances, see Pearce et al. (1974) and Worthington (1975). The main details of the method were worked out by Kuiper and Corsten; see the references in Worthington (1975).

*5.6. ANALYSIS OF COVARIANCE AND METHODS FOR MISSING VALUES

Among the approaches available for handling analysis of variance calculations when there are missing values is a method based on the analysis of covariance. It is thus convenient to discuss these together in one section. Comments will be kept brief.

Analysis of Covariance

Suppose that

$$\boldsymbol{y} = \boldsymbol{X}\boldsymbol{b} + \boldsymbol{Z}\boldsymbol{g} + \boldsymbol{e}$$

where the columns of $\boldsymbol{X}$ correspond to qualitative effects and represent the experimental design, whereas the columns of $\boldsymbol{Z}$ correspond to measured covariates. Then the full model may be fitted in the following two steps:

1. Carry out an analysis of variance for $\boldsymbol{y}$ and for the columns of $\boldsymbol{Z}$ in turn. The vector of residuals from the analysis of variance for $\boldsymbol{y}$ is

taken to be $\tilde{y}$, whereas the vector of residuals corresponding to any column of $\boldsymbol{X}$ becomes the corresponding column of $\tilde{\boldsymbol{X}}$.

2. Regress $\tilde{y}$ upon columns of $\tilde{\boldsymbol{X}}$.

It will be necessary to repeat these calculations for each of a sequence of submodels chosen to give the terms required for the analysis of variance table. (To fit a submodel, a subset of the columns of $\boldsymbol{X}$ is taken.) Calculations cycle through Steps 1 and 2 repeatedly, as the columns of $\boldsymbol{X}$ corresponding to each new factor or interaction are included in the model. For the theory underlying this method see Sections 2.14 and 4.9.

Methods for Missing Values

It sometimes happens that one or more observations y_i is lost for reasons that have nothing to do with the entries in the corresponding row of $\boldsymbol{X}$. In that case a least squares analysis can be carried out as for an unbalanced design. Complications of interpretation may arise because the individual entries in the analysis of variance table will depend on the order in which terms are fitted. However, if the proportion of missing values is small, any changes in the table due to the choice of a difference sequence of submodels will be of no consequence.

An alternative involves setting up one covariate for each missing observation. The covariate takes the value one when the observation to which it corresponds is missing; all other values are zero. Values of the dependent variate are set to zero (or otherwise chosen arbitrarily) for the missing observations. Analysis of variance calculations are then carried out as before.

A third method may be regarded as a particular case of the *expectation-maximization* (EM) algorithm discussed in Dempster et al. (1977). Initial estimates of the missing values are taken and the following two steps are repeated until convergence:

1. Fit the model. (This is the maximization step of the EM algorithm.)

2. Use the fitted value(s) from **1** as the new estimate(s) of the missing value(s). (This is the expectation step of the EM algorithm.)

To get an exact sequential analysis of variance table, the fitting procedures must, as before, be carried out for each submodel separately.

For a comparative discussion of these and other approaches to the analysis of designed experiments with missing observations, and brief historical details, see Jarrett (1978).

*5.7. THE ESTIMATION OF VARIANCE COMPONENTS

Consider a horticultural treatment where it is convenient to take a bunch of fruit on a tree as the treatment unit. The treatments, for example, might be (1) fruit left to mature normally, (2) plastic bags tied over the fruit, (3) plastic bags as in preceding situation but with holes in the bags to allow freer circulation of air. The variable of interest is the sugar content at harvest. It is assumed that what is done to one bunch has a negligible influence on other bunches on the tree. Assume an experimental setup where each of the I treatments appears just once on each tree used. The J trees used in the experiment are a random sample of all trees of that variety in the research orchard. The model assumed is

$$y_{ij} = \mu + \alpha_i + \tau_j + \varepsilon_{ij} \quad (i = 1, 2, 3; j = 1, 2) \tag{7.1}$$

where $\text{var}[\tau_j] = \sigma_\tau^2$, $\text{var}[\varepsilon_{ij}] = \sigma^2$, and it is assumed that τ_j and ε_{ij} are independent. This model implies that if there were no treatment differences, then the variance of the difference in sugar level for two bunches on the same tree would be $2\sigma^2$. The variance of the difference between bunches on different trees receiving the same treatment is $2(\sigma_\tau^2 + \sigma^2)$. The model accords with intuition by insisting that between tree variation is at least as great as within tree variation. The analysis of variance table follows:

	Sum of Squares	Degrees of Freedom	Expected Mean Square
Between trees	$\sum_i \sum_j (y_{.j} - \bar{y}_{..})^2$	$J - 1$	$\sigma^2 + I\sigma_\tau^2$
Between treatments	$\sum_i \sum_j (y_{i.} - \bar{y}_{..})^2$	$I - 1$	$\sigma^2 + J\left(\sum \alpha_i\right)^2/(I - 1)$
Residual	$\sum_i \sum_j (y_{ij} - \bar{y}_{i.} - \bar{y}_{.j} + \bar{y}_{..})^2$	$(I - 1)(J - 1)$	σ^2

The comparison of treatments does not involve σ_τ^2. However, knowledge of σ_τ^2 will be useful in estimating the reduction in precision from taking each separate bunch from a separate tree in another such experiment. The simplest way to estimate σ_τ^2 is to equate the expected mean square to the observed mean square in the first and third lines of the ANOVA table and then solve for σ_τ^2. This may lead, embarassingly, to a negative estimate. However, for models with equal subclass numbers these *analysis of variance* estimators appear as satisfactory overall as any other type of estimator that has been investigated. For unbalanced models they may, however, be far from optimal. The three main alternatives are *maximum likelihood* (ML), *restricted maximum likelihood* (REML), and *minimum variance quadratic unbiased estimation* (MINQUE). Other proposed methods can in most instances be regarded as modifications of one of these methods. In estimating the variance components, REML maximizes just that part of the likelihood that is free of the fixed effects. Provided that the non-negativity restrictions are not invoked, REML estimators are identical to the analysis of variance estimators for data with equal subclass numbers. See Harville (1977). The two types of maximum likelihood estimators (ML and REML) lead to heavy iterative computations when the model is at all complicated. MINQUE, because it does not require iteration, is a little more tractable. In addition to Harville (1977), see Searle (1978) for a comparative account of these and other methods in use. See also Kennedy and Gentle (1980).

The attempt to use variance components models with observational data is beset with more than usual difficulty. The various independence assumptions are unlikely to be satisfied, and the variances apply to the observed nonrandom sample rather than to any wider population.

*5.8. THE METHOD OF CONJUGATE GRADIENTS

Conjugate gradients offers an approach to general least squares calculation suited for use in the analysis of data from large experimental designs. The method is efficient when a substantial proportion of the elements of the model matrix $\boldsymbol{X}$ are zeros, and $\boldsymbol{X}'\boldsymbol{X}$ has a relatively small number of distinct eigenvalues. Possibilities for routine use of methods of this type in the analysis of experimental designs have so far been little explored. As yet there seems to be no similarly efficient method for calculating the variance-covariance matrix. There is no easy way to detect linear dependencies.

Details of the Algorithm

We will begin by developing a method for the solution of $\boldsymbol{Sb} = \boldsymbol{s}_y$, where $\boldsymbol{S}$ is a q by q positive definite (or positive semidefinite) matrix. In the present context $\boldsymbol{S} = \boldsymbol{X}'\boldsymbol{X}$, $\boldsymbol{s}_y = \boldsymbol{X}'\boldsymbol{y}$. An easy adaptation will then give a method that works directly with $\boldsymbol{X}$, so gaining maximum advantage from sparseness in that matrix. In this instance there is little point in beginning with an example of the calculations. Until some rationale has been given, they are likely to appear mysterious. Some readers may, however, wish to glance at the calculations at the end of this section before proceeding.

The reader whose main interest is in the formal derivation of the algorithm may now skip directly to eq. (8.6). Intervening material motivates the derivation and is relevant to understanding of the properties of the algorithm.

Suppose that the nonzero eigenvalues of $\boldsymbol{S}$ assume the distinct values $\lambda_1, \lambda_2, \ldots, \lambda_m$. Let $\boldsymbol{K}$ be a matrix whose columns form a complete set of eigenvectors corresponding to these eigenvalues. Define

$$f(\boldsymbol{S}) = (\lambda_1\boldsymbol{I} - \boldsymbol{S})(\lambda_2\boldsymbol{I} - \boldsymbol{S}) \cdots (\lambda_m\boldsymbol{I} - \boldsymbol{S}) \tag{8.1}$$

$$= c_0(\boldsymbol{I} + c_1\boldsymbol{S} + \cdots + c_m\boldsymbol{S}^m)$$

$$= c_0(\boldsymbol{I} - \boldsymbol{S}^-\boldsymbol{S}) \tag{8.2}$$

where

$$c_0 = \prod_{i=1}^{m} \lambda_i \neq 0$$

and

$$\boldsymbol{S}^- = -\left(c_1\boldsymbol{I} + c_2\boldsymbol{S} + \cdots + c_m\boldsymbol{S}^{m-1}\right) \tag{8.3}$$

Now observe that $f(\boldsymbol{S})\,\boldsymbol{S} = 0$. For any vector $\boldsymbol{x}$ may be written $\boldsymbol{x} = \boldsymbol{Ka}$, where the columns of $\boldsymbol{K}$ form a complete set of eigenvectors of $\boldsymbol{S}$. Permuting the factors of $f(\boldsymbol{S})\,\boldsymbol{S}$ as required to take each column of $\boldsymbol{K}$ with the corresponding factor from $f(\boldsymbol{S})\,\boldsymbol{S}$, it follows that $f(\boldsymbol{S})\,\boldsymbol{Sx} = 0$ for every $\boldsymbol{x}$, and hence that $f(\boldsymbol{S})\,\boldsymbol{S} = 0$. Then

$$\boldsymbol{0} = f(\boldsymbol{S})\,\boldsymbol{S} = c_0(\boldsymbol{I} - \boldsymbol{SS}^-)\boldsymbol{S}$$

whence $\boldsymbol{SS}^-\boldsymbol{S} = \boldsymbol{S}$. [When $\boldsymbol{S}$ is nonsingular, then $\boldsymbol{S}^- = \boldsymbol{S}^{-1}$, and in fact $f(\boldsymbol{S}) = 0$.] Then a solution to $\boldsymbol{Sb} = \boldsymbol{s}_y$ is

$$\boldsymbol{b} = \boldsymbol{S}^-\boldsymbol{s}_y = -\left(c_1\boldsymbol{I} + c_2\boldsymbol{S} + \cdots + c_m\boldsymbol{S}^{m-1}\right)\boldsymbol{s}_y \tag{8.4}$$

where $\boldsymbol{S}^-$ is defined as in eq. (8.3).

It may happen that later terms in the expansion for $\boldsymbol{S}^-\boldsymbol{s}_y$ can be reduced to expressions involving earlier terms. An extreme case is where $\boldsymbol{s}_y$ is an eigenvector of $\boldsymbol{S}$ so that $\boldsymbol{b} = \lambda^{-1}\boldsymbol{s}_y$. Thus generally

$$\boldsymbol{b} = -\left(c_1^*\boldsymbol{I} + c_2^*\boldsymbol{S} + \cdots + c_t^*\boldsymbol{S}^t\right)\boldsymbol{s}_y \quad \text{for } t < m \tag{8.5}$$

Conjugate gradients may be viewed as a method for finding the coefficients c_i^* in eq. (8.5).

Now it is a standard result in linear algebra that any vector belonging to the column space of $\boldsymbol{S}$ may be expressed uniquely as a sum of vectors, with each vector in the sum corresponding to just one nonzero eigenvalue of $\boldsymbol{S}$. Then it may be shown that $t = m - 1$ in eq. (8.5), and $c_i^* = c_i$ $(1 < i < m)$ if and only if no term in the sum is zero when $\boldsymbol{s}_y$ is represented in this manner. The proof depends on the fact that a polynomial of degree t can have no more than t roots.

The algorithm now proposed amounts to a method for finding the coefficients c_i^* $(i = 1, 2, \ldots)$ in the expansion (8.5). However, these coefficients will not be determined directly. Instead, observe that solving $\boldsymbol{Sb} = \boldsymbol{s}_y$ is equivalent to minimizing

$$g(\boldsymbol{b}) = \boldsymbol{r}'\boldsymbol{S}^{-1}\boldsymbol{r} \tag{8.6}$$

where $\boldsymbol{r} = \boldsymbol{Sb} - \boldsymbol{s}_y$. If $\boldsymbol{S}$ is singular, then $\boldsymbol{S}^{-1}$ should be replaced by $\boldsymbol{S}^{-}$. The vector $\boldsymbol{b}$ is determined iteratively, with the kth iterate allowing for the first k terms in eq. (8.5). The use of $\boldsymbol{S}^{-1}$ as weighting matrix ensures that it is $\boldsymbol{S}$ rather than $\boldsymbol{S}^2$ that appears at crucial points in the equations now obtained. Let $\boldsymbol{b}_0 = \boldsymbol{0}$, $\boldsymbol{r}_0 = \boldsymbol{s}_y$, and observe that

$$\begin{aligned} \boldsymbol{r}_{k+1} - \boldsymbol{r}_k &= (\boldsymbol{s}_y - \boldsymbol{Sb}_{k+1}) + (\boldsymbol{s}_y - \boldsymbol{Sb}_k) \\ &= \boldsymbol{S}(\boldsymbol{b}_k - \boldsymbol{b}_{k+1}) \\ &= -\boldsymbol{Sd}_k, \quad \text{where } \boldsymbol{b}_{k+1} = \boldsymbol{b}_k + \boldsymbol{d}_k \end{aligned}$$

Thus

$$\boldsymbol{r}_{k+1} = \boldsymbol{r}_k - \boldsymbol{Sd}_k \tag{8.7}$$

and rather than find $\boldsymbol{b}_{k+1}$ to minimize $g(\boldsymbol{b}_{k+1})$, one may find $\boldsymbol{d}_k$ to minimize $\boldsymbol{r}'_{k+1}\boldsymbol{S}^{-1}\boldsymbol{r}_{k+1}$.

Let $\boldsymbol{R}_k = [\boldsymbol{r}_0, \boldsymbol{r}_1, \ldots, \boldsymbol{r}_k]$, and take $\boldsymbol{d}_k = \boldsymbol{R}_k\boldsymbol{c}_k$. Then

$$\boldsymbol{r}_{k+1} = \boldsymbol{r}_k - \boldsymbol{SR}_k\boldsymbol{c}_k \tag{8.8}$$

This implies that, if for $i = 1, 2, \ldots, k$ the vector $\boldsymbol{r}_i$ is a linear combination of $\boldsymbol{S}^j\boldsymbol{r}_0$ $(j = 0, 1, \ldots, i)$, then $\boldsymbol{r}_{k+1}$ will be a linear combination of $\boldsymbol{S}^j\boldsymbol{r}_0$ $(j = 0, 1, \ldots, k+1)$. Thus use of eq. (8.7) is indeed equivalent to estimation of the coefficients c_i^* in eq. (8.5).

The normal equations for minimizing $\boldsymbol{r}'_{k+1}\boldsymbol{S}^{-1}\boldsymbol{r}_{k+1}$ are

$$\boldsymbol{R}'_k\boldsymbol{S}\boldsymbol{R}_k\boldsymbol{c}_k = \boldsymbol{R}'_k\boldsymbol{r}_k \tag{8.9}$$

Thus

$$\boldsymbol{R}'_k(\boldsymbol{r}_k - \boldsymbol{S}\boldsymbol{R}_k\boldsymbol{c}_k) = \boldsymbol{R}'_k\boldsymbol{r}_{k+1} = \boldsymbol{0} \tag{8.10}$$

A consequence of taking $\boldsymbol{d}_k = \boldsymbol{R}_k\boldsymbol{c}_k$ is that $\boldsymbol{r}'_i\boldsymbol{r}_j = 0$ for $i \neq j$. As this is possible for at most q nonzero vectors $\boldsymbol{r}_i$ (where it is assumed that $\boldsymbol{S}$ is $q \times q$), in the absence of the stronger result given earlier this would imply that $\boldsymbol{r}_i = \boldsymbol{0}$ for some $i < q$. In fact we know that $\boldsymbol{r}_t = \boldsymbol{0}$, where t is defined as in eq. (8.5). Rounding error means that in practice one has to test for $\boldsymbol{r}_i \simeq \boldsymbol{0}$.

Eq. (8.9) will not be used as it stands. In place of $\boldsymbol{d}_k = \boldsymbol{R}_k\boldsymbol{c}_k$ we take $\boldsymbol{d}_k = \boldsymbol{H}_k\boldsymbol{u}_k$, where $\boldsymbol{H}_k = [\boldsymbol{h}_0, \boldsymbol{h}_1, \ldots, \boldsymbol{h}_k]$ has the same column space as $\boldsymbol{R}_k$. The columns of $\boldsymbol{H}_k$ are chosen so that $\boldsymbol{H}'_k\boldsymbol{S}\boldsymbol{H}_k$ is a diagonal matrix, that is, $\boldsymbol{h}'_i\boldsymbol{S}\boldsymbol{h}_j = 0$ for $i \neq j$. They are further defined by eq. (8.15) below. Then eq. (8.9) is replaced by

$$\boldsymbol{H}'_k\boldsymbol{S}\boldsymbol{H}_k\boldsymbol{u}_k = \boldsymbol{H}'_k\boldsymbol{r}_k \tag{8.11}$$

As columns of $\boldsymbol{H}_k$ are linear combinations of those of $\boldsymbol{R}_k$, it follows from $\boldsymbol{r}'_i\boldsymbol{r}_j = 0$ for $i \neq j$ that $\boldsymbol{h}'_i\boldsymbol{r}_k = 0$ for $i < k$, and hence that

$$\boldsymbol{u}'_k = [0, \ldots, 0, u_k]; \quad \boldsymbol{b}_{k+1} = \boldsymbol{b}_k + \boldsymbol{d}_k = \boldsymbol{b}_k + \boldsymbol{h}_k u_k \tag{8.12}$$

Now $\boldsymbol{h}'_k\boldsymbol{S}\boldsymbol{h}_k = 0$ in eq. (8.11) would imply that calculations have terminated with $\boldsymbol{r}_k = \boldsymbol{0}$. Otherwise,

$$u_k = \frac{\boldsymbol{h}'_k\boldsymbol{r}_k}{\boldsymbol{h}'_k\boldsymbol{S}\boldsymbol{h}_k} \tag{8.13}$$

Anticipating a little, replace $k + 1$ by k in eq. (8.15) and premultiply by $\boldsymbol{r}'_k$ to show that $\boldsymbol{h}'_k\boldsymbol{r}_k$ may be replaced by $\boldsymbol{r}'_k\boldsymbol{r}_k$ in eq. (8.13).

Because $\boldsymbol{d}_k = \boldsymbol{H}_k u_k = \boldsymbol{h}_k u_k$ (by eq. 8.12), eq. (8.7) becomes

$$\boldsymbol{r}_{k+1} = \boldsymbol{r}_k - \boldsymbol{S}\boldsymbol{h}_k u_k \tag{8.14}$$

Now take

$$\boldsymbol{h}_{k+1} = \boldsymbol{r}_{k+1} + \boldsymbol{h}_k v_k \tag{8.15}$$

where v_k will be chosen so that $\boldsymbol{h}'_{k+1}\boldsymbol{S}\boldsymbol{h}_k = 0$, and an argument by induction will be used to show that $\boldsymbol{h}'_{k+1}\boldsymbol{S}\boldsymbol{h}_i = 0$ for $i < k$. Eq. (8.15) expresses $\boldsymbol{r}_{k+1}$ as a linear combination of columns of $\boldsymbol{H}_{k+1}$ and, substituting for $\boldsymbol{h}_k, \boldsymbol{h}_{k-1}, \ldots,$ in turn allows $\boldsymbol{h}_{k+1}$ to be expressed as a linear combination of columns of $\boldsymbol{R}_{k+1}$. Assume now that $\boldsymbol{h}'_k\boldsymbol{S}\boldsymbol{h}_i = 0$ for $i < k$. Then replacing each term in eq. (8.15) by its transpose and postmultiplying by $\boldsymbol{S}\boldsymbol{h}_i$ with $i < k$ yields

$$\boldsymbol{h}'_{k+1}\boldsymbol{S}\boldsymbol{h}_i = \boldsymbol{r}'_{k+1}\boldsymbol{S}\boldsymbol{h}_i \quad (i < k)$$

Now replace k by i (assumed $< k$) in eq. (8.14) and premultiply by $\boldsymbol{r}'_{k+1}$ to yield

$$u_i\boldsymbol{h}'_{k+1}\boldsymbol{S}\boldsymbol{h}_i = \boldsymbol{r}'_{k+1}\boldsymbol{r}_{i+1} - \boldsymbol{r}'_{k+1}\boldsymbol{r}_i = 0$$

Thus $\boldsymbol{h}'_{k+1}\boldsymbol{S}\boldsymbol{h}_i = 0$ for $i \leq k$. (Recall that $u_i = 0$ would imply $\boldsymbol{r}_i = \boldsymbol{h}_i = \boldsymbol{0}$.)

From eq. (8.14), taking the transpose and postmultiplying by $\boldsymbol{h}'_{k+1}$

$$\boldsymbol{r}'_{k+1}\boldsymbol{h}_{k+1} = \boldsymbol{r}'_k\boldsymbol{h}_{k+1} \tag{8.16}$$

Then premultiplying eq. (8.15) by $\boldsymbol{r}'_k$ and using eq. (8.16) together with $\boldsymbol{r}'_k\boldsymbol{h}_k = \boldsymbol{r}'_k\boldsymbol{r}_k$

$$v_k = \frac{\boldsymbol{r}'_{k+1}\boldsymbol{r}_{k+1}}{\boldsymbol{r}'_k\boldsymbol{r}_k}$$

Calculations may be started with $\boldsymbol{b}_0 = \boldsymbol{0}$, $\boldsymbol{h}_0 = \boldsymbol{r}_0 = \boldsymbol{s}_y$. Then for $k = 0, 1, \ldots$ one has (summarizing the preceding results)

$$u_k = \frac{\boldsymbol{r}'_k\boldsymbol{r}_k}{\boldsymbol{h}'_k\boldsymbol{S}\boldsymbol{h}_k}, \quad \boldsymbol{b}_{k+1} = \boldsymbol{b}_k + u_k\boldsymbol{h}_k, \quad \boldsymbol{r}_{k+1} = \boldsymbol{r}_k - \boldsymbol{S}\boldsymbol{h}_k u_k,$$

$$v_k = \frac{\boldsymbol{r}'_{k+1}\boldsymbol{r}_{k+1}}{\boldsymbol{r}'_k\boldsymbol{r}_k}, \quad \boldsymbol{h}_{k+1} = \boldsymbol{r}_{k+1} + v_k\boldsymbol{h}_k \tag{8.17}$$

For calculations based on $\boldsymbol{S}$ this is the favored approach from among several alternatives; see Reid (1971).

Calculations Based Directly on X

Suppose that $\boldsymbol{S} = \boldsymbol{X}'\boldsymbol{X}$, $\boldsymbol{s}_y = \boldsymbol{X}'\boldsymbol{y}$. Then simple modifications allow one to work directly with $\boldsymbol{X}$ and with $\boldsymbol{y}$, and take advantage of sparseness in $\boldsymbol{X}$.

Write

$$\boldsymbol{e}_k = \boldsymbol{y} - \boldsymbol{X}\boldsymbol{b}_k, \quad \boldsymbol{r}_k = \boldsymbol{X}'\boldsymbol{e}_k, \quad \boldsymbol{z}_k = \boldsymbol{X}\boldsymbol{h}_k$$

Then calculation of $\boldsymbol{e}_{k+1}$ from $\boldsymbol{e}_k$ replaces calculation of $\boldsymbol{r}_{k+1}$ from $\boldsymbol{r}_k$, and $\boldsymbol{r}_{k+1}$ is calculated as $\boldsymbol{X}'\boldsymbol{e}_{k+1}$. Also $\boldsymbol{h}_k'\boldsymbol{S}\boldsymbol{h}_k$ is calculated as $\boldsymbol{z}_k'\boldsymbol{z}_k$. The sequence of calculations is

$$\boldsymbol{e}_0 = \boldsymbol{y} - \boldsymbol{X}\boldsymbol{b}_0; \quad \boldsymbol{h}_0 = \boldsymbol{r}_0 = \boldsymbol{X}'\boldsymbol{e}_0 \tag{8.18}$$

Then from $k = 0, 1, \ldots$ until $\boldsymbol{r}_k = \boldsymbol{0}$ compute

$$\boldsymbol{z}_k = \boldsymbol{X}\boldsymbol{h}_k, \quad u_k = \frac{\boldsymbol{r}_k'\boldsymbol{r}_k}{\boldsymbol{z}_k'\boldsymbol{z}_k},$$

$$\boldsymbol{b}_{k+1} = \boldsymbol{b}_k + u_k\boldsymbol{h}_k, \quad \boldsymbol{e}_{k+1} = \boldsymbol{e}_k - u_k\boldsymbol{z}_k, \quad \boldsymbol{r}_{k+1} = \boldsymbol{X}'\boldsymbol{e}_{k+1},$$

$$v_k = \frac{\boldsymbol{r}_{k+1}'\boldsymbol{r}_{k+1}}{\boldsymbol{r}_k'\boldsymbol{r}_k}, \quad \boldsymbol{h}_{k+1} = \boldsymbol{r}_{k+1} + v_k\boldsymbol{h}_k \tag{8.19}$$

For this discussion a sequence of calculations has been taken which is easy to motivate and describe. ACM Algorithm 583, described in Paige and Saunders (1982), has better numerical properties. Paige and Saunders use an algorithm due to Golub and Kahan to reduce $\boldsymbol{X}$ to lower bidiagonal form. A further point is that the rate of convergence may be improved, leading to fewer iterations, if one works with a preconditioned version of $\boldsymbol{X}$ rather than with $\boldsymbol{X}$ itself.

Calculation of $\boldsymbol{S}^{-1}$

A reliable fast method has still to be found. One possible approach is to calculate numerical estimates of the derivatives of elements of $\boldsymbol{b}$ with respect to elements of $\boldsymbol{s}_y$.

Alternatively, assume a choice of $\boldsymbol{s}_y$ such that $t = m - 1$ in eq. (8.5). Then the u_i and v_i are at each step uniquely determined in eq. (8.17). From this information the coefficients c_i in eq. (8.4), and hence the generalized inverse $\boldsymbol{S}^-$, can be determined. However, it is not quite straightforward to ensure that $t = m - 1$, and this method is expensive computationally.

Example

Consider the one-way analysis of variance example given in Section 5.1, for which

$$\boldsymbol{X} = \begin{bmatrix} 1 & 1 & 0 & 0 \\ 1 & 1 & 0 & 0 \\ 1 & 0 & 1 & 0 \\ 1 & 0 & 1 & 0 \\ 1 & 0 & 0 & 1 \\ 1 & 0 & 0 & 1 \end{bmatrix}, \quad \boldsymbol{y} = \begin{bmatrix} 1.5 \\ 2.1 \\ 0.7 \\ 0.9 \\ 1.3 \\ 1.9 \end{bmatrix}$$

Take $\boldsymbol{b}_0 = \boldsymbol{0}$, $\boldsymbol{e}_0 = \boldsymbol{y}$, then

$$\boldsymbol{r}_0 = \boldsymbol{h}_0 = \boldsymbol{X}'\boldsymbol{e}_0 = \begin{bmatrix} 8.4 \\ 3.6 \\ 1.6 \\ 3.2 \end{bmatrix}, \quad \boldsymbol{z}_0 = \boldsymbol{X}\boldsymbol{h}_0 = \begin{bmatrix} 12 \\ 12 \\ 10 \\ 10 \\ 11.6 \\ 11.6 \end{bmatrix}$$

Thus $u_0 = \boldsymbol{r}_0'\boldsymbol{r}_0/\boldsymbol{z}_0'\boldsymbol{z}_0 = 96.32/757.12 = 0.1272$;

$$\boldsymbol{e}_1 = \boldsymbol{e}_0 - u_0\boldsymbol{z}_0 = \begin{bmatrix} 1.5 \\ 2.1 \\ 0.7 \\ 0.9 \\ 1.3 \\ 1.9 \end{bmatrix} - 0.1272\begin{bmatrix} 12 \\ 12 \\ 10 \\ 10 \\ 11.6 \\ 11.6 \end{bmatrix} = \begin{bmatrix} -0.0266 \\ 0.5734 \\ -0.5722 \\ -0.3722 \\ -0.1757 \\ 0.4243 \end{bmatrix}$$

$$\boldsymbol{r}_1 = \boldsymbol{X}'\boldsymbol{e}_1 = \begin{bmatrix} -0.1491 \\ 0.5467 \\ -0.9444 \\ 0.2486 \end{bmatrix}$$

Thus $v_0 = \boldsymbol{r}_1'\boldsymbol{r}_1/\boldsymbol{r}_0'\boldsymbol{r}_0 = 1.2748/96.32 = 0.0132$;

$$\boldsymbol{h}_1 = \boldsymbol{r}_1 + v_0\boldsymbol{h}_0 = \begin{bmatrix} -0.0379 \\ 0.5944 \\ -0.9232 \\ 0.2909 \end{bmatrix}, \quad \boldsymbol{z}_1 = \boldsymbol{X}\boldsymbol{h}_1 = \begin{bmatrix} 0.5565 \\ 0.5565 \\ -0.9611 \\ -0.9611 \\ 0.2530 \\ 0.2530 \end{bmatrix}$$

Thus $u_1 = \boldsymbol{r}_1'\boldsymbol{r}_1/\boldsymbol{z}_1'\boldsymbol{z}_1 = 1.2748/2.5948 = 0.4913$;

$$\boldsymbol{e}_2 = \boldsymbol{e}_1 - u_1\boldsymbol{z}_1 = \begin{bmatrix} -0.3 \\ 0.3 \\ -0.1 \\ 0.1 \\ -0.3 \\ 0.3 \end{bmatrix}$$

Thus $\boldsymbol{r}_2 = \boldsymbol{X}'\boldsymbol{e}_2 = \boldsymbol{0}$, and all that remains is to calculate $\boldsymbol{b}$. Then

$$\boldsymbol{b} = u_0\boldsymbol{h}_0 + u_1\boldsymbol{h}_1 = \begin{bmatrix} 1.05 \\ 0.75 \\ -0.25 \\ 0.55 \end{bmatrix}$$

The reader is invited to:

1. Verify that $\boldsymbol{b}$ is the minimum norm solution, that is, it minimizes $(b_0 + c)^2 + (b_1 - c)^2 + (b_2 - c)^2 + (b_3 - c)^2$ over all choices of c.
2. Repeat the calculations with a different choice of $\boldsymbol{s}_y$.
3. Repeat the calculations with a full rank parameterization.
4. Calculate the coefficients of the minimum polynomial, and hence calculate the eigenvalues of $\boldsymbol{X}'\boldsymbol{X}$.

The calculations are conveniently carried out using an interactive computer package/language that has facilities for matrix manipulation. **Minitab** (Ryan et al., 1981) and STATUS (Turner and Rogers, 1980) are suitable, from a number of possibilities.

5.9. FURTHER READING AND REFERENCES

The book or survey article that does justice to the topics treated cursorily in this chapter has still to be written. The apparent simplicity of the mathematical description of the models contrasts with the complexity of the practical issues that arise when a model has to be chosen to produce analyses and results in a manner that fairly represents the data.

Kennedy and Gentle (1980) discuss in greater detail the topics of Sections 5.3, 5.4, and 5.7 and give a number of additional references. For the material of Sections 5.5 and 5.8 it is necessary to consult the individual journal articles mentioned in the course of the discussion. For practical analysis, an excellent way to learn is to use some of the better computer programs written to handle analysis of data from designed experiments. The annotated computer outputs prepared by Searle and others (1978, 1979, 1980) are helpful both in understanding the output from individual computer packages and in making comparisons between packages.

On the method of conjugate gradients, see McIntosh (1982).

5.10. EXERCISES

1. Consider the model

$$y_{ij} = \mu + \alpha_i + \beta_j + \varepsilon_{ij}$$

where $\Sigma_{i=1}^{3}\alpha_i = 0$, $\Sigma_{j=1}^{2}\beta_j = 0$. Using the approach of equations (2.4) through (2.7) of Section 5.2, show that this model may be written

$$\boldsymbol{y} = \tilde{\boldsymbol{X}}_2\boldsymbol{\mu}_{(2)} + \boldsymbol{\varepsilon}$$

where

$$\tilde{X}_2 = \begin{bmatrix} 1 & -1 & -1 & -1 \\ 1 & -1 & -1 & 1 \\ 1 & 1 & 0 & -1 \\ 1 & 1 & 0 & 1 \\ 1 & 0 & 1 & -1 \\ 1 & 0 & 1 & 1 \end{bmatrix}, \quad \boldsymbol{\mu}_{(2)} = \begin{bmatrix} \mu \\ \alpha_2 \\ \alpha_3 \\ \beta_2 \end{bmatrix}$$

and

$$\boldsymbol{\mu}_{(1)} = \begin{bmatrix} \alpha_1 \\ \beta_1 \end{bmatrix} = \begin{bmatrix} 0 & -1 & -1 & 0 \\ 0 & 0 & 0 & -1 \end{bmatrix} \boldsymbol{\mu}_{(2)}$$

If $y' = [12 \;\; 14 \;\; 7 \;\; 2 \;\; 4 \;\; 9]$, complete the calculations.

2. (Continuation of Exercise 1)

Show how the matrices H_1 and H_2 of Section 5.2 may alternatively be chosen so that parameters corresponding to final factor levels and to interactions with final factor levels can readily be expressed in terms of earlier parameters. Show that in Exercise 1 this leads to

$$\boldsymbol{\mu}_{(2)} = \begin{bmatrix} \alpha_3 \\ \beta_2 \end{bmatrix} = \begin{bmatrix} 0 & -1 & -1 & 0 \\ 0 & 0 & 0 & -1 \end{bmatrix} \boldsymbol{\mu}_{(1)}$$

where $\boldsymbol{\mu}'_{(1)} = [\mu, \alpha_1, \alpha_2, \beta_1]$, and complete the calculations.

3. Use the method of eq. (2.3) to obtain the analysis of variance table and the parameter estimates for the data of Exercise 1.

4. Consider the model

$$y_{ijk} = \mu + \alpha_i + \beta_j + \tau_{ij} + e_{ijk} \quad (i = 1,2,3; j = 1,2; k = 1,2)$$

where the y_{ijk} are as follows:

	$j = 1$	$j = 2$
$i = 1$	10, 14	12, 16
$i = 2$	6, 8	1, 3
$i = 3$	1, 7	8, 10

Obtain the analysis of variance table and the parameter estimates using:

(i) The approach of eq. (2.3) to finding parameter estimates that satisfy the sigma restrictions.

(ii) The approach of eqs. (2.4) through (2.7) to finding parameter estimates that satisfy the sigma restrictions.

(iii) The parameterization using orthogonal contrasts (for both row and column effects) discussed in Section 5.1.

5. For the data of Exercise 4 determine fitted values:

   (i) After fitting μ.

   (ii) After fitting μ and α.

   (iii) After fitting μ, α, and β.

   Use the method of Section 5.4 to determine parameter estimates that satisfy the sigma restrictions.

6. For each of the sets of parameter estimates in Exercises 1 and 4 (i) through (iii) give the variance-covariance matrix. (Do not complete any numerical matrix calculations which may be necessary, except where these are trivial.)

7. For $r = 0, 1, \ldots, n-1$ the Chebyshev polynomials $\phi_r(x)$ may be defined as follows on the set of integers $x = 1, \ldots, n$:

$$\phi_0(x) = 1, \quad \phi_1(x) = x - \bar{x}, \quad \phi_{r+1}(x) = \phi_1(x)\phi_r(x) - \alpha_r\phi_{r-1}(x)$$

   where $\alpha_r = r^2(n^2 - r^2)/(16r^2 - 4)$. Assuming that

$$\sum_{x=1}^{n} \phi_i(x)\phi_j(x) = 0 \text{ if } i \neq j$$

   (see Plackett, 1960, for a proof) show that

$$\|\phi_r\|^2 = \alpha_r \|\phi_{r-1}\|^2$$

   where $\|\phi_r\|^2 = \sum_{x=1}^{n}\phi_r(x)^2$.

8. Let $x_1, x_2, \ldots, x_n$ be observations on the variate x. Let $\boldsymbol{X}$ be the matrix whose ith row is $(1, x_i, x_i^2, \ldots, x_i^p)$, where $p \leq n - 1$. Let $\boldsymbol{T}_p$ be the matrix that results from orthogonal reduction of $\boldsymbol{X}$ to upper triangular form. Now write

$$\begin{aligned}\hat{y} &= \boldsymbol{X}\boldsymbol{b} \\ &= \boldsymbol{X}\boldsymbol{T}_p^{-1}\boldsymbol{T}_p\boldsymbol{b} \\ &= \boldsymbol{X}_*\boldsymbol{b}_*\end{aligned}$$

   Show that this gives

$$\hat{y} = b_0^*\phi_0(x) + \cdots + b_p^*\phi_p(x)$$

   where the $\phi_i(x)$ are orthogonal polynomials in x.

9. Consider the model $y = Xm + e$ subject to the identifiability restrictions $Hm = 0$, that is, $\{Xm\} = \{Xm|Hm = 0\}$. Prove that the least squares estimate of m is given by $T_p m = t_y$, where for a suitable orthogonal matrix Q

$$Q\begin{bmatrix} X & y \\ H & 0 \end{bmatrix} = \begin{bmatrix} T_p & t_y \\ 0 & z \end{bmatrix}$$

Prove that

$$\begin{aligned}\operatorname{var}(t_y) &= I\sigma^2 - \left(HT_p^{-1}\right)' HT_p^{-1}\sigma^2 \\ &= Q_{11}Q_{11}'\sigma^2\end{aligned}$$

where Q_{11} is a suitable leading submatrix of Q. Deduce that

$$\operatorname{var}(m) = T_p^{-1}Q_{11}\left(T_p^{-1}Q_{11}\right)'\sigma^2$$

10. Consider the model $y = Xb + e$, subject to linear constraints that may be written, possible after a preliminary use of Gaussian elimination, as $H_1 b_{(1)} + H_2 b_{(2)} = d$. The vector b is made up by a suitable interleaving of elements of $b_{(1)}$ with elements of $b_{(2)}$, and X is made up by interleaving columns of $X_{(1)}$ and columns of $X_{(2)}$ in the same way; see eqs. (2.5) through (2.7). Let

$$y_* = y - X_1 H_1^{-1} d$$

$$X_2^* = X_2 - X_1 G$$

where $G = H_1^{-1}H_2$. Show that the least squares problem reduces to that of minimizing the sum of squares of elements of $y_* - X_2^* b_{(2)}$.

11. Consider the two-way additive model

$$y_{ij} = \mu + \alpha_i + \beta_j + \varepsilon_{ijk} \quad (i = 1,2,3; j = 1,2)$$

where k runs from 1 to n_{ij}. The cell numbers n_{ij} are

	$j = 1$	$j = 2$
$i = 1$	1	4
$i = 2$	8	2
$i = 3$	1	9

Determine, in terms of estimable functions of the parameters, the hypothesis that $R(\alpha|\mu) = 0$ tests.

***12.** (Continuation of Exercise 11)

Determine in terms of estimable functions the hypothesis that $R(\alpha|\mu) = 0$ tests when the model of Exercise 11 is modified to include an interaction term γ_{ij}.

***13.** (Continuation of Exercise 11; this requires access to a computer program for least squares regression calculations)

For the model of Exercise 11 suppose that the cell means $\bar{y}_{ij}$ are

	$j = 1$	$j = 2$
$i = 1$	12	14
$i = 2$	7	2
$i = 3$	4	9

Using any parameterization that you find convenient, determine fitted values

(i) After fitting μ.

(ii) After fitting μ and α.

(iii) After fitting μ, α, and β.

What happens if you now try to carry through the calculations used in Section 5.4 for obtaining parameter estimates?

***14.** Show that in the regression of $\boldsymbol{y}$ upon columns of $\boldsymbol{X}$ the effect of the linear constraint $\boldsymbol{l}'\boldsymbol{b} = 0$ upon the residual sum of squares is to increase it by an amount

$$\frac{(\boldsymbol{l}'\boldsymbol{b})^2}{(\boldsymbol{l}'\boldsymbol{S}^{-1}\boldsymbol{l})}$$

where $\boldsymbol{S} = \boldsymbol{X}'\boldsymbol{X}$.

[Hint: Let $\boldsymbol{C}$ be any nonsingular matrix with its final row equal to $\boldsymbol{l}'$. Then $\hat{\boldsymbol{y}} = \boldsymbol{X}\boldsymbol{b} = \boldsymbol{X}^*\boldsymbol{b}^*$; where $\boldsymbol{X}^* = \boldsymbol{X}\boldsymbol{C}^{-1}$, $\boldsymbol{b}^* = \boldsymbol{C}\boldsymbol{b}$, and the final element of $\boldsymbol{b}^*$ equals $\boldsymbol{l}'\boldsymbol{b}$. Then use the result of eq. (5.1) in chapter 2.]

Show that the increase in the residual sum of squares as a result of k linear restrictions $\boldsymbol{L}\boldsymbol{b} = \boldsymbol{0}$ is

$$\boldsymbol{b}'\boldsymbol{L}'\left[\boldsymbol{L}(\boldsymbol{X}'\boldsymbol{X})^{-1}\boldsymbol{L}'\right]^{-1}\boldsymbol{L}\boldsymbol{b}$$

(See Golub and Styan, 1973.)

***15.** Consider the fitted model

$$y = Xb + zg + e$$

where X is the model matrix for a specific experimental design (e.g., a complete factorial) and z gives values of a covariate. Computer programs for the experimental model and for simple linear regression are available. Show how the analysis may be handled by the combined use of these two programs. Extend to the case where there are several covariates, that is, zg is replaced by Zg, and multiple regression is used in place of simple linear regression. (See Sections 2.14 and 4.9. This approach will often provide a simpler and more pleasant form of output than would be available from analysis of a general linear least squares model.)

16. Show that the setup of Exercise 15 may be used with $z_j = 1$, $z_i = 0$, for $i \neq j$, to handle the least squares analysis of an experimental design in which the jth observation is missing. Give details. (Refer to Exercise 6, Chapter 4.)

17. Using the notation of Section 5.8, eq. (8.12) show that an alternative expression for

$$b = h_0 u_0 + h_1 u_1 + \cdots + h_t u_t$$

is

$$b = S^{\sim} s_y$$

where

$$S^{\sim} = \sum_{i=0}^{t} \left(h_i' S h_i\right)^{-1} h_i h_i'$$

Show that $S^{\sim}$ satisfies $S^{\sim} S S^{\sim} = S^{\sim}$. Show, however, that it is not generally true that $S S^{\sim} S = S$. (To find a counterexample, consider a model in which X consists only of a column of ones.)

18. Repeat the calculations for the example given in Section 5.8, but using a full rank parameterization (e.g., leave off the last column of X).

CHAPTER 6

Classical Multivariate Calculations

Estimation of parameters in the general multivariate linear model is a straightforward extension of the methods used for one dependent variate. The usefulness of the Cholesky decomposition extends to a variety of other multivariate calculations, including canonical correlation. A brief discussion of principal components is included, taking for granted algorithms for eigenvalue calculations.

6.1. SUMS OF CSSP MATRICES

The setup is that for one-way multivariate analysis of variance (MANOVA). On each of p variates $y_1, y_2, \ldots, y_p$ one has

$$\left.n_1 \text{ observations}\right\} \begin{array}{l}\text{Group 1 } (n_1 \text{ observations})\\ \text{CSSP matrix } \mathbf{S}_1\end{array}$$

$$\left.n_2 \text{ observations}\right\} \begin{array}{l}\text{Group 2 } (n_2 \text{ observations})\\ \text{CSSP matrix } \mathbf{S}_2\end{array}$$

$$\vdots$$

$$\left.n_k \text{ observations}\right\} \begin{array}{l}\text{Group } k \ (n_k \text{ observations})\\ \text{CSSP matrix } \mathbf{S}_k\end{array}$$

Let $\mathbf{S} = \mathbf{S}_1 + \mathbf{S}_2 + \cdots + \mathbf{S}_k$. Calculations involving $\mathbf{S}$ are usually best based, not directly on $\mathbf{S}$, but on the upper triangle matrix $\mathbf{T}$ such that $\mathbf{T}'\mathbf{T} = \mathbf{S}$. For example, $\boldsymbol{h}'\mathbf{S}^{-1}\boldsymbol{h}$ is best evaluated as the sum of squares of elements of $\mathbf{T}'^{-1}\boldsymbol{h}$.

A convenient starting point for calculation of $\mathbf{T}$ from the data values is provided by inserting k initial columns of indicator variates prior to the columns of values of $y_1, y_2, \ldots, y_p$. The value in the ith such column is 1 if the observation belongs to the ith group, and 0 otherwise. Consider an example in which there are just two variates, with three groups of four observations:

$$\mathbf{X} = \overbrace{\begin{bmatrix} 1 & 0 & 0 & 2 & 1 \\ 1 & 0 & 0 & 1 & 3 \\ 1 & 0 & 0 & 3 & 2 \\ 1 & 0 & 0 & 6 & 6 \\ 0 & 1 & 0 & 4 & 3 \\ 0 & 1 & 0 & 7 & 3 \\ 0 & 1 & 0 & 2 & 2 \\ 0 & 1 & 0 & 9 & 2 \\ 0 & 0 & 1 & 3 & 0 \\ 0 & 0 & 1 & 3 & 6 \\ 0 & 0 & 1 & 4 & 1 \\ 0 & 0 & 1 & 6 & 1 \end{bmatrix}}^{\text{Indicator Columns} \quad y_1 \quad y_2} \begin{matrix} \left.\vphantom{\begin{matrix}1\\1\\1\\1\end{matrix}}\right\} \text{Group 1, CSSP matrix } \mathbf{S}_1 \\ \left.\vphantom{\begin{matrix}1\\1\\1\\1\end{matrix}}\right\} \text{Group 2, CSSP matrix } \mathbf{S}_2 \\ \left.\vphantom{\begin{matrix}1\\1\\1\\1\end{matrix}}\right\} \text{Group 3, CSSP matrix } \mathbf{S}_3 \end{matrix}$$

Orthogonal reduction of $\mathbf{X}$ to upper triangular form, or formation of the Cholesky decomposition of $\mathbf{X}'\mathbf{X}$, yields the matrix

$$\boldsymbol{T} = \begin{bmatrix} 2 & 0 & 0 & 6 & 6 \\ & 2 & 0 & 11 & 5 \\ & & 2 & 8 & 4 \\ & & & 7 & 1 \\ & & & & 6 \end{bmatrix}$$

Deletion of the first three rows and columns yields

$$\mathsf{T} = \begin{bmatrix} 7 & 1 \\ & 6 \end{bmatrix}$$

As may readily be verified, $\mathsf{T}'\mathsf{T}$ is the sum of the separate CSSP matrices. In general, with k initial columns of indicator variates, the first k rows and columns of $\boldsymbol{T}$ must be deleted to give T.

Suppose that $\mathbf{X} = [\boldsymbol{X}, \boldsymbol{Y}]$, where $\boldsymbol{X}$ consists of the columns of $\boldsymbol{X}$ which correspond to indicator variates. Let

$$\boldsymbol{T} = \begin{bmatrix} \boldsymbol{T}_{11} & \boldsymbol{T}_{12} \\ & \boldsymbol{T}_{22} \end{bmatrix}$$

correspondingly. Thus $\mathbf{T} = T_{22}$. In Section 3.5 it was shown that

$$S_{22.1} = T_{22}'T_{22} = \mathbf{T}'\mathbf{T}$$

is the matrix of sums of squares and products of residuals from the regression of columns of Y upon columns of X. In the present case the residuals are deviations from the group means; it follows that $\mathbf{T}'\mathbf{T}$ is the sum of the individual CSSP matrices.

The indicator columns may be viewed as representing a factor with k levels. In place of the matrix X shown here one may thus use the X that corresponds to one of the alternative parameterizations discussed in Section 5.1.

6.2. MULTIVARIATE T^2 AND ONE-WAY MANOVA STATISTICS

Examples of some of the multivariate T^2 and ANOVA calculations will be given, at the same time noting the requisite formulae.

A Single-Sample Problem

Here the variates are written $d_1, d_2, \ldots, d_p$, rather than $y_1, y_2, \ldots, y_p$; they represent differences between the results of two sampling methods:

d_1	d_2	d_3
−0.2	1.6	1.3
8.2	11.1	1.1
−1.9	−2.2	0.9
4.4	6.2	2.5
1.5	4.6	2.0
2.1	2.7	0.3
1.7	1.6	1.8
−1.5	−0.2	3.0
2.3	6.9	3.4

Assuming that these observations are a random sample from a multivariate normal distribution with mean $\boldsymbol{\mu}$, the null hypothesis $\boldsymbol{\mu} = \mathbf{0}$ may be tested using the statistic

$$F = (\bar{\boldsymbol{d}} - \boldsymbol{0})'\mathbf{S}^{-1}(\bar{\boldsymbol{d}} - \boldsymbol{0}) \cdot \frac{n(n-p)}{p} \tag{2.1}$$

Degrees of freedom are p and $n - p$.

Here the upper triangle matrix such that $\mathbf{T}'\mathbf{T} = \mathbf{S}$ is

$$\mathbf{T} = \begin{bmatrix} 8.759 & 10.737 & -0.229 \\ & 4.063 & 1.850 \\ & & 2.203 \end{bmatrix}$$

Then $(\mathbf{T}'^{-1}\bar{\boldsymbol{d}})' = [0.211, 0.327, 0.569]$. The sum of squares of elements of this vector is 0.475, so that

$$F = 0.475 \times \frac{9 \times 6}{3} = 8.56$$

Hotelling's T^2 statistic is essentially the same as the foregoing F-statistic. It is

$$T^2 = n(n-1)\bar{\boldsymbol{d}}'\mathbf{S}^{-1}\bar{\boldsymbol{d}}$$

In the two-sample case the F-statistic takes the form

$$F = \frac{n_1 + n_2 - p - 1}{p}(\bar{y}_1 - \bar{y}_2)'\mathbf{S}^{-1}(\bar{y}_1 - \bar{y}_2)\frac{n_1 n_2}{n_1 + n_2} \tag{2.2}$$

with degrees of freedom p and $n_1 + n_2 - p - 1$. Also used is

$$D^2 = (n_1 + n_2 - 2)(\bar{y}_1 - \bar{y}_2)'\mathbf{S}^{-1}(\bar{y}_1 - \bar{y}_2) \tag{2.3}$$

This is the Mahalanobis measure of separation between the two groups. Then

$$T^2 = \frac{n_1 n_2}{n_1 + n_2} D^2$$

One-Way Multivariate ANOVA

Where the number k of groups is three or more, none of the statistics available for testing the null hypothesis that observations in all groups are from a multivariate normal distribution with a common mean μ has an F-distribution.

The matrix of between groups sums of squares and products about the mean is

$$\mathbf{A} = \sum_{i=1}^{k} n_i(\bar{y}_i - \bar{\bar{y}})(\bar{y}_i - \bar{\bar{y}})' \tag{2.4}$$

where $\bar{\bar{y}} = (\Sigma n_i)^{-1}\Sigma_{i=1}^{k} n_i \bar{y}_i$. If observations in all groups are from a distribution with the same mean μ and with the same variance-covariance matrix Σ, then writing $n = \Sigma_{i=1}^{k} n_i$,

$$E\left[(k-1)^{-1}\mathbf{A}\right] = E\left[(n-k)^{-1}\mathbf{S}\right] = \Sigma$$

This implies that $(n-k)(k-1)^{-1}\mathbf{AS}^{-1}$ should be approximately equal to the unit matrix. Hotelling's generalized T^2-statistic looks at the trace of a closely related matrix; one has

$$T^2 = (n-k)\,\text{trace}\,(\mathbf{AS}^{-1})$$

$$= (n-k)\sum_{i=1}^{k} n_i(\bar{y}_i - \bar{\bar{y}})'\mathbf{S}^{-1}(\bar{y}_i - \bar{\bar{y}})$$

Note that when $k = 2$

$$\sum_{i=1,2} n_i(\bar{y}_i - \bar{\bar{y}})'\mathbf{S}^{-1}(\bar{y}_1 - \bar{\bar{y}}) = \frac{n_1 n_2}{n_1 + n_2}(\bar{y}_1 - \bar{y}_2)'\mathbf{S}^{-1}(\bar{y}_1 - \bar{y}_2)$$

The F-distribution may be used to obtain approximate percentage points for $(n-k)^{-1}T^2$. Details are given in the next section. In addition comments are made on available alternatives to T^2.

6.3. THE MULTIVARIATE GENERAL LINEAR MODEL

The multivariate general linear model is a straightforward generalization of the univariate model considered in earlier chapters. In place of choosing $\boldsymbol{b}$ to minimize $\boldsymbol{e}'\boldsymbol{e}$ in $\boldsymbol{y} = \boldsymbol{Xb} + \boldsymbol{e}$, the problem is to choose $\boldsymbol{B}$ to minimize trace $(\boldsymbol{E}'\boldsymbol{E})$ in

$$\boldsymbol{Y} = \boldsymbol{XB} + \boldsymbol{E} \tag{3.1}$$

Here columns of $\boldsymbol{X}$ are numbered from 0 to m, and columns of $\boldsymbol{Y}$ (and of $\boldsymbol{B}$ and of $\boldsymbol{E}$) from $m+1$ to $m+p$. The columns of $\boldsymbol{B}$ may be obtained by carrying out p separate univariate regressions. In the first such regression the first column of $\boldsymbol{Y}$ is regressed upon columns of $\boldsymbol{X}$, and the vector of coefficients obtained is the first column of $\boldsymbol{B}$.

Now take $\mathbf{X} = [\boldsymbol{X}, \boldsymbol{Y}]$. Orthogonal reduction of $\mathbf{X}$ to upper triangular form, or formation of the Cholesky decomposition of $\mathbf{X}'\mathbf{X}$ yields

$$\boldsymbol{T} = \begin{bmatrix} \boldsymbol{T}_{XX} & \boldsymbol{T}_{XY} \\ \mathbf{0} & \boldsymbol{T}_{YY} \end{bmatrix} \begin{matrix} \}\ \text{Rows } 0 \text{ to } m \\ \}\ \text{Rows } m+1 \text{ to } m+p \end{matrix}$$

The matrix $\boldsymbol{B}$ of regression coefficients is obtained by solving column by column for $\boldsymbol{B}$ in

$$\boldsymbol{T}_{XX}\boldsymbol{B} = \boldsymbol{T}_{XY} \tag{3.2}$$

The matrix of sums of squares and products of residuals is $\boldsymbol{T}'_{YY}\boldsymbol{T}_{YY}$. Refer back to Section 3.5 for details of the argument.

Now let the columns of $\boldsymbol{X}$ be partitioned

$$\boldsymbol{X} = \left[\boldsymbol{X}_{(0)}, \boldsymbol{X}_{(1)}, \ldots, \boldsymbol{X}_{(\ell)}\right]$$

Let

$$\boldsymbol{T}_{XY} = \begin{bmatrix} \boldsymbol{T}_{0Y} \\ \boldsymbol{T}_{1Y} \\ \vdots \\ \boldsymbol{T}_{\ell Y} \end{bmatrix}$$

where $\boldsymbol{T}_{0Y}$ has the same number of rows as $\boldsymbol{X}_{(0)}$ has columns, $\boldsymbol{T}_{1Y}$ has the same number of rows as $\boldsymbol{X}_{(1)}$ has columns, and so on. The univariate ANOVA table generalizes to the MANOVA table:

		Degrees of Freedom
Due to columns of $\boldsymbol{X}_{(0)}$	$\boldsymbol{T}'_{0Y}\boldsymbol{T}_{0Y}$	k_0
Due to columns of $\boldsymbol{X}_{(1)}$ (adjusted for earlier columns)	$\boldsymbol{T}'_{1Y}\boldsymbol{T}_{1Y}$	k_1
....	...	...
....	...	...
Due to columns of $\boldsymbol{X}_{(\ell)}$ (adjusted for earlier columns)	$\boldsymbol{T}'_{\ell Y}\boldsymbol{T}_{\ell Y}$	k_ℓ
Residual	$\boldsymbol{T}'_{YY}\boldsymbol{T}_{YY}$	$n - \Sigma k_i$

In order to decide whether columns of $X_{(\ell)}$ should be included in the model, the CSSP matrix $\mathbf{A} = T'_{\ell Y}T_{\ell Y}$ with $\nu_1 = k_\ell$ degrees of freedom, may be compared with $S_{YY.X} = T'_{YY}T_{YY}$ with $\nu_2 = n - \Sigma k_i$ degrees of freedom. For later reference let G be such that

$$T'_{YY}G' = T'_{\ell Y} \tag{3.3}$$

Also let $\lambda_i (i = 1,\ldots,p)$ be the roots (or eigenvalues) in λ of $\det(\mathbf{A} - \lambda S_{YY.X}) = 0$, and let $\phi_i = \lambda_i(1 + \lambda_i)^{-1}$. Observe that ϕ_i is a root in ϕ of $\det[\mathbf{A} - \phi(\mathbf{A} + S_{YY.X})] = 0$.

Then alternative statistics on which the comparison between $\mathbf{A}$ and $S_{YY.X}$ may be based are:

1. $T^2 = (n - \Sigma k_i)$ trace $(\mathbf{A}S_{YY.X}^{-1})$
   $= (n - \Sigma k_i)$ trace $(G'G)$.
2. Roy's largest root criterion:

$$\lambda_{\max} = \max_{1 \le i \le p} \lambda_i, \quad \text{or equivalently} \quad \phi_{\max} = \max_{1 \le i \le p} \phi_i.$$

3. Pillai's statistic based on trace $[\mathbf{A}(\mathbf{A} + S_{YY.X})^{-1}]$.
4. Wilks' statistic:

$$W = \frac{\det S_{YY.X}}{\det(\mathbf{A} + S_{YY.X})}$$

Comments on the calculation of eigenvalues (the λ_i and the ϕ_i) will be left to Section 6.4. Here note that, although Pillai's statistic may be calculated as $\Sigma_{i=1}^{p}\phi_i$, and Wilks' statistic as $W = \Pi_{i=1}^{p}(1 + \lambda_i)^{-1}$, eigenvalue calculations lead to more arithmetic than is necessary in both cases. Let U be the matrix that results from orthogonal reduction to upper triangular form of

$$\begin{bmatrix} T_{\ell Y} \\ T_{YY} \end{bmatrix}$$

Then W may be calculated by taking the product of the diagonal elements of T_{YY}, dividing by the product of those of U, and squaring. Pillai's statistic may be calculated as trace$(H'H)$, where $HU = T_{\ell Y}$.

Critical values of W and $\lambda_{\max}$ are given in the *Biometrika Tables for Statisticians* (Pearson and Hartley, 1972, vol. 2, tables 47 and 48, with discussion on pp. 99–102). Hughes and Saw (1972) give an approximation for critical values for T^2 which works remarkably well. Let $\mathscr{T} = \nu_2^{-1}T^2$. Then take the distribution of $\mathscr{T}$ to be $\alpha\chi_f^2$, where

$$\alpha = \frac{(\nu_2 - 1)(\nu_1 + \nu_2 - p - 1)}{(\nu_2 - p)(\nu_2 - p - 1)(\nu_2 - p - 3)}$$

$$f = \frac{p\nu_1}{\alpha(\nu_2 - p - 1)}$$

Linear Dependencies between Columns of X

The discussion in Chapter 3 (Sections 3.4 through 3.8) carries over to the multivariate general linear model. The sums of squares and products matrices in the various rows of the MANOVA table are uniquely determined. But whenever $t_{ii} (i \leq m)$ is zero, all elements in the corresponding row of $\boldsymbol{B}$ in the model $\boldsymbol{Y} = \boldsymbol{XB} + \boldsymbol{E}$ are indeterminate.

6.4. EIGENVALUES, EIGENVECTORS, AND THE SVD

Suppose that $\mathbf{A}$ is a symmetric matrix. It is first necessary to have a method for finding the roots (or eigenvalues) $\lambda_1, \ldots, \lambda_p$ in λ, and corresponding eigenvectors $\boldsymbol{v}_1, \ldots, \boldsymbol{v}_p$, such that

$$(\mathbf{A} - \lambda \boldsymbol{I})\boldsymbol{v} = \mathbf{0} \tag{4.1}$$

Observe that $(\mathbf{A} - \lambda \boldsymbol{I})\boldsymbol{v} = \mathbf{0}$, for some vector $\boldsymbol{v}$, if and only if $\det(\boldsymbol{A} - \lambda \boldsymbol{I}) = 0$.

For a discussion of algorithms for handling such calculations the reader must look elsewhere; see especially Wilkinson and Reinsch (1971), Stewart (1973), and Nash (1979). Here the discussion will be restricted to showing how several standard problems either appear naturally in the form (4.1) or reduce to it. Section 10.9 gives a BASIC program that treats eigenvalue and eigenvector calculations for a symmetric matrix as a special case of calculations for the *singular value decomposition* (SVD). The SVD is described at the end of this section. The algorithm used is that described in Nash (1979), which is suitable for matrices of order less than about 10 or 15. For matrices of larger order it compares increasingly unfavorably with the use of Householder's tridiagonalization as described, for example, in Wilkinson and Reinsch (1971).

Principal Components

Let $y = [y_1, y_2, \ldots, y_p]'$ be a multivariate random variate, with $\mathrm{var}(y) = \Sigma$. In practice Σ is unlikely to be available; instead one has $\hat{\Sigma} = (n-1)^{-1}\mathbf{S}$, where $\mathbf{S}$ is a CSSP matrix. Let $\lambda_1, \ldots, \lambda_p$, with corresponding eigenvectors $v_1, \ldots, v_p$, be solutions in λ and v of

$$(\hat{\Sigma} - \lambda I)v = \mathbf{0} \tag{4.2}$$

If $\lambda_i \neq \lambda_j$, then it may be shown that $v_i'v_j = 0$. Even if some of the λ_i are equal, it is convenient to choose the vectors v_i so that $v_i'v_j = 0$ for $i \neq j$. In addition it will be assumed that $v_i'v_i = 1$.

Let $\Lambda = \mathrm{diag}(\lambda_1, \lambda_2, \ldots, \lambda_p)$. Suppose furthermore that the λ_i are ordered so that

$$\lambda_1 \geq \lambda_2 \geq \cdots \geq \lambda_p$$

Then

$$\hat{\Sigma}V = V\Lambda$$

where $V = [v_1, \ldots, v_p]$, so that

$$V'\hat{\Sigma}V = \Lambda \tag{4.3}$$

Now consider

$$f = V'y \tag{4.4}$$

Then, assuming $\mathrm{var}(y) = \hat{\Sigma}$,

$$\begin{aligned} \mathrm{var}(f) &= V'\hat{\Sigma}V \\ &= \Lambda \end{aligned}$$

that is, the elements of f are uncorrelated.

It may be shown that $\mathrm{var}(f_1)$ is a maximum among all linear combinations $\ell'y$ of elements of y, with $\ell'\ell = 1$; similarly $\mathrm{var}(f_2)$ is a maximum among all linear combinations that are uncorrelated with f_1, and so on. The new variates $f_1, f_2, \ldots, f_p$ are known as the *principal components*.

Singular Value Decomposition

Suppose that $\hat{\Sigma} = \mathbf{Z}'\mathbf{Z}$. Then an alternative to forming $\hat{\Sigma}$ and finding its eigenvalues and eigenvectors is to form the singular value decomposition of

Z. For this

$$U'\mathbf{Z}V = D \tag{4.5}$$

where U and V are orthogonal matrices, and D is a diagonal matrix. It is readily verified that the matrix V so formed will serve as the matrix V in eq. (4.3), and that $\Lambda = D^2$.

*6.5. THE GENERALIZED EIGENVALUE PROBLEM

A statistic suggested in Section 6.3 for assessing the statistical significance of a term in a multivariate analysis of variance table was the maximum of the roots (or eigenvalues) λ_i in the matrix equation

$$(\mathbf{A} - \lambda\mathbf{C})v = \mathbf{0} \tag{5.1}$$

where $\mathbf{A} = T'_{\ell Y}T_{\ell Y}$, $\mathbf{C} = S_{YY.X} = T'_{YY}T_{YY}$. This may be rewritten

$$[(1 + \lambda)\mathbf{A} - \lambda(\mathbf{A} + \mathbf{C})]\,v = \mathbf{0}$$

or

$$[\mathbf{A} - \phi(\mathbf{A} + \mathbf{C})]\,v = \mathbf{0} \tag{5.2}$$

where $\phi = \lambda(1 + \lambda)^{-1}$. Equations (5.1) and (5.2) are precisely equivalent, except when (5.2) $\phi = 1$, which requires an infinite value for λ in eq. (5.1). In all cases of interest in the present context $\mathbf{A}$ and $\mathbf{C}$ will be positive definite (or at least positive semidefinite) matrices, and it will be assumed that the Cholesky decomposition of $\mathbf{C}$ is available as a matrix T_{YY}. Then, provided all diagonal elements of T_{YY} are nonzero, eq. (5.1) may be rewritten

$$\left(T'^{-1}_{YY}\mathbf{A}T^{-1}_{YY} - \lambda I\right)v^* = \mathbf{0} \tag{5.3}$$

where $v^* = T_{YY}v$. The calculations are thus reduced to those for the standard form of eigenvalue problem.

Note that $T'^{-1}_{YY}\mathbf{A}T^{-1}_{YY}$ is obtained as $G'G$, where G is such that $T'_{YY}G' = T'_{\ell Y}$. Eigenvalues v_i in the original problem, if required, are found by solving for v_i in

$$T_{YY}v_i = v_i^*$$

If T_{YY} has any zero diagonal elements, omission of the corresponding variates (perhaps using the devices suggested in Section 3.7) will allow calculations based on eq. (5.3) to proceed. Possibilities for legitimate use of the results of such calculations depend on the particular application.

Use of eq. (5.2) is a little more complicated; it requires formation of the upper triangle matrix U such that $U'U = \mathbf{A} + \mathbf{C}$, probably as the orthogonal reduction to upper triangular form of

$$\begin{bmatrix} T_{\ell Y} \\ T_{YY} \end{bmatrix}$$

Subsequent calculations proceed without problem unless $\mathbf{A} + \mathbf{C}$ and hence U is singular, which implies the existence of a vector $d \neq 0$ such that $(\mathbf{A} + \mathbf{C})d = 0$. Then

$$0 = d'(\mathbf{A} + \mathbf{C})d = d'\mathbf{A}d + d'\mathbf{C}d$$

which implies, as both $\mathbf{A}$ and $\mathbf{C}$ are non-negative definite, that $\mathbf{A}d = \mathbf{C}d = 0$. This implies that wherever U has a zero on the diagonal, so does T_{YY}, and vice versa. Calculations proceed by solving for H in

$$U'H' = T'_{\ell Y} \tag{5.4}$$

Then eq. (5.2) becomes

$$(H'H - \phi I)v^{\#} = 0 \tag{5.5}$$

where $v^{\#} = Uv$. The ith row of H' in eq. (5.4) should be set to zero whenever the corresponding diagonal element of U is zero.

Canonical Correlations

For canonical correlations the view taken is that rows of $\mathbf{X} = [X, Y]$ are, apart from an initial column of ones, observations on a multivariate (and preferably multinormal) random variate

$$[x', y'] = [x_1, \ldots, x_m, y_1, \ldots, y_p]$$

Deletion of the first row and column from the matrix T of Section 6.3 yields

$$\mathbf{T} = \begin{bmatrix} \mathbf{T}_{XX} & \mathbf{T}_{XY} \\ 0 & T_{YY} \end{bmatrix}$$

Then $\mathbf{T}'\mathbf{T}$ is the CSSP matrix

$$\mathbf{S} = \begin{bmatrix} \mathbf{S}_{XX} & \mathbf{S}_{XY} \\ \mathbf{S}'_{XY} & \mathbf{S}_{YY} \end{bmatrix}$$

Assuming that $\boldsymbol{X}$ has n rows, the matrix

$$\hat{\Sigma} = (n-1)^{-1}\mathbf{S}$$

will be taken as an estimate of the variance-covariance matrix.

Now consider linear combinations

$$l_i = \boldsymbol{g}_i'\boldsymbol{x} \quad (i = 1,\ldots,m)$$

$$L_j = \boldsymbol{h}_j'\boldsymbol{y} \quad (j = 1,\ldots,p)$$

subject to the conditions:

1. l_i is uncorrelated with $l_1, l_2,\ldots,l_{i-1}$; and L_i is uncorrelated with $L_1, L_2,\ldots,L_{i-1}$.
2. The correlation of l_1 with L_1 is the maximum possible; and subject to Condition 1 that of l_2 with L_2 is the next largest possible, and so on for $i = 1, 2,\ldots,\min(m, p)$.

The correlation between l_i and L_i is the ith canonical correlation. The squares of the canonical correlations are the eigenvalues ϕ, and the linear combinations L_i are given by the eigenvectors $\boldsymbol{h}$, in

$$[\mathbf{A} - \phi(\mathbf{A} + \mathbf{C})]\boldsymbol{h} = \mathbf{0} \tag{5.6}$$

where $\mathbf{A} = \mathbf{T}'_{XY}\mathbf{T}_{XY}$, $\mathbf{C} = \boldsymbol{T}'_{YY}\boldsymbol{T}_{YY}$. A computationally simpler alternative is to solve

$$(\mathbf{A} - \lambda\mathbf{C})\boldsymbol{h} = \mathbf{0} \tag{5.7}$$

and take $\phi = \lambda(1+\lambda)^{-1}$.

Given $L_i = \boldsymbol{h}_i'\boldsymbol{y}$, the corresponding $l_i = \boldsymbol{g}_i'\boldsymbol{x}$ is obtained by solving for $\boldsymbol{g}_i$ in

$$\mathbf{T}_{XX}\boldsymbol{g}_i = \mathbf{T}_{XY}\boldsymbol{h}_i \tag{5.8}$$

A proof of these results is left as an exercise.

The size of the eigenvalue problem (5.6) or (5.7) is kept to a minimum if, in cases where the two sets contain different numbers of variates, the set containing the smaller number is taken last.

Where a diagonal element of $\mathbf{T}$ is zero, rows and columns which relate to the corresponding variate should be omitted in forming eq. (5.7). The linear relation may be determined as described in Section 3.9. If the linear relation gives a variate as a linear combination of earlier variates in the same set, its only use is to allow canonical variates to be written in completely general form (cf. section 3.9). Otherwise the linear relation will determine canonical variates $\boldsymbol{g}'\boldsymbol{x}$ and $\boldsymbol{h}'\boldsymbol{y}$ which are equal and so have correlation 1.0. One has

$$\mathbf{T}_{XX}\boldsymbol{g} - \mathbf{T}_{XY}\boldsymbol{h} = \mathbf{0}$$

$$-\boldsymbol{T}_{YY}\boldsymbol{h} = \mathbf{0}$$

It is then straightforward to show that any variates $\boldsymbol{h}'\boldsymbol{y}$ which satisfy these equations are uncorrelated with each other and with canonical variates obtained from solving eq. (5.7). The same is true for the corresponding variates $\boldsymbol{g}'\boldsymbol{x}$.

Equations (5.6) and (5.7) determine each $\boldsymbol{h}_i$ to within a common scale factor for all elements. Similarly for the $\boldsymbol{g}_i$. It makes sense to scale so that, as far as it can be estimated, the variance for all l_i and L_i is comparable. In the present case estimated variances are as follows:

$$\begin{aligned}(n-1)\text{var}(\boldsymbol{g}_i'\boldsymbol{x}) &= \boldsymbol{g}_i'\mathbf{S}_{XX}\boldsymbol{g}_i \\ &= \text{SS of elements of } \mathbf{T}_{XX}\boldsymbol{g}_i \qquad (5.9)\end{aligned}$$

$$\begin{aligned}(n-1)\text{var}(\boldsymbol{h}_i'\boldsymbol{y}) &= \boldsymbol{h}_i'\mathbf{S}_{YY}\boldsymbol{h}_i \\ &= \boldsymbol{h}_i'(\mathbf{T}_{XY}'\mathbf{T}_{XY} + \boldsymbol{T}_{YY}'\boldsymbol{T}_{YY})\boldsymbol{h}_i \qquad (5.10)\end{aligned}$$

that is, SS of elements of $\mathbf{T}_{XY}\boldsymbol{h}_i$ + SS of elements of $\boldsymbol{T}_{YY}\boldsymbol{h}_i$. One might for example scale so that the variance is in all cases 1.0, or perhaps $(n-1)^{-1}$. Some computer programs scale so that all l_i and L_i are of length one, and yet others so that in each case the maximum element is one.

Canonical Variate Analysis

Canonical variate analysis provides a perspective on the multivariate analysis of variance setup. The matrix $\boldsymbol{X}$ now consists of an initial column of ones and remaining columns of indicator variates that determine the group to which any particular row belongs in the manner of Section 5.1, eq. (1.8). The matrix $\mathbf{A} = \mathbf{T}'_{XY}\mathbf{T}_{XY}$ is then the matrix of between groups sums of squares and products; as in Section 6.2 it may be written

$$\mathbf{A} = \sum_{i=1}^{k} n_i (\bar{y}_i - \bar{\bar{y}})(\bar{y}_i - \bar{\bar{y}})'.$$

It is the matrix of due to regression sums of squares and products (adjusted for means) in the regression of columns of Y on columns of X. The matrix $\mathbf{C} = \mathbf{T}'_{YY}\mathbf{T}_{YY}$ is the matrix of within groups sums of squares and products. It is the matrix of sums of squares and products of residuals from the regression of columns of Y on columns of X.

For any linear combination $\mathbf{h}'\mathbf{y}$ of the y's, write BSS for the between groups sum of squares ($= \mathbf{h}'\mathbf{A}\mathbf{h}$). Write WSS for the within groups sum of squares ($= \mathbf{h}'\mathbf{C}\mathbf{h}$). Then the first canonical variate $L_1 = \mathbf{h}'_1 \mathbf{y}$ is chosen so that the BSS/WSS ratio is as large as possible. Subsequent canonical variates L_i are determined, in the order $i = 2, 3, \ldots$ by the rule that L_i is uncorrelated with $L_1, L_2, \ldots, L_{i-1}$; and that subject to this the BSS/WSS ratio is as large as possible. The vectors of coefficients $\mathbf{h}_i$ for the linear combinations are given by the solutions in $\mathbf{h}$ of the matrix equation

$$(\mathbf{A} - \lambda\mathbf{C})\mathbf{h} = \mathbf{0} \tag{5.11}$$

The eigenvalues λ in eq. (5.11) give the BSS/WSS ratios. See the exercises at the end of this chapter for a proof that the eigenvectors and eigenvalues in eq. (5.11) meet the required conditions. An example of the calculations is given in Section 6.8.

6.6. EXAMPLE—THE MULTIVARIATE GENERAL LINEAR MODEL

$$[X, Y] = \begin{bmatrix} 1 & 7 & 5 & 6 & | & 7 & 1 \\ 1 & 2 & -1 & 6 & | & -5 & 4 \\ 1 & 7 & 3 & 5 & | & 6 & 10 \\ 1 & -3 & 1 & 4 & | & 5 & 5 \\ 1 & 2 & -1 & 0 & | & 5 & -2 \\ 1 & 2 & 1 & 7 & | & -2 & 4 \\ 1 & -3 & -1 & 3 & | & 0 & -6 \\ 1 & 2 & 1 & 1 & | & 8 & 2 \\ 1 & 2 & 1 & 4 & | & 3 & 0 \end{bmatrix}$$

with column headings $x_1 \quad x_2 \quad x_3 \quad y_1 \quad y_2$

Orthogonal reduction of this matrix to upper triangular form yields

$$\begin{bmatrix} \boldsymbol{T}_{XX} & \boldsymbol{T}_{XY} \\ \boldsymbol{0} & \boldsymbol{T}_{YY} \end{bmatrix} = \left[\begin{array}{cccc|cc} 3 & 6 & 3 & 12 & 9 & 6 \\ & 10 & 4 & 2 & 4 & 6 \\ & & 4 & 2 & 6 & 2 \\ & & & 6 & -10 & 4 \\ \hline & & & & 2 & 10 \\ & & & & & \sqrt{10} \end{array}\right]$$

This allows the writing down of a multivariate analysis of variance table for the regression of columns of $\boldsymbol{Y}$ upon columns of $\boldsymbol{X}$, thus:

Due to constant term	$\begin{bmatrix} 9 \\ 6 \end{bmatrix}[9 \quad 6] = \begin{bmatrix} 81 & 54 \\ 54 & 36 \end{bmatrix}$
Due to x_1	$\begin{bmatrix} 4 \\ 6 \end{bmatrix}[4 \quad 6] = \begin{bmatrix} 16 & 24 \\ 24 & 36 \end{bmatrix}$
Due to x_2 (adj. for x_1)	$\begin{bmatrix} 6 \\ 2 \end{bmatrix}[6 \quad 2] = \begin{bmatrix} 36 & 12 \\ 12 & 4 \end{bmatrix}$
Due to x_3 (adj. for x_1, x_2)	$\begin{bmatrix} -10 \\ 4 \end{bmatrix}[-10 \quad 4] = \begin{bmatrix} 100 & -40 \\ -40 & 16 \end{bmatrix}$
Residual	$\begin{bmatrix} 2 & 0 \\ 10 & \sqrt{10} \end{bmatrix}\begin{bmatrix} 2 & 10 \\ 0 & \sqrt{10} \end{bmatrix} = \begin{bmatrix} 4 & 20 \\ 20 & 110 \end{bmatrix}$
Total about mean	$\begin{bmatrix} 156 & 16 \\ 16 & 166 \end{bmatrix}$

For a test whether x_3 should be included in the regression, form $\boldsymbol{G}'$ (or $\boldsymbol{g}$) such that

$$\begin{bmatrix} 2 & 0 \\ 10 & \sqrt{10} \end{bmatrix} \boldsymbol{G}' = \begin{bmatrix} -10 \\ 4 \end{bmatrix}$$

that is,

$$\boldsymbol{G}' = \begin{bmatrix} -5 \\ 17.076 \end{bmatrix}$$

The Hotelling statistic is

$$\begin{aligned} T^2 &= (9-4)\,\text{trace}\,(\boldsymbol{G'G}) \\ &= 5 \times (5^2 + 17.076^2) \\ &= 1582.9 \end{aligned}$$

Degrees of freedom are $\nu_2 = 9 - 4 = 5$ and $\nu_1 = 1$.

Eigenvalues of $\boldsymbol{G'G}$ are readily calculated from

$$\begin{vmatrix} 5^2 - \lambda & 5 \times 17.076 \\ 5 \times 17.076 & 17.076^2 - \lambda \end{vmatrix} = 0$$

Then $\lambda_{\max} = 5^2 + 17.076^2 = 316.6$. (The other root is 0.) It follows that

$$\phi_{\max} = \frac{316.6}{1 + 316.6} = 0.9969$$

$$W = (1 + 316.6)^{-1} = 0.00315$$

Alternatively, calculations may be based on the orthogonal reduction to upper triangular form of

$$\begin{bmatrix} -10 & 4 \\ 2 & 10 \\ 0 & \sqrt{10} \end{bmatrix}$$

This yields

$$\boldsymbol{U} = \begin{bmatrix} 10.198 & -1.9612 \\ 0 & 11.052 \end{bmatrix}$$

Solving

$$\boldsymbol{U'H'} = \begin{bmatrix} -10 \\ 4 \end{bmatrix}$$

yields

$$\boldsymbol{H'} = \begin{bmatrix} -0.9806 \\ 0.1879 \end{bmatrix}$$

Then

$$\phi_{\max} = 0.9806^2 + 0.1879^2 = 0.9969$$

$$W = \frac{(2\sqrt{10})^2}{(10.198 \times 11.052)^2}$$

$$= 0.00315$$

6.7. EXAMPLE—CANONICAL CORRELATION CALCULATIONS

Using data from Section 6.6, let (x_1, x_2, x_3) and (y_1, y_2) be the two sets of variates. From the upper triangle matrix given in Section 6.6, we pick out

$$\mathbf{T}_{XY} = \begin{bmatrix} 4 & 6 \\ 6 & 2 \\ -10 & 4 \end{bmatrix}, \quad \boldsymbol{T}_{YY} = \begin{bmatrix} 2 & 10 \\ 0 & \sqrt{10} \end{bmatrix}$$

Note that $\mathbf{T}_{XY}$ is obtained by deleting the first row from $\boldsymbol{T}_{XY}$.

Following eqs. (5.1) and (5.3), we first solve for $\boldsymbol{G}'$ in $\boldsymbol{T}'_{YY}\boldsymbol{G}' = \mathbf{T}'_{XY}$. This yields

$$\boldsymbol{G}' = \begin{bmatrix} 2 & 3 & -5 \\ -4.427 & -8.854 & 17.076 \end{bmatrix}$$

whence

$$\boldsymbol{G}'\boldsymbol{G} = \begin{bmatrix} 38.0 & -120.8 \\ -120.8 & 389.6 \end{bmatrix}$$

The eigenvalues λ are then found by solving

$$\begin{vmatrix} 38 - \lambda & -120.8 \\ -120.8 & 389.6 - \lambda \end{vmatrix} = 0$$

yielding $\lambda = 427.1, 0.4967$. Thus $\phi = \lambda(1 + \lambda)^{-1}$ has the values 0.9977, 0.3319.

Calculations for determining the eigenvectors $\boldsymbol{h}_1$ and $\boldsymbol{h}_2$ then proceed as follows:

1. For each eigenvalue λ find $\boldsymbol{v}^*$ such that

$$\begin{bmatrix} 38 - \lambda & -120.8 \\ -120.8 & 389.6 - \lambda \end{bmatrix} \boldsymbol{v}^* = \mathbf{0}$$

For simplicity take the initial element of $\boldsymbol{v}^*$ to be 1 in each case. Then

$$\boldsymbol{v}^* = \begin{bmatrix} 1 \\ -3.221 \end{bmatrix}, \quad \begin{bmatrix} 1 \\ 0.3105 \end{bmatrix}$$

2. Now solve for $\boldsymbol{h}$ in $\boldsymbol{T}_{YY}\boldsymbol{h} = \boldsymbol{v}^*$, giving

$$\boldsymbol{h} = \begin{bmatrix} 5.593 \\ -1.018 \end{bmatrix}, \quad \begin{bmatrix} 0.00912 \\ 0.09818 \end{bmatrix}$$

(A suitable form of standardization for $\boldsymbol{h}$ will be considered later.)

Calculations for obtaining the canonical variates for the x's will require

$$\mathbf{T}_{XX} = \begin{bmatrix} 10 & 4 & 2 \\ & 4 & 2 \\ & & 6 \end{bmatrix}, \quad \mathbf{T}_{XY} = \begin{bmatrix} 4 & 6 \\ 6 & 2 \\ -10 & 4 \end{bmatrix}$$

We begin by calculating

$$\mathbf{T}_{XY}\boldsymbol{h} = \begin{bmatrix} 16.264 \\ 31.522 \\ -60.0 \end{bmatrix}, \quad \begin{bmatrix} 0.6256 \\ 0.2511 \\ 0.3015 \end{bmatrix}$$

Solving $\mathbf{T}_{XX}\boldsymbol{g} = \mathbf{T}_{XY}\boldsymbol{h}$ then gives

$$\boldsymbol{g} = \begin{bmatrix} -1.526 \\ 12.88 \\ -10.0 \end{bmatrix}, \quad \begin{bmatrix} 0.03745 \\ 0.03765 \\ 0.05025 \end{bmatrix}$$

In order to make estimated variances $(n-1)^{-1}$ in each case, divide by scale factors; thus

$$\boldsymbol{g}_1 : (16.264^2 + 31.522^2 + 60^2)^{1/2} = 69.70 \quad (= \|\mathbf{T}_{XY}\boldsymbol{h}\|)$$

$$\boldsymbol{g}_2 : (0.6256^2 + 0.2511^2 + 0.3015^2)^{1/2} = 0.7385$$

$$\boldsymbol{h}_1 : (69.7^2 + 1 + 3.221^2)^{1/2} = 69.80 \quad \left(= \left(\|\mathbf{T}_{XY}\boldsymbol{h}\|^2 + \|\boldsymbol{v}^*\|^2\right)^{1/2}\right)$$

$$\boldsymbol{h}_2 : (0.7385^2 + 1 + 0.3105^2)^{1/2} = 1.281.$$

Results may be presented thus

l_i	L_i	$\rho = \sqrt{\phi}$
$-0.022x_1 + 0.185x_2 + 0.143x_3$	$0.080x_1 - 0.015x_2$	0.9988
$0.0507x_1 + 0.0510x_2 + 0.0680x_3$	$0.0071x_1 + 0.077x_2$	0.576

*6.8. EXAMPLE—CANONICAL VARIATE ANALYSIS

Consider

$$[X, Y] = \begin{bmatrix} 1 & 1 & 0 & 2 & 1 \\ 1 & 1 & 0 & 1 & 3 \\ 1 & 1 & 0 & 3 & 2 \\ 1 & 1 & 0 & 6 & 6 \\ 1 & 0 & 1 & 4 & 3 \\ 1 & 0 & 1 & 7 & 3 \\ 1 & 0 & 1 & 2 & 2 \\ 1 & 0 & 1 & 9 & 2 \\ 1 & 0 & 0 & 3 & 0 \\ 1 & 0 & 0 & 3 & 6 \\ 1 & 0 & 0 & 4 & 1 \\ 1 & 0 & 0 & 6 & 1 \end{bmatrix} \begin{matrix} \left.\vphantom{\begin{matrix}1\\1\\1\\1\end{matrix}}\right\} \text{Group 1, } n_1 = 4 \text{ observations} \\ \left.\vphantom{\begin{matrix}1\\1\\1\\1\end{matrix}}\right\} \text{Group 2, } n_2 = 4 \\ \left.\vphantom{\begin{matrix}1\\1\\1\\1\end{matrix}}\right\} \text{Group 3, } n_3 = 4 \end{matrix}$$

Orthogonal reduction of this matrix to upper triangular form yields

$$T = \left[\begin{array}{ccc|cc} 2\sqrt{3} & \frac{2}{\sqrt{3}} & \frac{2}{\sqrt{3}} & \frac{25}{\sqrt{3}} & 5\sqrt{3} \\ 0 & 2\sqrt{\frac{2}{3}} & -\sqrt{\frac{2}{3}} & -\frac{7}{\sqrt{6}} & \sqrt{\frac{3}{2}} \\ 0 & 0 & \sqrt{2} & \frac{3}{\sqrt{2}} & \frac{1}{\sqrt{2}} \\ \hline 0 & 0 & 0 & 7 & 1 \\ 0 & 0 & 0 & 0 & 6 \end{array}\right]$$

Deleting the first row and column, one has:

$$\begin{bmatrix} \mathbf{T}_{XX} & \mathbf{T}_{XY} \\ \mathbf{0} & \boldsymbol{T}_{YY} \end{bmatrix} = \left[\begin{array}{cc|cc} 2\sqrt{\frac{2}{3}} & -\sqrt{\frac{2}{3}} & -\frac{7}{\sqrt{6}} & \sqrt{\frac{3}{2}} \\ 0 & \sqrt{2} & \frac{3}{\sqrt{2}} & \frac{1}{\sqrt{2}} \\ \hline 0 & 0 & 7 & 1 \\ 0 & 0 & 0 & 6 \end{array}\right]$$

The matrix equation to be solved is

$$(\mathbf{A} - \lambda \mathbf{C})\boldsymbol{h} = \mathbf{0}$$

where $\mathbf{A} = \mathbf{T}'_{XY}\mathbf{T}_{XY}$, $\mathbf{C} = \boldsymbol{T}'_{YY}\boldsymbol{T}_{YY}$. Solving $\boldsymbol{T}'_{YY}\boldsymbol{G}' = \mathbf{T}'_{XY}$ yields

$$\boldsymbol{G}' = \begin{bmatrix} -0.4083 & 0.3030 \\ 0.2722 & 0.0673 \end{bmatrix}$$

$$\boldsymbol{G}'\boldsymbol{G} = \begin{bmatrix} 0.2585 & -0.0907 \\ -0.0907 & 0.0786 \end{bmatrix}$$

Setting $\det(\boldsymbol{G}'\boldsymbol{G} - \lambda \boldsymbol{I}) = 0$ yields $\lambda = 0.2963,\ 0.0408$. Solving $(\boldsymbol{G}'\boldsymbol{G} - \lambda \boldsymbol{I})\boldsymbol{v} = \mathbf{0}$ then gives

$$\boldsymbol{v} = \begin{bmatrix} 1 \\ -0.4176 \end{bmatrix}, \quad \begin{bmatrix} 1 \\ 2.400 \end{bmatrix}$$

$$\|\boldsymbol{v}\| = 1.0833, 2.600$$

Solving $\boldsymbol{T}_{YY}\boldsymbol{h} = \boldsymbol{v}$ yields

$$\boldsymbol{h} = \begin{bmatrix} 0.1528 \\ -0.0694 \end{bmatrix}, \quad \begin{bmatrix} 0.0857 \\ 0.4000 \end{bmatrix}$$

An estimate of $\mathrm{var}(\boldsymbol{h}'\boldsymbol{y})$ is

$$(n - k)^{-1}\boldsymbol{h}'\boldsymbol{T}'_{YY}\boldsymbol{T}_{YY}\boldsymbol{h} = (n - k)^{-1}\boldsymbol{v}'\boldsymbol{v}$$

where $k = 3$. Thus dividing elements in these vectors by $1.0833/\sqrt{9}$ and $2.6/\sqrt{9}$, respectively, one obtains the following variates:

Canonical Variate	BSS/RSS Ratio
$0.423y_1 - 0.192y_2$	$\lambda = 0.296$
$0.099y_1 + 0.462y_2$	$\lambda = 0.041$

A useful visual presentation of results is provided by a plot where the variates are the first two canonical variates, and the different points (identified by group) are obtained by evaluating them at individual data points. Plots of any third canonical variate against the first or second canonical variate may also be examined.

6.9. FURTHER READING AND REFERENCES

Harris (1975) gives an account of classical multivariate methods which emphasizes practical use and interpretation. The chapter by Finn in Enslein et al. (1977, pp. 203–264) gives details on the calculation of a wide variety of output information which may be required from a multivariate analysis. It is a useful supplement to the present chapter. For eigenvalue calculations, see Stewart (1973), Wilkinson and Reinsch (1971), Nash (1979) and Parlett (1980). Dongarra et al. (1979) give full discussion and documentation of their FORTRAN subroutine for the singular value decomposition. However, a preferred implementation is now ACM Algorithm 581, described in Chan (1982).

The thrust of current work in multivariate methods is to emphasize structure and description, rather than classical statistical inference. The book by Gnanadesikan (1977) is a good example of this changed emphasis. Gordon (1981) provides an attractively written introduction to methods of classification, with a bias toward methods based on notions of Euclidean geometry. Included are brief details on computer programs that implement classification algorithms.

Cluster analysis and related problems in the detection of pattern have a close connection with the problems of pattern recognition important to computer scientists. Chapter 19 of Hofstadter (1979) is an imaginative exploration of pattern recognition from a computer scientist's point of view.

6.10. EXERCISES

1. (i) The two variates x_1 and x_2 each have unit variance, and have correlation ρ. Determine the principal components.

(ii) Consider variates $x_1, x_2, \ldots, x_5$ all with unit variance.

The correlation between x_1 and x_2 is $\rho > 0$, that between x_3 and x_4 is 1.0, and all other correlations are zero. Determine the principal components.

2. Let λ_i $(i = 1, \ldots, p)$ be the eigenvalues, and $\boldsymbol{v}_i$ the corresponding eigenvectors, of $\mathbf{X}'\mathbf{X}$. Show that:

(i) $\mathbf{X}'\mathbf{X} = \sum_{i=1}^{p} \lambda_i \boldsymbol{v}_i \boldsymbol{v}_i'$.

(ii) $(\mathbf{X}'\mathbf{X})^{-1} = \sum_{i=1}^{p} \lambda_i^{-1} \boldsymbol{v}_i \boldsymbol{v}_i'$.

Show how the representation (ii) may be used to partition $\text{var}[b_i] = (\mathbf{X}'\mathbf{X})^{-1}\sigma^2$ into p components, one associated with each eigenvalue of

$\mathbf{X'X}$. How is this partition modified if the eigenvalues are not all distinct?

***3.** As in Section 3.11 define $\|\boldsymbol{A}\|_2$ to be the spectral norm of the square matrix $\boldsymbol{A}$. Prove that $\|\boldsymbol{A}\|_2$ is the maximum eigenvalue, and $\|\boldsymbol{A}^{-1}\|_2$ the minimum eigenvalue, of $\boldsymbol{A}$.

***4.** Consider the data matrix

$$[\boldsymbol{X}, \boldsymbol{Y}] = \begin{array}{c} \\ \end{array}\left[\begin{array}{cccc|cc} & x_1 & x_2 & x_3 & y_1 & y_2 \\ 1 & 2 & 8 & 6 & -1 & 4 \\ 1 & 1 & 0 & 2 & -2 & 2 \\ 1 & 2 & 5 & 5 & 7 & 8 \\ 1 & 0 & 4 & -2 & 7 & 4 \\ 1 & 1 & 0 & -2 & 0 & -2 \\ 1 & 1 & 3 & 3 & 0 & 3 \\ 1 & 0 & 1 & -3 & -5 & -4 \\ 1 & 1 & 3 & -1 & 2 & 3 \\ 1 & 1 & 3 & 1 & 1 & 0 \end{array}\right]$$

Determine the canonical correlations between the two sets of variates and the canonical variates involved.

***5.** Let

$$[\boldsymbol{X}, \boldsymbol{Y}] = \left[\begin{array}{cccc|cc} & x_1 & x_2 & x_3 & y_1 & y_2 \\ 1 & -1 & 0 & 1 & 0 & -\frac{1}{2} \\ 1 & 3 & 0 & 1 & 0 & 3\frac{1}{2} \\ 1 & 2 & -2 & -2 & -2 & 3\frac{1}{2} \\ 1 & -2 & -1 & 1 & 1 & -1 \\ 1 & -1 & 1 & -1 & -1 & -1 \\ 1 & 3 & 3 & 1 & 3 & 2 \\ 1 & 2 & 2 & 2 & 4 & 1\frac{1}{2} \\ 1 & -2 & -1 & -1 & -2 & -1 \\ 1 & 2 & 1 & 1 & 3 & 2 \end{array}\right]$$

(The variates are a rearrangement of those which appear in the example of Section 3.8.) Determine the canonical correlations and the canonical variates associated with them.

6. Let $\boldsymbol{T}$ be the matrix that results from orthogonal reduction to upper triangular form of $[\boldsymbol{X}, \boldsymbol{Y}]$, where $\boldsymbol{X}$ has an initial column of ones and remaining columns of indicator variates that determine the group to which any row belongs. Deletion of the first row and column of $\boldsymbol{T}$

yields

$$\mathbf{T} = \begin{bmatrix} \mathbf{T}_{XX} & \mathbf{T}_{XY} \\ \mathbf{0} & T_{YY} \end{bmatrix}$$

Prove that $\mathbf{T}'_{XY}\mathbf{T}_{XY}$ is the matrix of between groups sums of squares and products [e.g., see the definition in Section 6.5, in the discussion preceding eq. (5.11)].

*7. The Cauchy-Schwartz inequality states that if $\boldsymbol{g}$ and $\boldsymbol{h}$ are any real vectors, then

$$(\boldsymbol{g}'\boldsymbol{h})^2 \leq (\boldsymbol{g}'\boldsymbol{g})(\boldsymbol{h}'\boldsymbol{h})$$

with equality if and only if $\boldsymbol{g} = \alpha\boldsymbol{h}$ for some scalar α. Extend this result to show that, if $\mathbf{S}_{11}$ and $\mathbf{S}_{12}$ are any real matrices such that

$$\mathbf{S}_{11} = \mathbf{T}'_{11}\mathbf{T}_{11}, \quad \mathbf{S}_{12} = \mathbf{T}'_{11}\mathbf{T}_{12}$$

then

$$(\boldsymbol{g}'\mathbf{S}_{12}\boldsymbol{h})^2 \leq (\boldsymbol{g}'\mathbf{S}_{11}\boldsymbol{g})(\boldsymbol{h}'\mathbf{T}'_{12}\mathbf{T}_{12}\boldsymbol{h})$$

with equality if and only if $\mathbf{T}_{11}\boldsymbol{g} = \alpha\mathbf{T}_{12}\boldsymbol{h}$ for some scalar α. Deduce that, if $\boldsymbol{U}$ is upper triangular such that $\boldsymbol{U}'\boldsymbol{U} = \mathbf{S}_{22}$, and $\boldsymbol{G}$ is such that $\boldsymbol{G}\boldsymbol{U} = \mathbf{T}_{12}$, then

$$\sup_{\boldsymbol{g},\boldsymbol{h}} \frac{(\boldsymbol{g}'\mathbf{S}_{12}\boldsymbol{h})^2}{(\boldsymbol{g}'\mathbf{S}_{11}\boldsymbol{g})(\boldsymbol{h}'\mathbf{S}_{22}\boldsymbol{h})} = \sup_{\boldsymbol{v}} \frac{(\boldsymbol{G}\boldsymbol{v})'\boldsymbol{G}\boldsymbol{v}}{\boldsymbol{v}'\boldsymbol{v}}$$

8. Show that

$$\sup_{\boldsymbol{v}} \frac{\boldsymbol{v}'\boldsymbol{G}'\boldsymbol{G}\boldsymbol{v}}{\boldsymbol{v}'\boldsymbol{v}}$$

is the maximum eigenvector of $\boldsymbol{G}'\boldsymbol{G}$, and that the choice of $\boldsymbol{v}$ which yields this maximum is the corresponding eigenvector. Deduce the results of Section 6.5, eqs. (5.6) through (5.8).
[Hint: Write $\boldsymbol{z} = \boldsymbol{Q}\boldsymbol{v}$, where $\boldsymbol{Q}$ is orthogonal such that $\boldsymbol{Q}'\boldsymbol{G}'\boldsymbol{G}\boldsymbol{Q} = \operatorname{diag}(\lambda_1^2, \ldots, \lambda_p^2)$. Then

$$\frac{z_1^2\lambda_1^2 + \cdots + z_p^2\lambda_p^2}{z_1^2 + \cdots + z_p^2}$$

lies between the least and greatest of the λ_i^2.]

CHAPTER 7

Nonlinear Methods

Emphasis is on the use of general nonlinear methods for the solution of maximum likelihood equations. The examples considered are such that the likelihood can be shown to have a unique maximum in the interior of the parameter space. The maximum is then the solution of the equation or equations that result from equating to zero the derivative of the loglikelihood with respect to each parameter in turn. Use of the loglikelihood (which is a maximum if and only if the likelihood is a maximum) has the convenience that for independent observations it is the sum of the loglikelihoods for the individual observations.

The maximizing of the loglikelihood might alternatively be treated as a problem in optimization, for which general methods are available. Such methods will not be considered here.

Later sections of the chapter discuss loglinear and other models which Nelder and Wedderburn have characterized as *generalized linear models*.

7.1. NONLINEAR EQUATIONS IN ONE UNKNOWN

One or more roots $u = x$ are to be found for an equation $f(u) = 0$, where $f(x)$ is nonlinear in x. In most cases it is desirable to accompany details of the root or roots with a graph of $f(x)$ against x over a range of values of x surrounding the root. It makes sense to begin by drawing the graph, which will be used to set initial bounds on the root or roots.

The alternative methods are discussed very cursorily. The important concept of order of convergence is mentioned only in passing. For a more adequate discussion of the theory of nonlinear methods, and fuller discussions of details important in the implementation of general algorithms, see Dahlquist and Bjorck (1974, ch. 6) and Forsythe et al. (1977).

Calculations based on several alternative methods will be illustrated taking

$$f(u) = u - \lambda^{-1}\log(1 + au), \tag{1.1}$$

with $\lambda = 1.31$, $a = 1.93$. (The logarithm is to base e.) A root $u \neq 0$ is required. The problem is one that arose in the fitting of a mathematical model to explain how tiny wasps which lay their eggs in housefly pupae tend to avoid pupae which are already parasitized. For details see the note at the end of this section. Biological considerations make it likely that $0 < u < 1$. Observe that

$$f(1) = 1 - 1.31^{-1}\log(2.93) > 0$$

The approximation $\log(1 + au) \simeq au - \frac{1}{2}a^2u^2$ leads to $u = 0.33$; furthermore $f(0.33) < 0$. Hence $f(u) = 0$ has a root in the interval $0.33 < u < 1$. Where a method requires one initial estimate, $u_0 = 0.33$ will be taken.

The Biological Background

Assume that the number of visits any pupa receives from a female parasitic wasp follows a Poisson distribution with mean λ. The probability that a visit will lead to the laying of an egg is assumed to depend on the number of eggs already present in the pupa. If there are no previous eggs, this probability is taken as $\delta_0 = 1$; if there is one previous egg, it is $\delta_1 \leq 1$. (Carrying on in this way one might define $\delta_2, \delta_3, \ldots$.) Then λ may be estimated by taking $p_0 = e^{-\lambda}$, where p_0 is the sample proportion of pupae containing no eggs. Equating the sample proportion p_1 of pupae that contain just one egg to the corresponding theoretical proportion gives

$$p_1 = e^{-\lambda}\left[\lambda + \sum_{i=2}^{\infty} \frac{(1 - \delta_1)^{i-1}\lambda^i}{i!}\right]$$

$$= (1 - \delta_1)^{-1}(e^{-\lambda\delta_1} - e^{-\lambda}).$$

Rearrangement with $u = 1 - \delta_1$ and $a = p_1/p_0$ leads to eq. (1.1). Taking $p_0 = 0.27$ and $p_1 = 0.52$, leads to the particular values of λ and a which are given.

7.2. THE BISECTION METHOD, AND BEYOND

The bisection method is a systematic method for determining a sequence of points $x = m_i$ at which $f(x)$ is tabulated. Suppose that a_0 and b_0 are given such that $f(a_0)$ and $f(b_0)$ are of opposite sign, that is, $f(a_0)f(b_0) < 0$. Then, if $f(x)$ is continuous for $a_0 < x < b_0$, it follows that the function has at least one root in this interval.

Now take $m_1 = 0.5(a_0 + b_0)$. Then unless $f(m_1) = 0$, either

$$f(a_0)f(m_1) < 0; \quad \text{if so set } a_1 = a_0, b_1 = m_1$$

or

$$f(m_1)f(b_0) < 0; \quad \text{if so set } a_1 = m_1, b_1 = b_0$$

This gives an interval (a_1, b_1), containing a root u of $f(u) = 0$, which is half the length of the interval (a_0, b_0). Continuing in this way, one traps a root u such that $f(u) = 0$ between ever-closer bounds. In this interval starting values $a_0 = 0.33$ and $b_0 = 0.97$ will be used to demonstrate the use of this method to find a root of the equation

$$u - 1.31^{-1}\log(1 + 1.93u) = 0$$

which is eq. (1.1). Choosing $b_0 = a_0 + 0.64$ makes calculation of the successive midpoints simple! Following each value of a_i, the sign of $f(a_i)$ is given in brackets, and similarly for b_i and m_i. Successive steps are then:

a_i	b_i	m_i	$f(m_i)$
0.33(−)	0.97(+)	0.65(+)	0.029
0.33(−)	0.65(+)	0.49(−)	−0.018
0.49(−)	0.65(+)	0.57(+)	0.003
0.49(−)	0.57(+)	0.53(−)	−0.008
0.53(−)	0.57(+)	0.55(−)	−0.002
0.55(−)	0.57(+)	0.56(+)	0.0006
0.55(−)	0.56(+)	0.555(−)	−0.0008

Correct to two decimal places, $u = 0.56$. Convergence is slow but sure. The bisection method, and such modifications of it as *regula falsi*, which will be discussed shortly, are in practice a useful recourse in narrowing the search interval sufficiently that potentially much faster methods will be effective.

Regula falsi, or the *method of false position*, differs from the bisection method only in using linear interpolation to determine the point at which the current interval is divided. The point of division m_i is the best linear estimate of u such that $f(u) = 0$. Thus

$$m_i = a_i + h_i$$

where

$$h_i = -f(a_i)\frac{b_i - a_i}{f(b_i) - f(a_i)} \tag{2.1}$$

The Secant Method

The secant method has faster ultimate convergence than either the bisection method or *regula falsi*. If $u_{i-1} = a_i$ and $u_i = b_i$ are two successive approximands, then the secant method calculates $u_{i+1} = m_i$ as for *regula falsi* [eq. (2.1)] and takes this as the next approximand.

An Algorithm with Guaranteed Convergence

Brent (1973) gives details of an algorithm with guaranteed convergence that may be regarded as a distillation of the ideas discussed so far, with inverse quadratic interpolation added in for good measure. For a FORTRAN implementation and more detail than is given here, see Forsythe et al. (1977). Brent (1973) gives some information on historical antecedents, and **Algol** W and FORTRAN implementations. A BASIC implementation appears in Chapter 10 of this book.

At each step three points a_i, b_i, and c_i are carried. Here a_i is the previous iterate, b_i is the latest iterate, and c_i (which may coincide with a_i) is the most recent iterate such that b_i and c_i are on either side of the root. An inverse quadratic interpolation, or a linear interpolation if the current points are not distinct, is used to calculate a *first try* at the point to be taken as the next iterate. If this *first try* lies reasonably well within the current interval, it will be used as the next iterate. Otherwise, it is rejected in favor of the point given by the bisection method. Thus convergence is guaranteed, as with the bisection method and *regula falsi*.

Use of the BASIC version of this subroutine with eq. (1.1), with (0.33, 1.0) as the interval of search, gave convergence to two decimal place precision in one iteration. The algorithm is remarkably powerful and effective.

7.3. FUNCTIONAL ITERATION, AND THE NEWTON-RAPHSON METHOD

The common feature of the functional iteration and the Newton-Raphson method is that both calculate successive iterates from an equation of the form $u_i = \phi(u_{i-1})$, that is, the new iterate depends only on the iterate at the previous step. For *functional iteration* the equation whose root is required must appear explicitly in the form $u = \phi(u)$. The method may fail or work well, depending on which of the possibilities for writing the equation in this form is chosen. Convergence is assured, provided the starting value u_0 belongs to some suitable interval surrounding the root in which $|\phi'(x)| \leq c < 1$. The *Newton-Raphson* method uses the first-order (linear) Taylor series approximation to $f(x)$ around $x = u_i$. It will be left to the reader to demonstrate that, for a suitable choice of $\phi(\cdot)$, the method is equivalent to a functional iteration $u_{i+1} = \phi(u_i)$.

Functional Iteration—Rewriting $f(u) = 0$ as $u = \phi(u)$

Equation (1.1) is already in the required form, with

$$\phi(u) = 1.31^{-1}\log(1 + 1.93u).$$

It is readily verified that $\phi'(0.33) = 0.90$. Taken with the fact that $\phi''(x) < 0$ for x in the interval of interest, the omens for the use of the iteration $u_i = \phi(u_{i-1})$ seem good. A sequence of iterations appears in the first column of Table 7.1. The final three columns give calculations for *Aitken extrapolation*; this will be explained later.

A brief discussion of the theoretical properties of this method is desirable. Suppose that $|\phi'(\cdot)| \leq c < 1$ in an interval centered at a root u, that is, $u = \phi(u)$. Assuming that u_{i-1} lies in this interval, it then follows from the

Table 7.1. Functional Iteration for Finding a Root of $f(u) = u - 1.31^{-1}\log(1 + 1.93u)$, Together with Aitken Extrapolation

u_i	$u_{i+1} - u_i$	$u_{i+2} - 2u_{i+1} + u_i$	$\tilde{u}_i$
0.33	0.0462	−0.00571	0.704
0.376	0.0405	−0.00669	0.621
0.417	0.0338	−0.00668	0.587
0.450	0.0271	−0.00604	0.572
0.478	0.0211	−0.00508	0.565
0.499	0.0160	−0.00408	0.561
0.515	0.0119	−0.00316	0.560
0.527	0.00875	−0.00238	0.559
0.535	0.00636	−0.00177	0.558
0.542	0.00460	−0.00129	0.558
0.546	0.00331		
0.550			

Note: Aitken extrapolation calculations are given in columns 2 through 4. Iterates in the first column need to be calculated to six or seven decimal places in order to obtain the accuracy given in later columns.

mean value theorem that

$$u_i = \phi(u_{i-1}) = \phi(u) + (u_{i-1} - u)\phi'(z) \tag{3.1}$$

where z lies between u and u_{i-1}. Hence $|u_i - u| \le c|u_{i-1} - u|$. A further consequence of eq. (3.1) is that, when u_{i-1} is close to u,

$$u_i - u \simeq (u_{i-1} - u)\phi'(u).$$

If $\phi'(u) \neq 0$, this implies that the ratio of the distances of two successive iterates from the root is approximately constant. In such a case convergence is said to be linear. If $\phi'(u) = 0$, then eq. (3.1) implies that $|u_i - u|$ is, ultimately, very much smaller than $c|u_{i-1} - u|$, so that convergence is better than linear.

The *Aitken extrapolation* is intended for use when $\phi'(u) \neq 0$. One calculates $u_i = \phi(u_{i-1})$ as before, but now the sequence u_i is used to determine a second sequence $\tilde{u}_i$ which has a faster rate of convergence. For this set

$$\tilde{u}_i = u_i - \frac{(u_{i+1} - u_i)^2}{u_{i+2} - 2u_{i+1} + u_i} \tag{3.2}$$

The final three columns of Table 7.1 give the sequence of calculations for the example with which this chapter started.

If the original sequence converges linearly, a worthwhile further refinement is to use $\tilde{u}_i$ rather than u_i in calculating subsequent functional iterates. This leads to *Steffenson*'s method, for which

$$u_{i+1} = u_i - \frac{[\phi(u_i) - u_i]^2}{\phi[\phi(u_i)] - 2\phi(u_i) + u_i} \tag{3.3}$$

If this method is used, still staying with the same example, the successive iterates are

0.33 0.704 0.569 0.558 0.558

The equation used to demonstrate the calculations may, alternatively, be rewritten as $u = 1.31^{-1}(e^{\lambda u} - 1)$. The reader may care to investigate whether this form of equation is satisfactory for use as the basis of an iterative scheme.

The Newton-Raphson Method

The first-order Taylor series approximation to $f(x)$ around $x = u_i$ is

$$f(x) = f(u_i) + (x - u_i)f'(u_i)$$

Setting $f(x) = 0$ and replacing x by u_{i+1} gives

$$u_{i+1} = u_i - \frac{f(u_i)}{f'(u_i)} \tag{3.4}$$

It may be shown that the sequence u_i converges, provided $f'(u) \neq 0$ and u_0 is sufficiently close to the desired root. Checks should be applied that convergence is in fact occurring. See Dahlquist and Bjorck (1974, secs. 6.3, 6.5, and 6.7).

Taking $f(x)$ as in eq. (1.1) and

$$f'(x) = 1.0 - \frac{1.93}{1.31(1 + 1.93x)}$$

the sequence of iterates is

$$u_0 = 0.33 \qquad 0.792 \qquad 0.591 \qquad 0.559 \qquad 0.558 \qquad \ldots$$

Convergence is more rapid than for the secant method mentioned in Section 7.2. However $f'(\cdot)$ as well as $f(\cdot)$ must be evaluated at each step. There is a close connection between Newton-Raphson and the secant method. If in eq. (3.4), $f'(u_i)$ is replaced by the slope estimate $[f(u_i) - f(u_{i-1})]/(u_i - u_{i-1})$, the iterative scheme reduces to that for the secant method.

The Newton-Raphson method is readily extended to problems in finding the vector-valued root of a set of simultaneous equations. Section 7.8 shows how this may be done.

*7.4. NONLINEAR LEAST SQUARES

Consider the problem of minimizing

$$F(a) = \sum_{i=1}^{n} [g_i(a)]^2$$

Two chief methods are available. Under quite weak analytical assumptions [e.g., $g_i(a)$ everywhere differentiable] $F'(a) = 0$ at any minimum. If $F'(a) = 0$ at more than one point a, it is necessary to identify the particular root that makes $F(a)$ a minimum. Any of the methods discussed in Sections 7.1 through 7.3 may be used to solve for a.

Alternatively, one may write

$$g_i(a) \simeq g_i(a^{(0)}) + (a - a^{(0)})g_i'(a^{(0)})$$

where $a^{(0)}$ is an initial estimate of a. Then it is expected that a value of a which more nearly minimizes $F(a)$ will be obtained by minimizing

$$\sum_{i=1}^{n} \left[g_i(a^{(0)}) + (a - a^{(0)}) g_i'(a^{(0)}) \right]^2 = \sum_{i=1}^{n} \left[y_i - \Delta a^{(0)} x_i^2 \right]$$

where $y_i = -g_i(a^{(0)})$, $x_i = g_i'(a^{(0)})$, and $\Delta a^{(0)} = a - a^{(0)}$. Then take $a^{(1)} = a^{(0)} + \Delta a^{(0)}$ as the next approximation to a. This gives an iterative scheme in which each iteration is a least squares calculation. A possibility that must be checked is that successive iterates may converge to a local minimum.

In general, a will be replaced by a vector $\boldsymbol{a}$ whose elements are to be determined. The special case, when $g(\boldsymbol{a})$ is a monotonic function of a linear combination of elements of $\boldsymbol{a}$, is formally identical to the problem of fitting a generalized linear model with a normal distribution assumed for the dependent variate, as described in Section 7.9.

For a discussion of the nonlinear least squares problem, with reference to a FORTRAN algorithm which is described in detail in an accompanying article in the same journal issue, see Dennis et al. (1981).

7.5. SOLVING A SYSTEM OF NONLINEAR EQUATIONS—AN EXAMPLE

The *Bradley-Terry* model has been commonly used in analyzing data from taste panel experiments whereby each panelist is to compare only two out of some larger number of food treatments. The example provides a good introduction to the discussion of methods for the simultaneous solution of a set of nonlinear equations. The resulting model, which leads to maximum likelihood equations that can be solved using a simple iterative method, will be demonstrated in this section. Section 7.6 will take a somewhat similar approach to the estimation of fitted values in loglinear models. Both types of model might alternatively be fitted either directly, using the method of Newton-Raphson (Section 7.8) for solving systems of nonlinear equations, or by treating them as *generalized linear models* as defined in Nelder and Wedderburn (1972). Sections 7.9 and 7.10 will develop the theory of such generalized linear models.

A Taste Panel Experiment

Consider now results from an actual experiment. The four different types of fruit to be compared were from $k = 4$ different storage treatments. We will call the four types of fruit A, B, C, and D. Each of a number of taste panelists was offered fruit of two of the four types and asked, among other things, to decide which was firmer to the taste. Each of the six possible comparisons was repeated $n = 10$ times, with the following results:

	Compare with				
	A	B	C	D	
A firmer	—	1	1	0	(times out of 10)
B	9	—	3	3	
C	9	7	—	6	
D	10	7	4	—	

Define ν_{ij} to be the number of times (out of n) that the ith fruit was preferred to the jth fruit. Define the "score" a_i for the ith fruit to be the sum over all comparisons involving the ith fruit of the number of times [out of $n(k-1)$] that the ith fruit was preferred to one of the alternatives. Thus in the foregoing table $a_1 = 2$, $a_2 = 15$ $(= 9 + 3 + 3)$, $a_3 = 22$, $a_4 = 21$.

The Bradley-Terry Model

The *Bradley-Terry* model assumes that the ith fruit has some overall preference probability π_i, such that in a comparison between the ith fruit and the jth fruit, the probability that i will be preferred to j is

$$\pi_{ij} = \frac{\pi_i}{\pi_i + \pi_j} \tag{5.1}$$

Now define

$$\begin{aligned}\text{logit}(\pi_{ij}) &= \log\left(\frac{\pi_i}{\pi_j}\right)\\ &= \log(\pi_i) - \log(\pi_j)\\ &= \rho_i - \rho_j\end{aligned}$$

The ρ_i are unique to within addition of a constant to all k parameters. Use of the ρ_i in place of the π_i as the parameters gives a scale on which preferences are additive. The likelihood ℓ is the product of the binomial probabilities for the $\binom{k}{2}$ pairings which form the matrix of preferences. It is

$$\ell = C\prod_{i<j} \pi_{ij}^{\nu_{ij}}(1 - \pi_{ij})^{n-\nu_{ij}} = C\prod_{i \neq j} \frac{\pi_i^{\nu_{ij}}}{(\pi_i + \pi_j)^{\nu_{ij}}}$$

where C does not involve the unknown parameters. Note that $n - \nu_{ij}$ (with $i < j$) in the first expression becomes ν_{ji} in the second expression. Then

$$\log \ell = \sum_i a_i \log(\pi_i) - \sum_{i \neq j} \nu_{ij} \log(\pi_i + \pi_j) + C^*$$

where C^* does not involve the unknown parameters. Differentiating with respect to π_i gives equations that determine the maximum likelihood estimates p_i of the π_i:

$$\frac{a_i}{p_i} - \sum_j{}' \frac{n}{p_i + p_j} = 0 \quad (i = 1,\ldots,k) \tag{5.2}$$

where Σ' is used to denote summation over all $j \neq i$. Thus for each $i = 1,\ldots,k$

$$p_i = \frac{a_i}{n \sum_j{}' (p_i + p_j)^{-1}} \tag{5.3}$$

Initial estimates of the p_i are necessary; for example, they may all be taken equal to k^{-1}. A new set of p_i are obtained by taking the equality sign in eq. (5.3) as an assignment, successively for $i = 1,\ldots,k$. Each newly calculated p_i replaces the current value before proceeding. The whole cycle of calculations is repeated as necessary until iterates converge. If desired, the p_i can be rescaled at the end of each cycle so that they sum to 1.0.

For the example given at the beginning of this section, successive iterates were:

$p_1^{(0)} = 0.25$	$p_2^{(0)} = 0.25$	$p_3^{(0)} = 0.25$	$p_4^{(0)} = 0.25$
0.042	0.252	0.358	0.348
0.024	0.221	0.389	0.367
0.022	0.201	0.404	0.373
0.021	0.190	0.413	0.376
0.021	0.183	0.419	0.377
0.021	0.180	0.422	0.377
0.021	0.177	0.425	0.377
0.020	0.176	0.427	0.377

For better starting values, suggested by O. Dykstra, see David (1963).

Variance and covariance estimates may be obtained using the asymptotic maximum likelihood formulae given in Section 7.8. In order to use these, it is necessary to express one of the p_i's, perhaps p_k, in terms of the others. Variances and covariances are then obtained by taking the inverse of the $k-1$ by $k-1$ matrix whose (i, j)th element is $\partial^2 L/\partial p_i\, \partial p_j$, where $L = \log \ell$ is the loglikelihood.

In practice it may be more important to calculate variances and covariances for differences between treatments. (However, the scale implied by the p_i's is not necessarily the most suitable for this purpose, as will be shown shortly in this section.) This allows the use of t-tests, possibly modified so that the overall Type I error rate is at the chosen level (conventionally 5%), to be used in comparing each pair of treatments.

Further discussion of Bradley-Terry models may be found in Fienberg (1979).

When Simple Devices Suffice

The estimates p_i obtained using eq. (5.3) provide an intuitively appealing summary of experimental results. However, if each comparison is repeated the same number of times, it is best to base comparisons between treatments directly on the scores a_i. For as shown in David (1963), a_i is then less than a_j if and only if p_i is less than p_j. Moreover the standard error of the difference $a_j - a_i$ is, on the assumption of no difference between treatments, asymptotically equal to $\sqrt{(nk/2)}$.

7.6. ITERATIVE PROPORTIONAL SCALING—LOGLINEAR MODELS

This section will demonstrate the use of *iterative proportional scaling* in fitting loglinear models. Two-way tables illustrate the approach well enough, even though the calculations may appear trivial. The transition to fitting models for three-way and higher tables, where fitted values are not always available in closed form, is then straightforward. The second example taken is a three-way model, of a type such that fitted values are not available in closed form and the cycle of calculations must be repeated until acceptable accuracy has been obtained.

A Two-Way Table

R. O. Murray (1972) gives the incidence of deformity of the hip, as apparent in a radiographic examination, in samples from each of three groups of adult male Britishers aged between 17 and 21. The investigation was prompted by the suspicion that vigorous athletic activity is likely to lead to nascent hip damage. Sample *A* were from a country boarding school where considerable importance was given to sport. Sample *B* had attended a city school where more importance was given to intellectual concerns. Sample *C* came from industry; all had attended state schools where games were to a large extent voluntary. Results were

	Tilt Deformity		
	Present	Absent	Total
Sample *A* (sportsmen)	23	71	94
Sample *B* (intellectuals)	7	70	77
Sample *C* (from industry)	12	68	80
Total	42	209	251

Anyone who has taken an elementary course in statistics is likely to be familiar with the chi-square test for no association between school attended ($i = 1, 2, 3$) and presence or absence of the tilt deformity ($j = 1, 2$). The test assumes that in each row the individuals sampled fall independently into one or the other column. This assumption is, with data such as this, open to challenge. This is a point that will be taken up in due course, following the calculations.

The assumption of no association between row and column implies omission of the term that appears as an interaction in the loglinear model. Consider first the familiar multiplicative form of the model. If π_{ij} is the probability that an individual in the ith row will fall into column j, the hypothesis of interest specifies that

$$\pi_{ij} = \pi_{i.}\pi_{.j} \quad (i = 1, 2, 3; j = 1, 2) \tag{6.1}$$

for a suitable choice of $\pi_{i.}$ and $\pi_{.j}$. Iterative proportional scaling is only a little more complicated than the easy calculations needed for the standard method for obtaining fitted cell values in this model which is taught in every elementary course in statistics. The reader who has followed this almost trivial use of the iterative proportional scaling algorithm should be well placed to understand its use in more complicated models needed for multiway tables. Let the fitted value in the (i, j)th cell be m_{ij}. Then

$$\begin{aligned} \log(m_{ij}) &= \log(Np_{ij}) \\ &= \log N + \log(p_{i.}) + \log(p_{.j}) \end{aligned} \tag{6.2}$$

where $p_{i.}$ and $p_{.j}$ are the maximum likelihood estimates of $\pi_{i.}$ and $\pi_{.j}$, respectively. Equation (6.2) is formally identical to the equation that gives fitted values in terms of parameter estimates in a two-way analysis of variance model. (Note, however, that the fitted linear parameters do not satisfy any of the commonly adopted choices for identifiability conditions.) Despite the absence of a normally distributed dependent variate, eq. (6.2) carries with it the hope that some part of the theory associated with classical analysis of variance models will be applicable.

The notation now popular for multiway tables rewrites eq. (6.2) in the form

$$\begin{aligned} u_{ij} &= \log(m_{ij}) \\ &= u + u_{1(i)} + u_{2(j)} \end{aligned} \tag{6.3}$$

Here it is usual to insist that the parameters satisfy the analysis of variance sigma conditions:

$$\sum_{i=1}^{I} u_{1(i)} = \sum_{j-1}^{J} u_{2(j)} = 0 \tag{6.4}$$

where for the example $I = 3$ and $J = 2$.

The *iterative proportional scaling* algorithm works with the cell frequencies as they stand, after the style of eq. (6.1). In this it is to be contrasted with Nelder and Wedderburn's *generalized linear model* approach which is the subject of Sections 7.7, 7.9, and 7.10 and which works with the linearized form of the model given in eqs. (6.2) and (6.3).

The *no association* model of eq. (6.3) specifies parameters corresponding to both the row and the column margins. Let m_{i+} be the sum of the fitted values m_{ij}, and let n_{i+} be the sum of the observed values n_{ij}, in the ith row. Sums of fitted and observed values in each column are similarly defined as m_{+j} and n_{+j}, respectively. It may then be shown that the maximum likelihood estimates of the parameters are such that

$$m_{i+} = n_{i+} \quad (i = 1, 2, 3)$$

$$m_{+j} = n_{+j} \quad (j = 1, 2)$$

subject to the further condition that $\log(m_{ij})$ is given in terms of parameters u, $u_{1(i)}$, and $u_{2(j)}$, as in eq. (6.3).

The *iterative proportional scaling* algorithm starts with all entries $m_{ij}^{(0)}$ in the table equal to 1.0, or otherwise chosen so that $u_{ij}^{(0)} = \log(m_{ij}^{(0)})$ is of the form given in eq. (6.3). Taking each margin in turn, table entries $m_{ij}^{(0)}$ are scaled so that marginal totals of fitted and observed values are the same. As will become apparent, all that is necessary in this simple case is to scale first elements in each row, and then elements in each column, according to the procedure described. The algorithm readily generalizes to more complicated cases, where there are three or more margins in which the totals for fitted values must be made to agree with totals for observed values. In the context of general multiway tables eq. (6.3) specifies a *direct* model, in which calculations converge after one cycle. In general, scaling so that totals agree in one margin may upset the relationships established in other margins of the table, and the cycle of calculations must be repeated until the successive iterates converge. By way of giving some rationale for the algorithm, observe that if

$$u_{ij}^{(\,)} = u + u_{1(i)}^{(\,)} + u_{2(j)}^{(\,)}$$

is of the form given by eq. (6.3), then so is

$$u_{ij}^{(*)} = u_{ij}^{(\,)} + \log(n_{i+}) - \log\left(m_{i+}^{(\,)}\right)$$

and similarly for the scaling used to adjust the j margin.

Calculations for a Two-Way Table—Two Examples

Calculations are given both for a two-way table with artificial data designed to make the arithmetic easy and for the tilt deformity data. The artificial data, with marginal totals in brackets, is

	$j = 1$	$j = 2$	Total
$i = 1$	6	8	(14)
$i = 2$	5	15	(20)
$i = 3$	9	7	(16)
Total	(20)	(30)	(50)

In the calculations in Table 7.2 the table margin currently under consideration is identified by placing in it, in brackets, the relevant totals of observed values. Following these are given the corresponding totals for the current fitted values.

From the fitted values in Table 7.2 it is an easy matter to calculate the parameters of eq. (6.3), with the usual analysis of variance sigma conditions.

Thus

$$u = u_{..}$$

$$u_{1(i)} = u_{i.} - u_{..}$$

$$u_{2(j)} = u_{.j} - u_{..} \tag{6.5}$$

In addition one may wish to calculate multiplicative versions of these parameter estimates:

$$t = \exp(u_{..}), \quad t_{1(i)} = \exp(u_{1(i)}), \quad t_{2(j)} = \exp(u_{2(j)})$$

Consider the parameter estimates corresponding to the i margin. Whereas the loglinear parameters have been chosen so that they sum to zero, the multiplicative versions multiply to 1.0, and similarly for the j margin.

A test whether the model is adequate may be based on the familiar *Pearson chi-square* statistic. This equals the sum, over all cells, of the quantity $(n_{ij} - m_{ij})^2/m_{ij}$. An alternative suggested by maximum likeli-

'able 7.2. Iterative Proportional Scaling Used to Find Fitted Values in Two-Way Tables, ssuming No Association between Rows and Columns

Artificial Data			Tilt Deformity Example		
1	1	(14) 2 $[\frac{14}{2} = 7]$	1	1	(94) 2 $[\frac{94}{2} = 47]$
1	1	(20) 2 $[\frac{20}{2} = 10]$	1	1	(77) 2 $[\frac{77}{2} = 38.5]$
1	1	(16) 2 $[\frac{16}{2} = 8]$	1	1	(80) 2 $[\frac{80}{2} = 40]$
↓	↓		↓	↓	
7	7		47	47	
10	10		38.5	38.5	
8	8		40	40	
20)	(30)		(42)	(209)	
25	25		125.5	125.5	
↓$[\frac{20}{25}]$	↓$[\frac{30}{25}]$		↓$[\frac{42}{125.5}]$	↓$[\frac{209}{125.5}]$	
.6	8.4	(14) 14.0	15.73	78.27	(94) 94.00
.0	12.0	(20) 20.0	12.88	64.12	(77) 77.00
.4	9.6	(16) 16.0	13.39	66.61	(80) 80.00

e: The table margin currently under scrutiny is identified by placing in it, in parentheses, the relevant totals of erved values. The multiplier for table entries in that row appears in square brackets.

hood theory is known either as G^2 or (using the terminology of Nelder and Wedderburn's generalized linear models) as the *deviance*, D. It equals twice the amount by which the loglikelihood is reduced from that for the full model which has one parameter for each cell in the table, multiplied by a scale factor ϕ which equals one for loglinear models in which it is assumed that items have entered independently into the cells of any row of the table. It is calculated by summing the quantity $2n_{ij}\log(n_{ij}/m_{ij})$ over all cells of the table. Provided the n_{ij} do not differ unduly from the m_{ij}, the chi-square statistic and the deviance (D or G^2) will be approximately equal. For the tilt deformity example, the chi-square statistic is 7.5, while the deviance is 7.7. In either case the comparison is with the critical value of a chi-square statistic with two degrees of freedom.

Here it is desirable to interject a warning: if items have not entered independently into the cells of the table, then inferences based on either statistic will be in doubt. Thus there will almost certainly have been instances, in the tilt deformity data, where brothers have appeared in the same row. As brothers might be expected to show a similar proneness (or lack of it) to tilt deformity, they are likely also to appear in the same column. The effect can be illustrated by considering the extreme case where two brothers are present from each family. If brothers always fall into the same column of the table, then the observational unit should be a pair of brothers. The counting of each brother as an independent observational unit makes each n_{ij} twice as large as it should be, so that both the chi-square statistic and the deviance are too large by a factor of two.

A Three-Way Table

Now consider a three-way table with fictitious data. The numbers have been chosen to illustrate the possibilities for misinterpretation which arise when a three-way table is collapsed into a two-way table. Individuals (male or female) are classified as porridge eaters (P) or other (NP), and as sleep-walkers (SW) or other (NSW):

	Males ($k = 1$)			Females ($k = 2$)			Add over k	
	SW	NSW		SW	NSW		($ij+$ marginal val	
P	4	21	(25)	34	14	(48)	(38)	(3
NP	16	59	(75)	26	6	(32)	(42)	(6
	(20)	(80)		(60)	(20)			

From such a table it is clear that there is a strong association between sex and porridge eating, and between sex and sleepwalking. Males are in both cases much less likely to be involved than females. For each of $k = 1$ and $k = 2$ the table margins show the ratios P : NP and SW : NSW in a way that differs little from the proportions within the body of the table. These are the *ik* margins (marginal totals n_{i+k}):

(25)	(48)
(75)	(32)

The *jk* margins (marginal totals n_{+jk}) are

(20)	(80)	(60)	(20)

The *ij* marginal totals on the right show an apparent association between porridge eating and sleepwalking which results from the associations between porridge eating and sex, and between sleepwalking and sex.

The fitted model required is one for which

$$m_{i+k} = n_{i+k} \quad (i, k = 1, 2) \tag{6.6}$$

$$m_{+jk} = n_{+jk} \quad (j, k = 1, 2) \tag{6.7}$$

In addition the log (fitted values) u_{ijk} must be of the form

$$u_{ijk} = u + u_{1(i)} + u_{2(j)} + u_{3(k)} + u_{12(ik)} + u_{23(jk)}$$

These values can be determined by fitting separately the two-way tables for $k = 1$ and for $k = 2$. Notice in passing that fitting the *ik* margins incidentally fits the *i* margins and the *k* margins. When the *jk* margins are also fitted this incidentally fits the *j* margins.

However, in order to provide an example of calculations that cannot be handled so easily, consider the model that includes all first-order interactions. In addition to the conditions given in eqs. (6.6) and (6.7)

$$m_{ij+} = n_{ij+} \quad (i, j = 1, 2) \tag{6.8}$$

For this model

$$u_{ijk} = u + u_{1(i)} + u_{2(j)} + u_{3(k)} + u_{23(jk)} + u_{13(ik)} + u_{12(ij)}$$

As before we place 1.0 in each cell initially. The first cycle of calculations is shown in Table 7.3. The *jk* marginal totals are first adjusted, then the *ik* marginal totals, and then the *ij* marginal totals. This cycle of calculations must be repeated several times.

Table 7.3. Use of Iterative Proportional Scaling to Calculate the Fitted Values for the Three-Way Table Given in the Text

1	1		1	1			
1	1		1	1			
(20)	(80)		(60)	(20)			
2	2		2	2			
↓ × 10.0	↓ × 40.0		↓ × 30.0	↓ × 10.0	[$\frac{20}{2}$ = 10.0, $\frac{80}{2}$ = 40.0, etc.]		
10	40	(25) 50.0	30	10	(48) 40.0		
10	40	(75) 50.0	30	10	(32) 40.0		
↓	↓		↓	↓			
5	20		36	12		(38) 41	(35) 32
15	60		24	8		(42) 39	(65) 68
↓	↓		↓	↓			
4.63	21.87		33.37	13.13			
16.15	57.35		25.85	7.65			
(20)	(80)		(60)	(20)			
20.78	79.22		59.22	20.78			

Note: The first cycle of calculations is shown. The *jk* marginal totals are first adjusted, then the *ik* marginal totals, and then the *ij* marginal totals. This cycle of calculations must be repeated several times.

Calculations repeat the pattern given in Table 7.3 as often as is necessary to obtain convergence. Six cycles were needed to give fitted values accurate to one decimal place:

3.7	21.3		34.3	13.7
16.3	58.7		25.7	6.3

It is left to the reader to calculate the loglinear parameter estimates and their multiplicative counterparts, using the approach shown in eq. (6.5).

Model Tests and Variance Estimates

Both the chi-square statistic and the deviance (or G^2-statistic) generalize in the obvious way to use in three-way and higher tables. In the analysis just carried out one might wish to test whether the $u_{12(ij)}$ term can reasonably be omitted from the model which includes all first-order interactions. The relevant deviances are 0.062 for the model with all first-order interactions (jk, ik, ij), on one degree of freedom; and 1.44 for the model that omits the ij interaction, on two degrees of freedom. Degrees of freedom are calculated just as for the corresponding analysis of variance model.

Methods for calculating variance estimates are discussed in Bishop, Fienberg, and Holland (1975) and in Fienberg (1980). Unless the model is *direct*, that is, except where the iterative proportional scaling algorithm will theoretically give convergence in a finite number of steps, there would appear to be no completely satisfactory alternative to calculating variance and covariance estimates as for iteratively reweighted least squares.

Nelder and Wedderburn's generalized linear model approach applies iteratively reweighted least squares to a particular class of problems, which includes loglinear models. This approach to computing parameter estimates in loglinear models is much slower, and requires more space for storage of arrays, than does iterative proportional scaling. It makes good sense to use iterative proportional scaling for the initial screening of alternative models and perhaps for getting initial parameter estimates in those models that appear most interesting, even if iteratively reweighted least squares is finally to be used in order to obtain estimates of variances and covariances.

*7.7. LOGLINEAR MODELS—THEORETICAL CONSIDERATIONS

Loglinear models arise in a variety of ways. An example that has little connection with multiway tables will now be presented, by way of introducing some important theoretical points.

A Simple Loglinear Model with Poisson Errors

The expected number λ_i of male poroporo moths (*Sceliodes cordalis*) captured using a pheromone (in this context a male sex attractant) at times t_i after release is, as a first approximation, described using the model

$$\eta_i = \alpha + \beta t_i \tag{7.1}$$

where $\eta_i = \log(\lambda_i)$. The following numbers were obtained in a mark-release study with these moths:

Time, t_i (days)	1	2	3	4	5	6
Number of moths, n_i	17	8	2	2	4	0

As a first approximation, it is assumed that the observed count n_i at time t_i has a Poisson distribution with mean λ_i. (In fact it is likely that the observed count at one time will influence the mean level at the next. In this instance the model that assumes independent Poisson errors nevertheless gives an adequate account of the data.)

Calculations for this example, using the Newton-Raphson method, are given in the next section.

The Equivalence of Poisson with Multinomial Assumptions

What may seem surprising is that, for purposes of inference, it makes no difference to estimates based on the likelihood function whether the joint distribution of the n_i $(i = 1, 2, \ldots, 6)$ in the example just given is that of six independent Poisson distributions corresponding to the six categories, or whether it is multinomial with total $N = 17 + 8 + 2 + 2 + 4 + 0 = 33$. Large sample variance and covariance estimates, based on asymptotic distributions, are the same in the two cases. A slight complication is that if one analyzes assuming Poisson errors when the model is multinomial, variances and covariances associated with the total N must be ignored. The equivalence holds whenever the loglinear model used includes a constant term [such as α in eq. (7.1)], this ensures that

$$\sum_i \lambda_i = \sum_i n_i = N$$

which is also the sum of the expected values for the multinomial distribution. The ease of working with models with independent Poisson errors carries over with minor modification to a wide class of loglinear models with multinomial errors.

In the context of multiway tables there will be one fixed multinomial total for each fixed marginal total. In the porridge eating-sleepwalking example of the previous section, the fixed marginal total is that for sex. There are two multinomial distributions, one for each sex with four categories each. The model must include one parameter for each such fixed marginal total. The simplest case to consider (and the only case discussed here) is that where the fixed margins do not overlap. There is then the same equivalence between an analysis that assumes Poisson errors over all cells of the table and one that makes appropriate multinomial assumptions. For further discussion, see Haberman (1974), Palmgren (1981).

The Loglikelihood for a Model with Poisson Errors

Suppose that y_i $(i = 1, \ldots, n)$ follow independent Poisson distributions, with means λ_i given by a loglinear model whose parameters are to be estimated. Parameter estimates will be chosen to maximize the loglikelihood. This may be written

$$L = \sum_i y_i \log \lambda_i - \sum_i \lambda_i + \sum_i c(y_i) \qquad (7.2)$$

for an appropriate choice of the function $c(\cdot)$. The maximum of L over all possible models is given by the full model in which the λ_i or $\log(\lambda_i)$ are themselves the parameters to be estimated. For this model the estimate of λ_i is $\tilde{\lambda}_i = y_i$ $(i = 1, \ldots, n)$, and

$$\tilde{L} = \sum_i y_i \log y_i - \sum_i y_i + \sum_i c(y_i) \qquad (7.3)$$

Let $\hat{\lambda}_i$ be the estimate of λ_i in the current model, and write $\hat{L}$ for the loglikelihood for this model. A statistic that corresponds to the residual sum of squares in a normal errors model is the deviance, which was mentioned earlier in Section 7.6 and is discussed in more detail in Section 7.9. Here, assuming that the model is one for which $\Sigma_i \hat{\lambda}_i = \Sigma_i y_i = N$, it equals

$$-2\phi(\hat{L} - \tilde{L}) = 2\sum_i y_i \log\left(\frac{y_i}{\hat{\lambda}_i}\right)$$

where $\phi = 1$. Assuming the current model, the deviance follows asymptotically a chi-square distribution with degrees of freedom equal to $n - q$, where q is the number of parameters which have been estimated.

The Comparison with the Normal Errors Model

The corresponding statistic for a model in which the y_i follows a normal distribution with mean μ_i and variance σ^2, independently for $i = 1,\ldots,n$, is

$$D = -2\phi(\hat{L} - \tilde{L})$$

$$= \sum_i (y_i - \hat{\mu}_i)^2$$

where $\phi = \sigma^2$. In this case the deviance D is the familiar residual sum of squares. Notice that D is defined so as to be independent of the unknown scale parameter ϕ. The usual estimate of ϕ is $s^2 = D/(n - q)$, where, as before, q is the number of parameters estimated. Use of this estimator is based on considerations other than maximum likelihood.

*7.8. THE NEWTON-RAPHSON METHOD FOR SEVERAL UNKNOWNS

Assuming Poisson errors, the loglikelihood for the simple growth model given in Section 7.7 is

$$L = -\sum_i \lambda_i + \sum_i n_i \log(\lambda_i) + c^*$$

where c^* does not involve the unknown parameters, and $\log(\lambda_i) = \alpha + \beta t_i$ $(i = 1, 2, 3)$. (See Table 7.4 in the next section.) Substituting a for α and b for β and equating to zero the derivative of L with respect to a and b in turn yields (writing f_0 for the derivative with respect to a, and f_1 for the derivative with respect to b)

$$f_0 = \frac{\partial L}{\partial a} = -\sum_i \hat{\lambda}_i + \sum_i n_i = 0$$

$$f_1 = \frac{\partial L}{\partial b} = -\sum_i \hat{\lambda}_i t_i + \sum_i n_i t_i = 0$$

where $\hat{\lambda}_i = \exp(a + bt_i)$. The Newton-Raphson method is suitable for solving such systems of nonlinear equations.

The Method of Newton-Raphson

Let $\boldsymbol{b}$ be the vector of parameter estimates to be determined, and suppose that a set of equations $f_i = 0$, where $f_i = f_i(\boldsymbol{b})$, is to be solved. In this book the main interest is in examples such as that just considered, where f_i is the derivative of a loglikelihood with respect to the ith parameter. Let $\boldsymbol{f} = \boldsymbol{f}(\boldsymbol{b})$ be the vector whose ith element is f_i, and let $\boldsymbol{J} = \boldsymbol{J}(\boldsymbol{b})$ be the matrix whose (i, j)th element J_{ij} is the derivative of f_i with respect to the jth parameter, evaluated at $\boldsymbol{b}$. Calculations proceed iteratively. Assume that, following the kth step, the approximation $\boldsymbol{b}^{(k)}$ has been obtained to the vector $\boldsymbol{b}$ such that $\boldsymbol{f}(\boldsymbol{b}) = 0$. Then writing

$$\boldsymbol{f}^{(k)} = \boldsymbol{f}(\boldsymbol{b}^{(k)}) \quad \text{and} \quad \boldsymbol{J}^{(k)} = \boldsymbol{J}(\boldsymbol{b}^{(k)})$$

the first-order Taylor series approximation to $\boldsymbol{f}$ in the neighborhood of $\boldsymbol{f}^{(k)}$ is

$$\boldsymbol{f} \simeq \boldsymbol{f}^{(k)} + \boldsymbol{J}^{(k)}(\boldsymbol{b} - \boldsymbol{b}^{(k)}) \tag{8.1}$$

Setting $\boldsymbol{f} = 0$ gives

$$-\boldsymbol{J}^{(k)}\Delta\boldsymbol{b}^{(k)} \simeq \boldsymbol{f}^{(k)} \tag{8.2}$$

where $\Delta\boldsymbol{b}^{(k)} = \boldsymbol{b} - \boldsymbol{b}^{(k)}$. Then providing $\boldsymbol{b}$ is sufficiently close to $\boldsymbol{b}^{(k)}$, taking

$$\boldsymbol{b}^{(k+1)} = \boldsymbol{b}^{(k)} + \Delta\boldsymbol{b}^{(k)} \tag{8.3}$$

will yield an improved approximation.

For the example with which this section started

$$J_{00} = \frac{\partial^2 L}{\partial a^2} = -\sum_i \exp(a + bt_i) = -\sum_i \lambda_i$$

$$J_{01} = \frac{\partial^2 L}{\partial a\,\partial b} = -\sum_i t_i \exp(a + bt_i) = -\sum_i t_i\lambda_i$$

$$J_{11} = \frac{\partial^2 L}{\partial b^2} = -\sum_i t_i^2 \exp(a + bt_i) = -\sum_i t_i^2\lambda_i$$

Variance and Covariance Estimates for Maximum Likelihood Estimates

For the discussion that now follows, it is assumed that $\boldsymbol{f}$ is the vector of first derivatives of a loglikelihood. Then $\boldsymbol{J}$ is the matrix of second derivatives, otherwise known as the *Hessian* and written as $\boldsymbol{H}$. It then follows from asymptotic maximum likelihood theory, under suitable regularity conditions, that the elements of $-\boldsymbol{J}^{-1}$ provide asymptotic variance and covariance estimates. Calculations are conveniently handled by first calculating the Cholesky matrix $\boldsymbol{T}$ such that $\boldsymbol{T}'\boldsymbol{T} = -\boldsymbol{J}$, followed by calculation of $\boldsymbol{T}^{-1}$ and then $-\boldsymbol{J}^{-1}$. Successful evaluation of $\boldsymbol{T}$, with all diagonal elements nonzero, establishes that $-\boldsymbol{J}$ is positive definite and hence that the solution to the Newton-Raphson equations gives at least a local minimum.

Alternatively the elements of $\boldsymbol{D}^{-1}$, where

$$\boldsymbol{D}(\boldsymbol{b}) = E\left[\boldsymbol{f}(\boldsymbol{b})\boldsymbol{f}(\boldsymbol{b})'\right] \tag{8.4}$$

evaluated at the maximum likelihood estimate of $\boldsymbol{b}$, may be used for this purpose. Moreover eq. (8.2) might be replaced by

$$\boldsymbol{D}^{(k)}\Delta\boldsymbol{b}^{(k)} = \boldsymbol{f}^{(k)} \tag{8.5}$$

This is equivalent to replacing the second derivatives in eq. (8.2) by their expected values. In general, this may be hazardous. If the model does not fit well, $\boldsymbol{D}^{(k)}$ may differ widely from $\boldsymbol{J}^{(k)}$, to the extent that iterations will not now converge.

For conditions under which the asymptotic maximum likelihood theory applies, see Cox and Hinkley (1974, sec. 9.1). Special care is needed if one of the parameters to be estimated is used in specifying the boundaries for the parameter space. General results regarding the adequacy of the asymptotic approximation in any particular case are not usually available, so that the obtained variance estimates must be treated as guides only.

Example of the Calculations

For the moth capture example of Section 7.7, starting values are conveniently obtained from a straight-line regression of $\log(n_i)$ upon t_i. For purposes of this starting value calculation $n_6 = 0$ is replaced by 0.5. This leads to initial values $a^{(0)} = 3.136$, $b^{(0)} = -0.5632$. The first step of the calculations then proceeds thus:

t_i	1	2	3	4	5	6	
n_i	17	8	2	2	4	0	$\left(\sum_i n_i = 33\right)$
$a^{(0)} + b^{(0)}t_i$	2.573	2.010	1.446	0.883	0.320	−0.243	
$\lambda_i^{(0)}$	13.102	7.460	4.248	2.419	1.377	0.784	

$$\left(\sum_i \lambda_i^{(0)} = 29.390\right)$$

Thus

$$f_0^{(0)} = \frac{\partial L}{\partial a} = -\sum_i \lambda_i^{(0)} + \sum_i n_i = 33 - 29.390 = 3.610$$

$$f_1^{(0)} = \frac{\partial L}{\partial b} = -\sum_i \lambda_i^{(0)} t_i + \sum_i n_i t_i = 67 - 62.031 = 4.969$$

Also

$$-J_{00}^{(0)} = \sum_i \lambda_i^{(0)} = 29.390$$

$$-J_{01}^{(0)} = \sum_i t_i \lambda_i^{(0)} = 62.031$$

$$-J_{11}^{(0)} = \sum_i t_i^2 \lambda_i^{(0)} = 182.527$$

Then solving $-\boldsymbol{J}^{(0)}\Delta\boldsymbol{b}^{(0)} = \boldsymbol{f}^{(0)}$ gives

$$\Delta\boldsymbol{b}^{(0)} = \begin{bmatrix} 0.231 \\ -0.0514 \end{bmatrix}$$

Hence

$$a^{(1)} = 3.136 + 0.231 = 3.367$$

$$b^{(1)} = -0.5632 - 0.0514 = -0.6146$$

Following the next iteration

$$a^{(2)} = 3.367 - 0.017 = 3.350$$

$$b^{(2)} = -0.6146 + 0.0039 = -0.6107$$

A third iteration gives

$$a^{(3)} = 3.350 + 0.00004$$

$$b^{(3)} = -0.6107 + 0.00007$$

Thus results are accurate to at least three significant decimal digits. Also

$$-\boldsymbol{J} = \begin{bmatrix} 33 & 67 \\ 67 & 189.69 \end{bmatrix}$$

has Cholesky decomposition

$$\boldsymbol{T} = \begin{bmatrix} 5.745 & 11.663 \\ 0 & 7.325 \end{bmatrix} \quad \text{with} \quad \boldsymbol{T}^{-1} = \begin{bmatrix} 0.1741 & -0.2772 \\ 0 & 0.1365 \end{bmatrix}$$

The fact that it has been possible to form $\boldsymbol{T}$ with positive diagonal elements implies that $-\boldsymbol{J}$ is positive definite, and hence that parameter values give at least a local maximum. As the likelihood can be readily shown to be convex, this is a global maximum. Then

$$-\boldsymbol{J}^{-1} = \begin{bmatrix} 0.107 & -0.0378 \\ -0.0378 & 0.0186 \end{bmatrix}$$

Thus SE[a] $\simeq$ 0.33, SE[b] $\simeq$ 0.14.

For this example the assumption of independent errors is dubious. Capture of moths at one time will reduce the number available for capture the next time.

Modified Newton Methods

Convergence is guaranteed for the Newton-Raphson method only if the initial approximation is good enough. Convergence in general can be ensured for a wide class of functions, provided eq. (8.3) is replaced by

$$\boldsymbol{b}^{(k+1)} = \boldsymbol{b}^{(k)} + \delta\,\Delta\boldsymbol{b}^{(k)} \tag{8.6}$$

where δ is chosen so that the set of values $(\boldsymbol{f}^{(k)})'\boldsymbol{f}^{(k)}$ form a strictly monotone decreasing sequence. For details, see Stoer and Bulirsch (1980, sec. 5.4).

In many applications the repeated reevaluation of the matrix $\boldsymbol{J}^{(k)}$ of derivatives will be unacceptably expensive. Methods are available that replace $\boldsymbol{J}^{(k)}$ by a simply calculated approximation, but still in such a way that a suitable choice of δ in eq. (8.6) will ensure convergence. The *delta* method calculates

$$J_{ij}^{(k)} = \frac{f_i^{(k)}(\boldsymbol{b} + \delta\boldsymbol{e}_i) - f_i^{(k)}(\boldsymbol{b})}{\delta}$$

where $\boldsymbol{e}_i$ has one in its ith position and zeros elsewhere. Choosing δ so that a good approximation to the derivative is obtained is not quite straightforward. A further reduction in the number of steps may be achieved by calculating an adequate approximation to $\boldsymbol{J}^{(k)}$ from $\boldsymbol{J}^{(k-1)}$. Stoer and Bulirsch (1980) discuss a method, due to Broyden, which calculates $\boldsymbol{J}^{(k)}$ by a simple *rank 1* update from $\boldsymbol{J}^{(k-1)}$.

Computer Programs

The writing of computer programs for solving general systems of nonlinear equations is a task for the expert. Anyone who has access to the Numerical Algorithms Group (Numerical Algorithms Group, 1981) or International Mathematical and Statistical Subroutine Library (IMSL, 1982), will be advised to begin by investigating the subroutines available there. Otherwise, the Association for Computing Machinery Transactions on Mathematical Software should be checked for the most recent subroutines published there.

*7.9. GENERALIZED LINEAR MODELS

Generalized linear models is the name which Nelder and Wedderburn (1972) use to describe a class of models which includes *loglinear models* and *logit* and *probit* models as used in bioassay. The *Bradley-Terry* model (Section 7.5) is a generalized linear model.

There are obvious conceptual economies from bringing a variety of different models under a single theory. This is exploited to good effect in the statistical package/language GLIM (Baker and Nelder, 1978), which is designed to cater for any model in this class. The Newton-Raphson procedure for solving the maximum likelihood equations works well for all the commonly used models.

The Exponential Family of Distributions

Table 7.4 shows how the loglikelihood for a single observation from a normal, binomial, or Poisson distribution may in each case be written

$$L = \phi^{-1}[\theta y - d(\theta)] + c(\phi, y) \tag{9.1}$$

where θ is a function of $\mu = E[y]$. Distributions whose loglikelihood can be written in this way comprise the family. Another member of this family likely to be encountered in the context of generalized linear models is the gamma distribution.

Definition

The definition given here follows, in general outline, Baker and Nelder (1978). See also Nelder and Wedderburn (1972). The definition is conveniently given in two parts:

i. The model has a dependent variate y (corresponding to a single element of $\mathbf{y}$) whose distribution is of exponential form.

Table 7.4. General Form and Particular Cases of the Loglikelihood for a Single Observation y from a Distribution Belonging to the Exponential Family

General Form

$\log \ell = \phi^{-1}[\theta y - d(\theta)] + c(\phi, y)$
$E[y] = d'(\theta)$; $\text{var}[y] = \phi d''(\theta)$

Normal

$\log \ell = \sigma^{-2}(\mu y - \frac{1}{2}\mu^2) + c(\phi, y)$
$[\phi = \sigma^2; \theta = \mu; d(\theta) = \frac{1}{2}\mu^2]$

Binomial

$\log \ell = \theta y - n \log(1 + e^{\theta}) + c(y, n)$
where $\theta = \log[\pi/(1 - \pi)]$ and n is the number of Bernoulli trials
$[\phi = 1; d(\theta) = n \log(1 + e^{\theta})]$

Poisson

$\log \ell = \theta y - e^{\theta} + c(y)$
where $\theta = \log \mu$
$[\phi = 1; d(\theta) = e^{\theta}]$

Note: As elsewhere, log( ) is logarithm to base e.

ii. Taking $\boldsymbol{x}' = [x_0, x_1, \dots, x_p]$ to be a vector whose elements are the explanatory variates (i.e., a single row of the matrix $\boldsymbol{X}$ that performs the role of the model matrix), one has $E[y] = \mu$, where

$$\eta = g(\mu) = \boldsymbol{x}'\boldsymbol{\beta} \tag{9.2}$$

Here $g(\cdot)$ is the *link function*, $\boldsymbol{\beta}$ is the vector of coefficients to be estimated, and $\eta = \boldsymbol{x}'\boldsymbol{\beta}$ is the *linear predictor*. It is assumed that $g(\cdot)$ is monotonic.

For a loglinear model the link function is $\log(\cdot)$, that is, $\eta = g(\mu) = \log(\mu) = \boldsymbol{x}'\boldsymbol{\beta}$. For the Bradley-Terry model the link function is the *logit* of μ/n, that is, $\eta = g(\mu) = \log(\pi/(1 - \pi))$. In these two cases the link function is the inverse function to the function $d'(\cdot)$ in eq. (9.1), and $\eta = \theta$. Choice of the link function as the inverse function to $d'(\cdot)$ has the attraction that the elements of $\boldsymbol{X}'\boldsymbol{y}$ are then *sufficient statistics*.

Expressions for the Mean and Variance of y

Differentiating eq. (9.1) with respect to θ,

$$\frac{\partial L}{\partial \theta} = \phi^{-1}[y - d'(\theta)] \tag{9.3}$$

Hence as (see, e.g., Cox and Hinkley, 1974),

$$E\left[\frac{\partial L}{\partial \theta}\right] = \frac{\partial}{\partial \theta} E[L] = 0$$

it follows that $\mu = E[y] = d'(\theta)$.

From eq. (9.3)

$$E\left[\left(\frac{\partial L}{\partial \theta}\right)^2\right] = \phi^{-2}\text{var}[y]$$

Then as

$$E\left[\left(\frac{\partial L}{\partial \theta}\right)^2\right] = -E\left[\frac{\partial^2 L}{\partial \theta^2}\right] = \phi^{-1}d''(\theta)$$

(see again Cox and Hinkley, 1974), it follows that

$$\text{var}[y] = \phi d''(\theta) = \phi \frac{d\mu}{d\theta}$$

In the sequel an approximate expression for var$[g(y)]$ will be needed. For this take

$$\text{var}[g(y)] \simeq [g'(\mu)]^2 \text{var}[y]$$

$$= \phi[g'(\mu)]^2 \frac{d\mu}{d\theta} \tag{9.4}$$

Getting an Initial Approximation for the Estimate b of β

The transformed values $g_i = g(y_i)$ follow the linear model

$$g_i = \mathbf{x}_i'\boldsymbol{\beta} + \xi_i \tag{9.5}$$

The distribution of ξ_i is that of $g(y_i - \mu_i)$, where it is y_i whose distribution has been specified. Least squares, or perhaps weighted least squares, may thus be used to minimize $\Sigma_i \xi_i^2$ or $\Sigma_i w_i \xi_i^2$ in eq. (9.5). It is safest to take the weights w_i all equal to 1.0 for these initial calculations. Alternatively, w_i might be chosen so that w_i^{-1} is proportional to the value obtained for $\text{var}[g(y_i)]$ in eq. (9.4) when μ_i is set equal to y_i.

Maximizing the Loglikelihood

The expression to be maximized is

$$\phi L = \sum_i \theta_i y_i - \sum_i d(\theta_i) + c^*(y) \tag{9.6}$$

The derivative of $\theta_i y_i - d(\theta_i)$ with respect to θ_i is $y_i - d'(\theta_i) = y_i - \mu_i$. Recall that θ_i is a function of μ_i. Furthermore μ_i is a function of η_i; this follows because the function $g(\cdot)$ giving η_i in terms of μ_i is assumed monotonic. The derivative of ϕL with respect to β_j is thus

$$\sum_i (y_i - \mu_i) \frac{d\theta_i}{d\mu_i} \frac{d\mu_i}{d\eta_i} x_{ij} \quad (j = 0, 1, \ldots, p)$$

Replacing μ_i by $\hat{\mu}_i$ and η_i by $\hat{\eta}_i = g(\hat{\mu}_i) = \boldsymbol{x}_i'\boldsymbol{b}$ and equating to zero, for all j, gives

$$\boldsymbol{X}'\boldsymbol{\Psi}\hat{\boldsymbol{\mu}} = \boldsymbol{X}'\boldsymbol{\Psi}\boldsymbol{y} \tag{9.7}$$

where

$$\boldsymbol{\Psi} = \text{diag}\left[\frac{d\hat{\theta}_i}{d\hat{\mu}_i} \frac{d\hat{\mu}_i}{d\hat{\eta}_i}\right]$$

and $\hat{\boldsymbol{\mu}}$ is such that $\boldsymbol{g}(\hat{\boldsymbol{\mu}}) = \hat{\boldsymbol{\eta}} = \boldsymbol{X}\boldsymbol{b}$.

A Newton-Raphson Procedure for Solving the Likelihood Equations

Suppose now that $\hat{\boldsymbol{\mu}}$ satisfies eq. (9.7) and that an initial approximation $\boldsymbol{m}^{(0)}$ to $\hat{\boldsymbol{\mu}}$ is available. The first-order Taylor series approximation to $\hat{\boldsymbol{\eta}}$ [with elements $g_i(\hat{\mu}_i)$] in terms of $\boldsymbol{\eta}^{(0)}$ [with elements $g_i^{(0)}(m_i)$] is

$$\hat{\boldsymbol{\eta}} = \boldsymbol{\eta}^{(0)} + \boldsymbol{D}(\hat{\boldsymbol{\mu}} - \boldsymbol{m}^{(0)}) \tag{9.8}$$

where $\boldsymbol{D} = \text{diag}[d\eta_i/d\mu_i]$, evaluated at $\boldsymbol{\mu} = \boldsymbol{m}^{(0)}$. Then in eq. (9.8) set $\hat{\boldsymbol{\eta}} = \boldsymbol{X}\boldsymbol{b}^{(1)}$, solve for $\hat{\boldsymbol{\mu}}$, and substitute in eq. (9.7) to obtain

$$\boldsymbol{X}'\boldsymbol{\Psi}\left[\boldsymbol{m}^{(0)} + \boldsymbol{D}^{-1}(\boldsymbol{X}\boldsymbol{b}^{(1)} - \boldsymbol{\eta}^{(0)})\right] = \boldsymbol{X}'\boldsymbol{\Psi}\boldsymbol{y}$$

Rearranging terms, this may be written

$$\boldsymbol{X}'\boldsymbol{W}^{(0)}\boldsymbol{X}\boldsymbol{b}^{(1)} = \boldsymbol{X}'\boldsymbol{W}^{(0)}\boldsymbol{u}^{(0)} \tag{9.9}$$

where $\boldsymbol{W}^{(0)} = \boldsymbol{\Psi}\boldsymbol{D}^{-1} = \text{diag}[(d\theta_i/d\mu_i)(d\mu_i/d\eta_i)^2]$, and $\boldsymbol{u}^{(0)}$ has elements

$$u_i^{(0)} = \eta_i^{(0)} + \frac{d\eta_i}{d\mu_i}\left[y_i - m_i^{(0)}\right]$$

The ith diagonal element of $\boldsymbol{W}^{(0)}$ is $\{\text{var}[y_i(d\eta_i/d\mu_i)]\}^{-1}\phi$. (Note again that derivatives such as $d\eta_i/d\mu_i$ are evaluated at the current estimates of the parameters.) Equation (9.9) is the equation for a weighted least squares regression of $\boldsymbol{u}^{(0)}$ upon columns of $\boldsymbol{X}$, with $\boldsymbol{W}^{(0)}$ as the matrix of weights.

Iterations continue, using

$$\boldsymbol{X}'\boldsymbol{W}^{(k)}\boldsymbol{X}\boldsymbol{b}^{(k+1)} = \boldsymbol{X}'\boldsymbol{W}^{(k)}\boldsymbol{u}^{(k)} \tag{9.10}$$

for $k = 1, 2, \ldots$ until, as it is hoped, the sequence of approximations $\boldsymbol{b}^{(0)}$, $\boldsymbol{b}^{(1)}, \ldots$ converges to $\boldsymbol{b}$. If the successive iterates converge, then the vector $\boldsymbol{m} = \hat{\boldsymbol{\mu}}$ such that $\boldsymbol{g}(\boldsymbol{m}) = \boldsymbol{X}\boldsymbol{b}$ is a solution for eq. (9.7). In a number of important cases it can be shown that the likelihood is a concave function of the linear parameters, and hence that any solution is unique. For the case of a log link with Poisson errors, see Haberman (1974). In particular, maximum likelihood estimates exist and are unique if all cells entries are nonzero. Logit models with binomial errors are covered by the same theory. See also Nelder and Wedderburn (1972).

Asymptotic variance and covariance estimates of $\boldsymbol{b}$ are given by the elements of the matrix $(\boldsymbol{X}'\boldsymbol{W}\boldsymbol{X})^{-1}$.

The Deviance

The *deviance* is defined to be $-2\phi(\hat{L} - \tilde{L})$, where $\hat{L}$ is the *loglikelihood* for the model under consideration, and $\tilde{L}$ is the *loglikelihood* for the full model, with as many parameters fitted as there are observations. The *loglikelihood ratio statistic* $-2(\hat{L} - \tilde{L})$ is a *scaled* version of the deviance, with scale factor ϕ. Refer back to Section 7.7 (discussion surrounding eqs. (7.2) and (7.3)) for details in the particular case of loglinear models.

Example

Consider the example (number n_i of moths captured at time t_i) for which calculations were handled in Section 7.8 by a direct use of the method of Newton-Raphson. As in Section 7.8 a straight-line regression of $\log(n_i)$ upon t_i (with $n_6 = 0$ replaced by 0.5) gives starting values $a^{(0)} = 3.136$, $b^{(0)} = -0.5632$. Data, together with the sequence of calculations for the first iteration, are:

t_i	1	2	3	4	5	6
n_i	17	8	2	2	4	0
$\eta_i^{(0)} = a^{(0)} + b^{(0)}t_i$	2.573	2.010	1.446	0.883	0.320	−0.243
$m_i^{(0)}$	13.102	7.460	4.248	2.419	1.377	0.784
$n_i - m_i^{(0)}$	3.898	0.540	−2.248	−0.419	2.623	−0.784
$u_i^{(0)}$	2.870	2.082	0.917	0.710	2.225	−1.243

For this example $g(\mu) = \log(\mu)$ is the link function, $g'(\mu) = \mu^{-1}$; the weight for the ith observation is proportional to the current estimate of μ_i; and the values $u_i^{(0)}$ of the modified dependent variable which appear among the data given here are calculated using

$$u_i^{(0)} = \eta_i^{(0)} + m_i^{(0)^{-1}}\left[n_i - m_i^{(0)}\right]$$

The regression of $u_i^{(0)}$ upon t_i with weights $m_i^{(0)}$ gives $a^{(1)} = 3.367$, $b^{(1)} = -0.6185$. Calculations for the second iteration then give

$\eta_i^{(1)}$	2.752	2.138	1.523	0.909	0.294	−0.320
$m_i^{(1)}$	15.682	8.482	4.588	2.482	1.342	0.726
$u_i^{(1)}$	2.837	2.081	0.959	0.715	2.274	−1.320

Then $a^{(2)} = 3.351$, and $b^{(2)} = -0.6109$. A third iteration gives $a^{(3)} = 3.350$ and $b^{(3)} = -0.6107$, leading to fitted values $\hat{\eta}_i$ and $m_i = \hat{\mu}_i$ as follows

$\hat{\eta}_i$	2.752	2.138	1.523	0.909	0.294	−0.320
m_i	15.682	8.482	4.588	2.482	1.342	0.726

The deviance is (see Exercise 10 at the end of this chapter)

$$2\sum_i n_i \log\left(\frac{n_i}{m_i}\right) = 6.97$$

[Take $n_6 \log(n_6/m_6) = \lim_{n\to 0} n \log(n/m_6) = 0$.] Asymptotic variance and covariance estimates may be calculated from

$$\boldsymbol{X'WX} = \begin{bmatrix} \sum m_i & \sum m_i t_i \\ \sum m_i t_i & \sum m_i t_i^2 \end{bmatrix} = \begin{bmatrix} 33.014 & 67.014 \\ 67.014 & 189.75 \end{bmatrix}$$

The Cholesky decomposition of this matrix is

$$\boldsymbol{T} = \begin{bmatrix} 5.746 & 11.663 \\ 0 & 7.329 \end{bmatrix}, \quad \boldsymbol{T}^{-1} = \begin{bmatrix} 0.174 & -0.2770 \\ 0 & 0.1364 \end{bmatrix}$$

Variances and covariances are given by the elements of $\boldsymbol{T}^{-1}\boldsymbol{T}'^{-1}$. Thus SE[$a$] = 0.33, SE[$b$] = 0.14, and cov[$a$, b] = 0.038.

*7.10. COMMENTS ON FITTING GENERALIZED LINEAR MODELS

Attention has already been drawn to the conceptual economies which are achieved by bringing a variety of different models under the one theory. In principle they can all be fitted by the iteratively reweighted least squares approach of the previous section. There is rarely any problem with convergence, at least with binomial and Poisson data. Other computational methods may, however, give much faster convergence for particular models. Iterative proportional scaling is the method that will be preferred in the fitting of loglinear models for multiway tables when all cells are filled.

Linear Dependencies in the Columns of X

Unless $\boldsymbol{W}$ has one or more zero diagonal elements, a diagonal zero in forming the Cholesky decomposition of $\boldsymbol{X}'\boldsymbol{W}\boldsymbol{X}$ will indicate overparameterization, that is, the corresponding column of $\boldsymbol{X}$ is a linear combination of earlier columns. A zero in the ith diagonal position of $\boldsymbol{W}$ can arise only if $g'(m_i)$ is regarded as infinite.

Distributions for Counts

Models with errors assumed to be binomial or Poisson are perhaps the most commonly used of all models with non-normal errors. The assumption that data follow a binomial or Poisson distribution must be used with some care. In the context of multiway tables it frequently breaks down because items have not entered independently into the cells of the table. Counts of the numbers of rotten apples on trays of 24 each are unlikely to follow a binomial distribution because the apples on a tray will not rot independently of one another. The likely effect will be to increase the variance above what one would expect for a binomial distribution. Nevertheless, the true (but unknown) distribution will have a variance that tails away to zero at the two ends of the scale (0% and 100%) in much the same way as does a binomial variance. An adequate analysis may be obtained by assuming that the unknown variance is some constant multiple ϕ of the binomial variance.

This motivates taking the loglikelihood to be just as for a binomial distribution, except that the scale parameter ϕ is no longer equal to 1.0. If the data are sufficiently extensive, and the model involves several factors, the scale parameter may be estimated from the mean deviance associated with high-order interaction terms.

If there is genuine replication (several independent trials of the experiment) then the deviance associated with between-replicates variation may be divided by degrees of freedom to give what may be used as an error mean square term in calculating standard errors. More generally, it may be necessary to have regard to more than one source of variation. Care is then needed in selecting an appropriate numerator mean deviance for use in an approximate test to determine whether a particular effect is statistically significant.

For comments on the effects of nonindependence in the analysis of multiway tables obtained from survey data, and suggestions for taking account of it, see Holt, Scott and Ewings (1980).

An Analysis of Deviance Table?

With unbalanced models that have a normal error structure, and with all models that have a non-normal error structure, the deviance associated with any term will differ depending on what other model terms have been previously fitted. In particularly favorable cases the deviance will change only slightly with a change in the order in which terms are fitted, and deviance information can be summarized in an ANOVA-like analysis of deviance table which pays no regard to the order of terms. This may be obvious because (to take the most easily recognized case) the variances for all cells are similar. Or one may try the effect of fitting terms in a variety of sequential orders.

In general (with all unbalanced models) a decision on whether to include a particular model term will require calculation of the amount by which the mean deviance changes when that term is omitted from a model that includes all other terms under consideration.

These issues are further discussed in McCullagh and Nelder (1983), Section 2.3.2.

Theoretical Approximations

Inference usually requires the assumption that the deviance is approximately distributed as chi-square. If the scale parameter ϕ has to be estimated from the data, use of an F-distribution for the scaled deviance is preferable, just as in the normal errors case. (Both numerator and denominator have approximate chi-square distributions.) There do not, as yet, appear to be available rules which might help in deciding whether such an F approximation is adequate.

For use with categorical data of the Pearson chi-square statistic (which will very nearly equal the deviance if the model fits well) a commonly used rule is that all expected frequencies must be at least 1.0, with no more than 20% of the expected frequencies less than 5. Lawal (1980) suggests that use of the asymptotic chi-square approximation will be satisfactory, if the minimum expectation is three or more, and gives a method that may be used in other cases. See also Lawal and Upton (1980). Rules with specific relevance to the deviance (for Poisson or binomial data) are given in Smith et al. (1981), together with suggested modifications to the chi-square approximation in cases where the unmodified approximation performs poorly. It is well to consider how results such as those just mentioned ought to influence the form of output presented by computer programs.

Standard errors of parameter estimates rely on the use of linear approximations that are not always satisfactory. Thus suppose that a model with logit link and binomial error term is used to compare a sample proportion of 0.2 for a control with rather smaller sample proportions for treatments. Let $z_0 = \text{logit}(p_0)$, where $p_0 = 0.2$, and $z = \text{logit}(p) = \log[p/(1-p)]$. The generalized linear model analysis assumes that the variance of a sample proportion p is given with adequate accuracy using the approximation $\text{var}[\text{logit}(p)] \simeq 1/(n\hat{p}(1-\hat{p}))$, where $\hat{p}$ is the corresponding fitted value. In order to compare treatment with control on a logit scale, use is in effect made of the formula $\text{var}[z_0 - z] = \text{var}[z_0] + \text{var}[z]$. This is satisfactory provided that the variance does not change drastically over the interval $[z, z_0]$. The following table shows how the approximation may break down:

p	logit(p) ($= z$)	var(z)	$s_d = \text{SE}(z_0 - z)$	$(z_0 - z)/s_d$
0.2	−1.38	$6.25/n$		
0.10	−2.20	$11.11/n$	$4.17/\sqrt{n}$	$0.199\sqrt{n}$
0.04	−3.18	$26.04/n$	$5.68/\sqrt{n}$	$0.315\sqrt{n}$
0.02	−3.89	$51.02/n$	$7.57/\sqrt{n}$	$0.330\sqrt{n}$
0.01	−4.60	$101.01/n$	$10.36/\sqrt{n}$	$0.310\sqrt{n}$

The t-statistic in the final column decreases when p decreases from 0.02 to 0.01, even though the increased difference between p and p_0 is stronger evidence for a difference between treatment and control. The comparison would be better made by replacing p by $\arcsin(\sqrt{p})$, when (assuming the angle is in radians) $\text{var}[\arcsin(\sqrt{p})] \simeq 0.25/n$, whatever the value of p.

*7.11. FURTHER GENERALIZATIONS OF GENERALIZED LINEAR MODELS

The *regression models for ordinal data* discussed in McCullagh (1980) may be seen as an extension of generalized linear models. Included are models for use with multiway tables when the classification corresponding to the dependent variate (the *dependent margin*) consists of a set of ordered categories. (See the example that follows.) In the past it has been common to use loglinear models in such cases. The cumulative totals in the dependent margin of the equivalent loglinear model appear, in a regression model, as values of a dependent variate. Regression models for ordinal data are thus conceptually easier than loglinear models. The complication of the calculations needed to fit them will cease to be a barrier to their use, once computer programs are developed that require only a few simple instructions in order to proceed. Consider now a simple example.

Regression Models for Ordinal Data—An Example

In an experiment that examined the effect of a fungicide on size of apples results from one block were:

	Size category			
	Small	Small/Medium	Medium/Large	Large
Number n_{1j} (controls)	28	107	74	35
Number n_{2j} (sprayed)	12	58	93	66

Now in each row form cumulative totals $z_{ij} = \sum_{k=1}^{j} n_{ik}$, and $\ell_{ij} = \log(z_{ij}/(n_{i+} - z_{ij}))$, where $n_{i+} = z_{i4}$ is the sum for row i ($i = 1$ or $i = 2$). This gives

	Size category			
	Small	Small/Medium	Medium/Large	Large
z_{1j}	28	135	209	244
ℓ_{1j}	−2.04	0.21	1.79	—
z_{2j}	12	70	163	229
ℓ_{2j}	−2.90	−0.82	0.90	—
$\ell_{1j} - \ell_{2j}$	0.86	1.03	0.89	

The effect of the spray has been to shift the three cutoff points separating the four categories by an average of $(0.86 + 1.03 + 0.89)/3 = 0.93$. The data are consistent with a shift of 0.93 at all three cutoff points.

Assume now that in each row $E[n_{ij}] = \mu_{ij}$. Let $\gamma_{ij} = \Sigma_{k=1}^{j}\mu_{ik}$, for $j = 1, 2, \ldots, J$. Then for these data it seems reasonable to assume the model

$$\eta_{ij} = \log\left(\frac{\gamma_{ij}}{n_{i+} - \gamma_{ij}}\right) = \alpha_i + \zeta_j, \quad (j = 1, 2, \ldots, J - 1)$$

where $\gamma_{iJ} = n_{i+}$. This is linear in α_i and ζ_j. This differs from the generalized linear models discussed earlier because the link function does not relate the means μ_{ij} of the observed values of the dependent variate directly to the linear predictor $\eta_{ij} = \alpha_i + \zeta_j$. Instead, setting

$$\boldsymbol{\mu}' = [\mu_{11}, \mu_{12}, \mu_{13}, \mu_{14}, \mu_{21}, \mu_{22}, \mu_{23}, \mu_{24}]$$

and similarly for $\boldsymbol{\gamma}$ and $\boldsymbol{\eta}$, one has $\boldsymbol{\mu} = \boldsymbol{C}\boldsymbol{\gamma}$, with $\boldsymbol{\gamma} = \boldsymbol{h}(\boldsymbol{\eta})$ and $\boldsymbol{\eta} = \boldsymbol{X}\boldsymbol{\beta}$.

A further refinement is the introduction of a different scale factor for the two rows of the table. In the preceding table $\ell_{13} - \ell_{11} = 3.83$, whereas $\ell_{23} - \ell_{21} = 3.80$, so that the distance between the two cutpoints is very similar in the two rows. However results for a second spray (a third row of data) gave $n_{31} = 5$, $n_{32} = 51$, $n_{33} = 87$, and $n_{34} = 67$. It is readily shown that $\ell_{33} - \ell_{31} = 4.47$, suggesting that the distribution is more spread out. This situation can be handled by a model in which $\eta_i = \boldsymbol{x}_i'\boldsymbol{\beta} \cdot \rho_i$, where $\log(\rho_i) = \boldsymbol{z}_i'\boldsymbol{\tau}$, is a linear combination of the ith row of a matrix $\boldsymbol{Z}$ whose columns are usually (but not necessarily) a subset of the columns of $\boldsymbol{X}$.

Regression Models for Ordinal Data—Theory

From this point on μ_i will be used to refer to element i of $\boldsymbol{\mu}$, and similarly for other vectors. The vector of observed counts will be written $\boldsymbol{y}$ in place of $\boldsymbol{n}$. With this change of notation, $\boldsymbol{h}(\boldsymbol{\eta})$ has elements $h(\eta_i)$, where taking the relevant row total as n

$$h(\eta_i) = \frac{\exp(\eta_i)}{1 + \exp(\eta_i)} n$$

which is the inverse function to the logit.

For data such as in the example the analysis will usually assume, at least for purposes of obtaining parameter estimates, that the n_{ij} follow a joint multinomial distribution in each row. As in the case of loglinear models this is asymptotically equivalent to assuming that the n_{ij} are distributed independently as Poisson.

As in eq. (9.6) the expression to be maximized is

$$\phi L = \sum_i \theta_i y_i - \sum_i d(\theta_i) + c^*(y)$$

where the elements y_i of $\boldsymbol{y}$ are cell frequencies. As before the derivative of $\theta_i y_i - d(\theta_i)$ with respect to θ_i is $y_i - d'(\theta_i) = y_i - \mu_i$. The derivative of ϕL with respect to β_j is then

$$\sum_i (y_i - \mu_i)\frac{d\theta_i}{d\mu_i}\left[\sum_k \frac{d\mu_i}{d\eta_k} x_{kj}\right] = \sum_i (y_i - \mu_i)\frac{d\theta_i}{d\mu_i}\left[\sum_k c_{ik}\frac{d\gamma_k}{d\eta_k} x_{kj}\right]$$

This is the jth element of

$$\boldsymbol{X'HC'\Xi y} - \boldsymbol{X'HC'\Xi\mu}$$

where $\boldsymbol{\Xi} = \text{diag}[d\theta_i/d\mu_i]$, $\boldsymbol{H} = \text{diag}[d\gamma_k/d\eta_k]$. The maximum likelihood equations are thus

$$\boldsymbol{X'HC'\Xi m} = \boldsymbol{X'HC'\Xi y} \tag{11.1}$$

where $\boldsymbol{m} = \boldsymbol{Ch}(\boldsymbol{Xb})$. This should be compared with eq. (9.7).

Now suppose that $\boldsymbol{m}^{(0)}$ is an initial approximation to $\boldsymbol{m}$. The first-order Taylor series approximation to $\boldsymbol{m}$ about the point $\boldsymbol{m}^{(0)}$ is

$$\boldsymbol{m} = \boldsymbol{m}^{(0)} + \boldsymbol{CHX}(\boldsymbol{b} - \boldsymbol{b}^{(0)})$$

(This is a change from the approach of Section 7.9 where use was made of the first-order Taylor series approximation to $\boldsymbol{\eta}$.) Substituting in eq. (11.1), one has

$$\boldsymbol{X}^{*\prime}\boldsymbol{\Xi X}^*\boldsymbol{b} = \boldsymbol{X}^{*\prime}\boldsymbol{\Xi}\left[(\boldsymbol{y} - \boldsymbol{m}^{(0)}) + \boldsymbol{X}^*\boldsymbol{b}^{(0)}\right] \tag{11.2}$$

where $\boldsymbol{X}^* = \boldsymbol{CHX}$. These are the normal equations for the regression of $\boldsymbol{u}^{(0)} = \boldsymbol{y} - \boldsymbol{m}^{(0)} + \boldsymbol{X}^*\boldsymbol{b}^{(0)}$ upon columns of $\boldsymbol{X}^*$, with weights given by the diagonal matrix $\boldsymbol{\Xi}$. This yields a new set of coefficients $\boldsymbol{b}^{(1)}$ and new vectors $\boldsymbol{\eta}^{(1)}$, $\boldsymbol{\gamma}^{(1)}$, $\boldsymbol{m}^{(1)}$, and $\boldsymbol{u}^{(1)}$ which form the basis for the next iteration.

If an allowance is made for a scale parameter, then set

$$\boldsymbol{H}_1 = \text{diag}\left[\frac{d\gamma_k}{d\eta_k}\rho_k\right], \quad \boldsymbol{H}_2 = \text{diag}\left[\frac{d\gamma_k}{d\eta_k}\eta_k\right]$$

$\tilde{X} = [CH_1X, CH_2Z]$, and $\tilde{b}' = [b', t']$ where t is the estimate of τ. Then eq. (11.2) becomes

$$\tilde{X}'\Xi\tilde{X}\tilde{b} = \tilde{X}'\Xi[(y - m^{(0)}) + \tilde{X}\tilde{b}^{(0)}] \tag{11.3}$$

It is possible to set up the necessary sequence of calculations in the third version of the statistical package GLIM. Some clues on how this may be done are provided in Thompson and Baker (1981). However, the effort involved, particularly if there is a variety of models to investigate and some of these include scale parameters, makes GLIM3 unsatisfactory for routine use. In the fourth version of GLIM there will be explicit provision for fitting such models, based on the use of eq. (11.2) or eq. (11.3). In GENSTAT, and in several other statistical packages, it is possible to handle the calculations by carrying out a series of matrix manipulations. Calculations using GENSTAT are less elegant and efficient than one would like because it is not possible to set up band or upper triangular matrices and take advantage of the special form of these matrices in subsequent calculations.

Models for Capture-Recapture Experiments

Cormack (1981) shows how various models for capture-recapture experiments on open populations may be formulated as generalized linear models and fitted using GLIM.

7.12. FURTHER READING AND REFERENCES

For a lucid summary of elementary aspects of the fitting of nonlinear models, see Dahlquist and Bjorck (1974). See also Forsythe et al. (1977). Stoer and Bulirsch (1980) give a more advanced treatment. An excellent brief discussion of loglinear models is that of Fienberg (1980). Plackett (1981) is a concise but remarkably wide-ranging treatment of models for categorical data, with numerous practical examples and with detailed references. For an introduction to the ideas associated with generalized linear models see Part I of the manual for the statistical package/language GLIM (Baker and Nelder, 1978). Pregibon (1981) discusses diagnostics for the detection of unusually large or influential data points in logistic regression. He shows how the results may be extended to other generalized linear models.

7.13. EXERCISES

1. Find the root(s) in u of $u = \cos u$ using:

(i) Functional iteration (Section 7.3).

(ii) Steffenson's method (Section 7.3).

(iii) The method of Newton-Raphson.

2. Show that if $(u_i - u) = c(u_{i-1} - u)$, then

$$u = u_i - \frac{(u_{i+1} - u_i)^2}{u_{i+2} - 2u_{i+1} + u_i}$$

Use this to give a rationale for Aitken extrapolation.

3. Investigate whether the functional iteration $u_{i+1} = \phi(u_i)$ can be used to solve the equation $u = \phi(u)$ with $\phi(u) = \lambda^{-1}(e^{\lambda u} - 1)$, where $\lambda = 1.31$.

4. Show that successive iterates using the Newton-Raphson method to solve $f(u) = 0$ are given by an iteration of the form $u_{i+1} = \phi(u_i)$ in which, provided $f'(u) \neq 0$, $\phi'(u) = 0$. Comment on the ultimate rate of convergence of this method. [Consider the Taylor series expansion of $\phi(u_i)$ around $u_i = u$.]

***5.** Repeat the calculations of Section 7.5 to six or seven decimal digit precision. At the same time calculate the Aitken extrapolates separately for each of p_1, p_2, p_3, and p_4, beginning with the second cycle of iterations. Is use of the Aitken extrapolates worthwhile in this case? [It will be desirable to use a computer or programmable calculator.]

6. Prove that, if $n_i - m_i$ is small relative to m_i and $\Sigma_i(n_i - m_i) = 0$, then

$$2\sum_i n_i \log\left(\frac{n_i}{m_i}\right) \simeq \sum_i \frac{(n_i - m_i)^2}{m_i}$$

and that the two expressions differ only in the term $(n_i - m_i)^3/m_i^2$ and higher-order terms in $(n_i - m_i)/m_i$.

7. Prove that, assuming $\Sigma_i(n_i - m_i) = 0$,

$$\sum_i \frac{(n_i - m_i)^2}{m_i} = \sum_i \frac{n_i^2}{m_i} - \sum_i n_i$$

***8.** Prove that the chi-square statistic for testing for no association in an $I \times 2$ contingency table may be calculated as

$$\frac{n_{++}}{n_{+1}} \sum_{i=1}^{I} \frac{n_{i1}^2}{n_{i+}} + \frac{n_{++}}{n_{+2}} \sum_{i=1}^{I} \frac{n_{i2}^2}{n_{i+}} - n_{++}$$

How can this formula be modified to reduce loss of precision from the final subtraction of n_{++}?

***9.** In a generalized linear model with a binomially distributed dependent variate y_i, replace y_i by $y_i' = wy_i$, and n_i by $n_i' = wn_i$, for each i. Prove that parameter estimates are unchanged. How do variance estimates change?

***10.** Prove that for a generalized linear model with Poisson distributed dependent variate the deviance may be written

$$2\sum_i \left[y_i \log\left(\frac{y_i}{m_i}\right) - (y_i - m_i) \right]$$

where m_i is the estimate of the Poisson parameter associated with the ith point in the fitted model. Show that, if the link function is $g(\mu) = \log(\mu)$, this reduces to

$$2\sum_i y_i \log\left(\frac{y_i}{m_i}\right)$$

[Hint: Consider the derivative of the loglikelihood with respect to β_0.]

***11.** The following table shows numbers of fruit without/with rots according to an experiment where fruit were stored individually at one of three temperatures:

	Storage temperature		
	Low	Medium	High
No rots found	82	32	63
Rots present	146	167	199

Use iterative proportional scaling to obtain fitted values for the model which assumes no association between storage temperature and absence or presence of rot. Determine parameter estimates satisfying the sigma restrictions in the loglinear model.

***12.** Consider the maximum likelihood equations giving the parameter estimates in a Nelder and Wedderburn generalized linear model:

$$\boldsymbol{X}'\boldsymbol{\Psi}\boldsymbol{m} = \boldsymbol{X}'\boldsymbol{\Psi}\boldsymbol{y}$$

where

$$\boldsymbol{\Psi} = \text{diag}\left[\frac{d\hat{\theta}_i}{dm_i}\frac{dm_i}{d\hat{\eta}_i}\right]$$

and $\boldsymbol{m}$ is such that $\boldsymbol{g}(\boldsymbol{m}) = \hat{\boldsymbol{\eta}} = \boldsymbol{X}\boldsymbol{b}$ [see eq. (9.7)]. Derive an iterative scheme for solving these equations based on the first-order Taylor series expansion for $\boldsymbol{m}$ in terms of $\boldsymbol{m}^{(0)}$.

13. Consider the solution of eq. (11.1):

$$\boldsymbol{X}'\boldsymbol{H}\boldsymbol{C}'\boldsymbol{\Xi}\boldsymbol{m} = \boldsymbol{X}'\boldsymbol{H}\boldsymbol{C}'\boldsymbol{\Xi}\boldsymbol{y}$$

where $\boldsymbol{m} = \boldsymbol{C}\boldsymbol{h}(\boldsymbol{X}\boldsymbol{b})$ by the repeated use of eq. (11.2). Show how calculations may be based around the following operations:

(i) Matrix multiplication.

(ii) Orthogonal reduction of a matrix to upper triangular form.

(iii) Solution of an upper triangular system of equations.

What special forms of matrix multiplication are needed if the multiplications are to be handled efficiently? What additional matrix operations are needed in order to determine the variances and covariances of parameter estimates?

CHAPTER 8

Further Variations on a Linear Model Theme

In the generalized linear models discussed in Chapter 7 the mean of the dependent variate followed, after transformation, a linear model. Here are considered further problems in which linear combinations of parameters appear. Splines, discussed in Section 8.1, can be fitted using linear least squares techniques when the knots are fixed. Later sections discuss robust variants of least squares and time series methods.

8.1. SPLINES

In view of their pleasant linear properties it would have been appropriate to include splines somewhere in the first six chapters. But, as no convenient slot could be found for a discussion which is too brief to merit a chapter of its own, they appear here. Splines approximate each of a set of suitably chosen subsections of a curve by a polynomial, commonly a cubic. At join points (knots) the constraint is imposed that derivatives should agree, conventionally up to a degree one less than that of the polynomial. This ensures a smooth transition from one section of the curve to the next. Cubic splines are a mathematical approximation to the curves that engineering draftspersons form by passing a flexible ruler through some fixed set of points. The name is derived from the flexible wooden strips used to construct flexible curves in laying out railway lines or in shipbuilding.

Once suitably transformed, a dependent variate can often be expressed satisfactorily as a linear or quadratic or cubic function of an explanatory variate x. Quadratic curves model responses that rise to a peak and then fall (or vice versa) as values of the explanatory variate x are varied. A cubic may be appropriate in modeling a response that rises, levels out, and then rises again. However, cubics have what is for this purpose a regrettable tendency to dip rather than level out before rising again. The use of any higher-degree polynomial than a cubic or quadratic is to be discouraged. High-degree polynomials can be expected to twist up and down between the data points in a manner that makes their use hazardous even for interpolating within the range of the data. Low-degree splines retain the simplicity and pleasant behavior of low-degree polynomials in each local region of the curve. For any fixed locations of the knots or join points, splines can be fitted by linear methods.

A least squares method for the fitting of spline curves will be illustrated for the case of the linear spline or broken stick. Suppose that the knots are at the points $d_1 < d_2 < \cdots < d_\ell$. In addition the two points $d_0 = d_1$ and $d_{\ell+1} = d_\ell$ will be required. Then for $i = 1, \dots, \ell$ define

$$
\begin{aligned}
s_i(x) &= 0, \quad x < d_{i-1} \text{ or } x > d_{i+1} \\
&= \frac{x - d_{i-1}}{d_i - d_{i-1}}, \quad d_{i-1} \le x < d_i \\
&= \frac{d_{i+1} - x}{d_{i+1} - d_i}, \quad d_i \le x < d_{i+1}
\end{aligned}
$$

Now consider

$$f(x) = b_1 s_1(x) + b_2 s_2(x) + \cdots + b_\ell s_\ell(x)$$

Then $f(d_i) = b_i$, and for $d_i < x < d_{i+1}$

$$f(x) = \frac{(d_{i+1} - x) b_i + (x - d_i) b_{i+1}}{d_{i+1} - d_i} \tag{1.1}$$

The functions $s_1(x), \dots, s_\ell(x)$ form a basis for the set of linear spline functions with knots at $d_1, d_2, \dots, d_\ell$.

Example

Consider the set of data points

x:	1	2	3	4	5	6	7	8	9	10
y:	4	12	18	28	37	35	32	30	25	22

A linear spline is to be fitted with knots at $x = 1, 5, 10$. As noted earlier, double knots will be assumed at $x = 1$ and $x = 10$. Then

$$s_1(x) = \frac{5 - x}{4}, \quad \text{for } 1 \le x \le 5, \text{ and is zero otherwise}$$

$$\begin{aligned} s_2(x) &= 0, \quad \text{for } x < 1 \text{ or } x > 10 \\ &= \frac{x - 1}{4}, \quad \text{for } 1 \le x < 5 \\ &= \frac{10 - x}{5}, \quad \text{for } 5 \le x < 10 \end{aligned}$$

$$s_3(x) = \frac{x - 5}{5}, \quad \text{for } 5 \le x < 10, \quad \text{and is zero otherwise}$$

The fitted model is then $\hat{y} = \boldsymbol{Xb}$, where $\boldsymbol{X}$ has three columns with values those of $s_1(x)$, $s_2(x)$, and $s_3(x)$, respectively. Thus

$$\boldsymbol{X} = \begin{bmatrix} 1.0 & 0 & 0 \\ 0.75 & 0.25 & 0 \\ 0.5 & 0.5 & 0 \\ 0.25 & 0.75 & 0 \\ 0 & 1.0 & 0 \\ 0 & 0.8 & 0.2 \\ 0 & 0.6 & 0.4 \\ 0 & 0.4 & 0.6 \\ 0 & 0.2 & 0.8 \\ 0 & 0 & 1.0 \end{bmatrix}, \quad y = \begin{bmatrix} 4 \\ 12 \\ 18 \\ 28 \\ 37 \\ 35 \\ 32 \\ 30 \\ 25 \\ 22 \end{bmatrix}$$

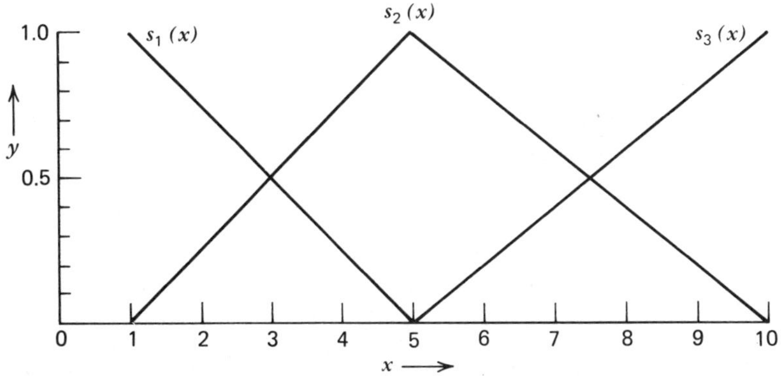

Figure 8.1a. Linear B-splines used as a basis for the set of all linear splines, with knots as shown.

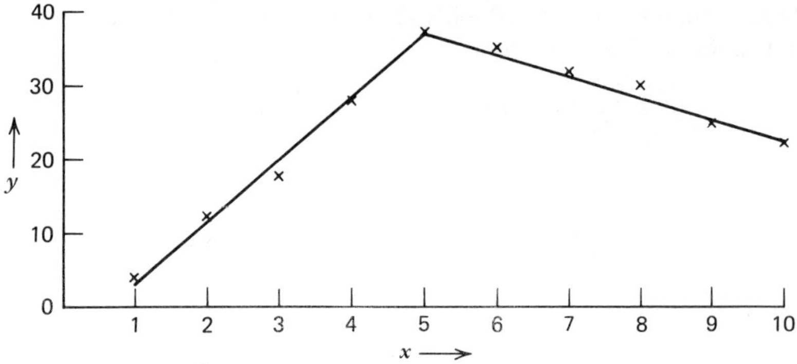

Figure 8.1*b*. Data points, together with the fitted linear spline.

The fitted least squares equation is

$$\hat{y} = 3.11s_1(x) + 37.08s_2(x) + 22.79s_3(x)$$

Thus

$$\begin{aligned} \hat{y} &= -5.38 + 8.49x, \quad 1 \le x < 5 \\ &= 51.37 - 2.86x, \quad 5 \le x < 10. \end{aligned}$$

Figure 8.1 gives graphs of $s_1(x)$, $s_2(x)$, and $s_3(x)$ and the fitted linear spline.

An Algorithm for Calculating B-Splines

The approach to higher-degree splines is very similar. The set of splines of any given degree form a linear space. The B-splines (due to Schoenberg) of degree k provide, in a form that is well suited to practical computation, a basis for the space spanned by splines of that degree. As before it is assumed that there are knots at $d_1 < d_2 < \cdots < d_\ell$. For splines of degree k (kth degree polynomials) it will be assumed that the initial and final knot points are each repeated $k + 1$ times. Then define

$$B_{j,0}(x) = \begin{cases} 1, & d_j \le x \le d_{j+1} \\ 0, & \text{otherwise} \end{cases}$$

$$B_{j,k}(x) = \frac{x - d_j}{d_{j+k} - d_j} B_{j,k-1}(x) + \frac{d_{j+k+1} - x}{d_{j+k+1} - d_{j+1}} B_{j+1,k-1}(x) \quad (1.2)$$

If $d_{j+k} = d_j$, the convention is that any term in which $d_{j+k} - d_j$ appears in the denominator is zero. It is readily shown that the basis element $B_{j,k}(x)$ has positive support on the interval $d_j < x < d_{j+k+1}$. Note that when $k = 1$

$$B_{j-1,1}(x) = s_j(x)$$

As defined here $B_{j,k}(x)$ coincides, on its support, with a polynomial of degree k. (In De Boor, 1978, k is one more than the degree of the polynomial.)

Some further flexibility in the form of the approximating curve can be achieved by allowing multiple knots at points located within the interval of interest. Thus a double knot in a cubic spline means that the two curves joining at that point need agree only up to the first derivative. For details, see De Boor (1978).

*8.2. ROBUST VARIANTS OF LEAST SQUARES

Consider the following ordered set of weights of 15 apples in a sample from the same tree and subject to the same treatment in an experiment designed to test possibilities for changing the shape of apples:

36 90 95 120 125 126 131 132 153 154

162 163 163 171 213

The smallest of the numbers stands out as lying well below the main group of measurements. This is even more evident on a logarithmic scale, which is very often the appropriate scale to use with biological measurement data. The numbers are then

3.58 4.50 4.55 4.79 4.83 4.84 4.88 4.88 5.03 5.04

5.09 5.09 5.09 5.14 5.36

The sample kurtosis is $k = \Sigma(x - \bar{x})^4/(ns^4) = 6.0$, where s is the sample standard deviation. Provided that normality can be assumed for the main distribution, this statistic gives a convenient test for one or more outliers. See Barnett and Lewis (1978, table XIVb, p. 312). The calculated value $k = 6.0$ is significant at a level of less than 1%. The point most distant from the mean, here 3.58, is the *outlier*. Once this point is removed from the data, k is no longer statistically significant (and the sample standard deviation is approximately halved).

Data where a small proportion of the values appear to be incompatible with the assumption of a normal distribution, and are thus identified as outliers, is common in practice. In this situation the power of classical normal theory methods may be dramatically reduced, so that they are now far from optimal. A simple and attractive model that has been used in theoretical investigations assumes that a fraction $1 - \eta$ of the observations, where typically $0.01 \leq \eta \leq 0.05$, is from a normal distribution with mean μ and variance σ^2 [i.e., is distributed as $N(\mu, \sigma^2)$], whereas the remaining fraction η of the observations is from a $N(\mu, d^2\sigma^2)$ distribution with $d \gg 1$. A popular choice has been $d = 3$; see Hampel (1974) and other work reported in Barnett and Lewis (1978). If $\eta = 0.05$ (i.e., 5% contamination), the efficiency of the mean as an estimator is reduced, for large samples, by a factor of around 1.4. The effect on the sampling variance of the variance estimator s^2 is spectacular; for large samples it is increased by a factor of 6.5 relative to that for a $N(\mu, \sigma^2)$ distribution. A 5% contamination from a $N(\mu, 9\sigma^2)$ distribution would be detectable only in a sample of several hundred observations.

Attention will be restricted to the *M-estimators*, so called because they behave like, or actually are, maximum likelihood estimators for a distribution with longer tails than the normal. Consider first the estimation of a location parameter ℓ, using Hampel's version of an M-estimator proposed by Huber. The estimate $\hat{\ell}$ of the location parameter ℓ is chosen to minimize

$$\sum_i \rho\left(\frac{x_i - \ell}{d}\right) \tag{2.1}$$

where

$$\rho(z) = \frac{z^2}{2}, \quad |z| < k$$

$$\rho(z) = k|z| - k^2/2, \quad |z| > k$$

Hampel proposed taking $d = \text{median}\{|x_i - \text{median}(x_i)|/0.6745\}$. The divisor 0.6745 is suggested because d is then approximately equal to the standard deviation if n is large and the distribution is normal. It is suggested that k be taken equal to about 1.5. The sample median may be taken as the initial estimate of ℓ. Differentiating eq. (2.1) with respect to z the equation to be solved may be written

$$\sum_i \psi\left(\frac{x_i - \ell}{d}\right) = 0 \tag{2.2}$$

where

$$\psi(z) = z, \quad |z| < k$$

$$\psi(z) = k\,\text{sgn}(z), \quad |z| > k$$

Because the function $\rho(\cdot)$ is convex, eq. (2.1) has a unique minimum. The estimate $\hat{\ell}$ has the property that it is the mean of the set of numbers that results when observations located more than a distance kd away from $\hat{\ell}$ are replaced by $\hat{\ell} + kd\,\mathrm{sgn}(x - \hat{\ell})$. This suggests a simple iterative procedure for finding $\hat{\ell}$, which will be demonstrated using the data set of log(weight) for 15 apples given at the beginning of this section. As an initial estimate $\hat{\ell}$, take the median of the observations; for the example this is 4.88. Then set

$$d = \mathrm{median}\left\{\frac{|x_i - \mathrm{median}(x_i)|}{0.6745}\right\} = \frac{0.21}{0.6745} = 0.3113$$

Any observation x_i more than $1.5d \simeq 0.47$ away from 4.88 is then replaced by $4.88 + 0.47\,\mathrm{sgn}(x_i - \hat{\ell})$. Thus 3.58 becomes 4.41, and 5.36 becomes 5.35. The new estimate $\hat{\ell}$ is then taken as the mean of these numbers, which is 4.90. Now return to the original data. The only number that differs from 4.90 by more than 0.47 is 3.58; this is now replaced by $4.90 - 0.47 = 4.43$. The mean of this new set of points is, to two decimal places, again 4.90. Hence $\hat{\ell} = 4.90$.

Robust Regression

Suppose now $\hat{\ell} = \boldsymbol{X}'\boldsymbol{b}$, where as before $\boldsymbol{b}$ is to be chosen so that the expression (2.1) is minimized. Then eq. (2.2) becomes

$$\sum_i x_{ij}\psi\left(\frac{y_i - \boldsymbol{x}_i'\boldsymbol{b}}{d}\right) = 0 \quad (j = 0,\ldots,p) \tag{2.3}$$

This may be written

$$\boldsymbol{X}'\boldsymbol{W}\boldsymbol{X}\boldsymbol{b} = \boldsymbol{X}'\boldsymbol{W}\boldsymbol{y} \tag{2.4}$$

where, setting $r_i = y_i - \boldsymbol{x}_i'\boldsymbol{b}$, $\boldsymbol{W} = \mathrm{diag}[\psi(r_i/d)/r_i]$. Suppose that an initial estimate $\boldsymbol{b}^{(0)}$ of $\boldsymbol{b}$ is available. Then subtracting $\boldsymbol{X}'\boldsymbol{W}\boldsymbol{X}\boldsymbol{b}^{(0)} = \boldsymbol{X}'\boldsymbol{W}\hat{\boldsymbol{y}}^{(0)}$ from both sides of eq. (2.4) and replacing $\boldsymbol{b}$ by $\boldsymbol{b}^{(1)}$ yields

$$\boldsymbol{X}'\boldsymbol{W}\boldsymbol{X}(\boldsymbol{b}^{(1)} - \boldsymbol{b}^{(0)}) = \boldsymbol{X}'\boldsymbol{W}(\boldsymbol{y} - \hat{\boldsymbol{y}}^{(0)}) \tag{2.5}$$

Thus, given an estimate of d, the following method may be attempted:

1. Calculate the vector $\boldsymbol{r}^{(0)}$ of residuals and hence the matrix $\boldsymbol{W}^{(0)}$ of weights.

2. Use eq. (2.5) to determine $\boldsymbol{b}^{(1)} - \boldsymbol{b}^{(0)}$ and hence $\boldsymbol{b}^{(1)}$.
3. If the new estimate of $\boldsymbol{b}$ is to within the limits of precision unchanged, terminate the calculations. Otherwise, return to Step 1 and repeat the cycle of calculations.

This leaves the problem of finding an estimate for d. An attractive initial estimate of d, if it were not expensive to calculate, would be the least absolute deviations (LAD or L_1) estimate. In the one-dimensional case this is the same as the use of the median. The estimator d of scale suggested in connection with eq. (2.1) generalizes in the regression context to the median of the absolute values of the nonzero residuals from a least absolute deviations regression, divided by 0.6745.

The more satisfactory approach is one due to Huber, which allows the scale factor d to be estimated simultaneously with $\boldsymbol{b}$. Huber's method has the important advantage over some other similar proposals that, because estimates are chosen to minimize a function convex in $\boldsymbol{b}$ and d, a unique solution is known to exist. Moreover some information, still not very adequate, is available on the theoretical properties of the method. Assume a function $\rho(\cdot)$ such that $\rho(0) = 0$ which is convex in $\boldsymbol{b}$. A suitable choice is that of eq. (2.1). Then $\boldsymbol{b}$ and d are chosen to minimize the function

$$g(\boldsymbol{b}, d) = n^{-1}\sum_i \rho\left(\frac{r_i}{d}\right)d + ad \tag{2.6}$$

where $r_i = y_i - \boldsymbol{x}_i'\boldsymbol{b}$, and $a > 0$ is chosen in a manner that will be considered shortly. Observe that, for given d, the minimum for $\boldsymbol{b}$ is the same as before. It may be shown (see Huber, 1981, p. 178) that $g(\cdot)$ is convex and thus that the solution $(\boldsymbol{b}, d)$ is unique. [Note that the present discussion uses $\rho(\cdot)$ where Huber uses $\rho_0(\cdot)$.]

For the function $g(\cdot)$ to be minimized eq. (2.3) must be satisfied and in addition (differentiating with respect to d)

$$\sum_i \chi\left(\frac{r_i}{d}\right) = na \tag{2.7}$$

where $\chi(z) = z\psi(z) - \rho(z)$. For the choice of $\rho(z)$ in eq. (2.1)

$$\chi(z) = [\psi(z)]^2$$

For consistency with the classical estimates for the normal model, take

$$a = \frac{n-p}{n}E[\chi(Z)]$$

where Z is distributed as normal with mean zero and variance one. The only modification needed from the algorithm described earlier is that, following calculation of a new value $\boldsymbol{b}^{(k+1)}$ for $\boldsymbol{b}$, a new estimate $d^{(k+1)}$ must be obtained for d, using

$$(d^{(k+1)})^2 = \frac{1}{na}\sum_i \chi\left(\frac{r_i}{d^{(k)}}\right)(d^{(k)})^2 \tag{2.8}$$

Huber (1981, p. 180) shows that, if $\rho(z)/z$ is convex for $z < 0$ and concave for $z > 0$, then $g(\boldsymbol{b}^{(k)}, d^{(k+1)}) < g(\boldsymbol{b}^{(k)}, d^{(k)})$, unless this function is already a minimum.

An Algorithm Requiring Only the Calculation of $X'X$

A simpler algorithm is available that more than compensates for any increase in the number of iterations by requiring reduced computation at each individual step. Following the kth step, let $\boldsymbol{r}^*$ be the vector whose elements are the pseudoresiduals

$$r_i^* = \psi\left(r_i^{(k)}/d^{(k)}\right)d^{(k)}$$

Then an improved value of $\boldsymbol{b}$ is $\boldsymbol{b}^{(k+1)} = \boldsymbol{b}^{(k)} + \Delta\boldsymbol{b}^{(k)}$, where $\Delta\boldsymbol{b}^{(k)}$ satisfies

$$\boldsymbol{X}'\boldsymbol{X}\Delta\boldsymbol{b}^{(k)} = \boldsymbol{X}'\boldsymbol{r}^* \tag{2.9}$$

See Huber (1981) for a proof that, for functions $\rho(\cdot)$ which satisfy $0 \le \rho''(\cdot) \le 1$,

$$g\left(\boldsymbol{b}^{(k+1)}, d^{(k)}\right) \le g\left(\boldsymbol{b}^{(k)}, d^{(k)}\right)$$

Unless eq. (2.3) is already satisfied, the inequality is strict. See Dutter (1977) for detailed comparisons of the performance of several variants of this and the first algorithm in calculations with sets of real data.

F-Tests in Robust Regression

The form of classical analysis of variance tables and F-tests can be preserved, provided that there are at least five times as many observations as parameters to be fitted. Observations y_i are replaced by pseudoobservations $\tilde{y}_i$, calculated as follows. An analysis of variance table is then formed, and F-tests are carried out, in the usual way. See Huber (1981, sec. 7.10) for a full discussion. Let

$$\tilde{r}_i = \frac{K\psi(r_i/d)d}{n^{-1}\sum_i \psi'(r_i/d)} \tag{2.10}$$

where

$$K = 1 + \frac{q}{n} \frac{\operatorname{var}[\psi']}{\left(E[\psi']\right)^2}$$

(Here q is the number of parameters to be estimated.) Suitable estimates of $E[\psi']$ and $\operatorname{var}[\psi']$ are, for example,

$$E[\psi'] \simeq n^{-1} \sum_i \psi'(r_i) = m$$

$$\operatorname{var}[\psi'] \simeq n^{-1} \sum_i [\psi'(r_i) - m]^2$$

Now let $\hat{\boldsymbol{y}}$ be the vector of fitted values for the largest model under consideration, and set

$$\tilde{\boldsymbol{y}} = \hat{\boldsymbol{y}} + \tilde{\boldsymbol{r}}$$

where $\tilde{\boldsymbol{r}}$ has elements $\tilde{r}_i$ calculated according to eq. (2.10). From this point on the fitting of smaller models and F-tests, and so on, proceed as in classical statistical theory, based on the pseudoobservations $\tilde{y}_i$. The residual sum of squares for the full model is the sum of squares of the $\tilde{r}_i$.

A Modification for Leverage Points

Let h_i be the ith diagonal element of $\boldsymbol{X}(\boldsymbol{X}'\boldsymbol{X})^{-1}\boldsymbol{X}'$, as in Section 4.11. Points for which h_i is large (e.g., more than 0.5 or $2p/n$, whichever is the larger) will have a major influence in determining the fitted equation, unless they lie close to the fitted regression plane. In order to dampen the effect of points with large leverage, Huber (1981) proposes to replace the expression to be minimized [eq. (2.6)] by

$$g^*(\boldsymbol{b}, d) = n^{-1} \sum_i \left[\rho\left(\frac{r_i}{d(1 - h_i)} \right)(1 - h_i)d + a(1 - h_i)d \right] \quad (2.11)$$

Descending M-Estimators

A more extreme approach to robust estimation is represented by descending M-estimators that reject completely all points whose residuals are larger than some threshold in absolute magnitude. Thus consider Tukey's biweight:

$$\rho(z) = z^2\left(1 - \left(\frac{z}{k}\right)^2\right)^2 \quad |z| < k$$

$$= 0 \quad |z| \geq k$$

where d is chosen as before and k is approximately 4. Care is needed to ensure that any estimate obtained corresponds to a global minimum.

8.3. TIME SERIES—TIME DOMAIN MODELS

Time series observations take their particular character from their ordering in time or in some other linear dimension that performs an equivalent role. The distinguishing features of time series data are that (1) neighboring observations are statistically dependent and (2) independent replicate series are usually unavailable.

The series will be represented as $\{Y_t\}$. A consecutive set of observations $y_1, y_2, \ldots, y_n$ is often called a *realization*. An important concept is stationarity (in the *wide* or *weak* sense). Define $\mu_t = (E[Y_t])$ to be the mean, and σ_t^2 (defined below as γ_0) to be the variance. Then for stationarity, μ_t and σ_t^2 are both independent of t, and so also is the autocovariance $\gamma_s = \text{cov}[Y_t, Y_{t-s}]$ between two observations which are s units apart. Define $\rho_s = \gamma_s/\gamma_0$ to be the autocorrelation of order s, with sample value $r_s = c_s/c_0$. The definition for the sample statistic c_s corresponding to γ_s will be considered shortly.

Most methods in common use for analyzing and modeling time series assume that the observed series can be obtained by a few simple algebraic steps from a stationary series. Thus it may be assumed that removal of a trend and/or cyclic component leaves a stationary series. Or, following the methods advocated by Box and Jenkins (1976, and elsewhere) it may be assumed that the given series consists of the successive partial sums (that is, integrands) of the terms of a stationary series. The underlying stationary series can then be recovered by taking the differences of successive terms. On occasion it may be appropriate to take second or higher differences.

Models for stationary series on which Box and Jenkins rely heavily are the *autoregressive* (AR) model, the *moving average* (MA) model, and the *autoregressive moving average* (ARMA) model which is obtained by adding together components of the first two types. If differencing is necessary in order to recover an underlying ARMA series, one has an autoregressive integrated moving average (ARIMA) model.

An *autoregressive* (AR) model (of order p) takes the form

$$Y_t = \beta_1 Y_{t-1} + \beta_2 Y_{t-2} + \cdots + \beta_p Y_{t-p} + \varepsilon_t \tag{3.1}$$

where the ε_t are assumed to be independently and identically distributed with common variance. More generally, Y_t may be assumed to have mean μ, so that in eq. (3.1) Y_t should be replaced by $Y_t - \mu$. Estimation for autoregressive models can usually be handled satisfactorily, though with some loss of information from the first few data points, using least squares regression methods. If Y_t is to be uncorrelated with future disturbance terms ε_{t+i} for $i > 0$ and the process is to be stationary, then all roots of

$$1 + \beta_1 z + \beta_2 z^2 + \cdots + \beta_p z^p = 0 \tag{3.2}$$

must lie outside the unit circle.

The kth-order *moving average* (MA) model is

$$Y_t = \mu + \varepsilon_t + \eta_1 \varepsilon_{t-1} + \eta_2 \varepsilon_{t-2} + \cdots + \eta_k \varepsilon_{t-k} \tag{3.3}$$

where the ε_{t-i} are independently and identically distributed with mean zero and variance σ^2. Parameter estimation in the moving average model, and in the mixed autoregressive moving average (ARMA) model requires use of a nonlinear minimization procedure.

With moving average models (or the moving average component of an ARMA model) it is again necessary to impose what Box and Jenkins (1976) call *invertibility* conditions, similar in form to the condition of eq. (3.2). The condition is a *predictability* requirement which ensures that Y_t depends only on past events.

Conditional Least Squares Estimates in Autoregressive Models

Parameter estimates $b_0 = \hat{\mu}, b_1, \ldots, b_p$ are to be obtained in

$$\tilde{y}_t = b_1 \tilde{y}_{t-1} + b_2 \tilde{y}_{t-2} + \cdots + b_p \tilde{y}_{t-p} + e_t$$

where $\tilde{y}_t = y_t - \mu$. The direct use of least squares methods usually gives satisfactory estimates. For these take $\boldsymbol{y} = \boldsymbol{X}\boldsymbol{b} + \boldsymbol{e}$, with

$$X = \begin{bmatrix} 1 & y_p & y_{p-1} & \cdots & y_1 \\ 1 & y_{p+1} & y_p & \cdots & y_2 \\ \cdot & \cdot & & \cdots & \cdot \\ \cdot & \cdot & & \cdots & \cdot \\ 1 & y_{n-1} & y_{n-2} & \cdots & y_{n-p} \end{bmatrix}, \quad \boldsymbol{y} = \begin{bmatrix} y_{p+1} \\ y_{p+2} \\ \cdot \\ \cdot \\ y_n \end{bmatrix}$$

If $\varepsilon_1, \varepsilon_2, \ldots, \varepsilon_n$ are identically distributed as normal with mean zero and variance σ^2, the least squares estimates of the β's are the maximum likelihood estimates given $Y_1 = y_1$, $Y_2 = y_2, \ldots, Y_p = y_p$. Any information contained in $y_1, y_2, \ldots, y_p$ is discarded. Unconditional maximum likelihood estimates are more complicated to obtain.

An alternative approach is to use the *Yule-Walker* equations, obtained by multiplying eq. (3.1) by Y_{t-i} and taking expectations, successively, for $i = 1, 2, \ldots, p$. When sample autocorrelations are substituted for their theoretical counterparts, the Yule-Walker equations become

$$\begin{bmatrix} 1 & r_1 & r_2 & \cdots & r_{p-1} \\ r_1 & 1 & r_1 & \cdots & r_{p-2} \\ r_2 & r_1 & 1 & \cdots & r_{p-2} \\ \vdots & \vdots & \vdots & & \vdots \\ r_{p-1} & r_{p-2} & r_{p-3} & \cdots & 1 \end{bmatrix} \begin{bmatrix} b_1 \\ b_2 \\ b_3 \\ \vdots \\ b_p \end{bmatrix} = \begin{bmatrix} r_1 \\ r_2 \\ r_3 \\ \vdots \\ r_p \end{bmatrix}$$

Suppose the sample autocovariances and autocorrelations are defined as

$$c_k = n^{-1} \sum_{i=1}^{n-k} (y_i - \bar{y})(y_{i+k} - \bar{y})$$

$$r_k = \frac{c_k}{c_0}$$

This use of the Yule-Walker equations is equivalent to a least squares regression of y^* upon columns of X^*, where setting

$$\mathbf{y} = \begin{bmatrix} y_1 - \bar{y} \\ y_2 - \bar{y} \\ \vdots \\ y_n - \bar{y} \end{bmatrix}, \quad y^* = \begin{bmatrix} \mathbf{y} \\ 0 \\ \vdots \\ 0 \\ 0 \end{bmatrix}$$

column i of X^* $(i = 1, \ldots, p)$ consists of i zeros followed by the elements of $\mathbf{y}$ followed by $p - i$ trailing zeros. It follows that with this definition of c_k and r_k the matrix of coefficients of the Yule-Walker equations is positive definite, or at least positive semidefinite.

Unconditional Least Squares Estimates in ARMA Processes

The method now discussed may be used for fitting the parameters of any *autoregressive* (AR), *moving average* (MA), or *mixed autoregressive moving average* (ARMA) process. Consider as an example a very short series of 10 numbers which as it happens were simulated to follow the model

$$y_t - 10 = 0.6(y_{t-1} - 10) + \varepsilon_t - 0.8\varepsilon_{t-1}, \quad \text{where} \quad \text{var}[\varepsilon_t] = 2.5$$

The series obtained was

12.9 11.9 5.2 13.5 9.1 15.4 6.1 9.8 12.5 11.7

This series is far too short for satisfactory estimation of the parameters. But it will serve for present purposes to demonstrate how to calculate a sum of squares of residuals e_t associated with a particular choice of parameter estimates. Assuming the process is normal (i.e., Gaussian) the parameter estimates that minimize the sum of squares of these estimated residuals are close to maximum likelihood estimates. In practice, e_t can be neglected for $t < 1 - \ell$, say, where ℓ is a little larger than the order p of the autoregressive component.

The calculation of residuals will be demonstrated for $m = 10.8$ (the mean of the 10 numbers), $b = 0.5$, $h = 0.75$, in the model which, when fitted, appears as

$$y_t - m = b(y_{t-1} - m) + e_t - he_{t-1} \tag{3.4}$$

The reference manual for the statistical package **Minitab** (Ryan et al., 1981) suggests that it is usually satisfactory to start with all parameters except m set equal to 0.1. In our case we are only concerned to illustrate the calculations, and the starting values $b = 0.5$ and $h = 0.75$ have been chosen because they provide convenient numbers with which to work. These are, incidentally, far from the best-fitting estimates. Furthermore a model with two linear parameters (b and d) is more complicated than is warranted for such a short data series.

Calculation of the residuals relies on two chief features of ARMA models. The first to note is that later values of residuals (e_i) are little affected by the residuals assigned to points observed a sufficient distance back in the series. The second feature is that a stationary time series *looks the same* whether the observations are run forward or backward in time. In principle any model that predicts later observations from those earlier in the series may equally well be used for prediction (postdiction?) backward in time. The covariance structure is the same whether the observations are ordered from first to last or from last to first. Thus in place of eq. (3.4) which took the form

$$y_t - m = b(y_{t-1} - m) + e_t - he_{t-1}$$

we might equally well have

$$y_{t-1} - m = b(y_t - m) + d_{t-1} - hd_t \tag{3.5}$$

where the d_t are estimates of independently and identically distributed residuals with the same variance as the ε_t. The forward and backward series correspond to different realizations of the same process, with different residuals in the two cases. Thus eq. (3.5) can be used, starting from time $t = n$ and taking $d_n = 0$ to calculate $d_{n-1}, d_{n-2}, \ldots, d_1$. Hence, taking $d_i = 0$ for $i < 1$, estimates can be obtained for $y_0, y_{-1}, \ldots, y_{-\ell}$, where $y_{-\ell} \simeq m$. Equation (3.4) can then be used to calculate forward, obtaining $e_{-\ell+1}, e_{-\ell+2}, \ldots, e_n$. One could then back calculate again using eq. (3.5). This will not be necessary (except perhaps as a check) unless either b is close to 1.0 or the series is so short (as in the present example) that the fitting of any model is hazardous!

First subtract $m = 10.8$ from each number in the series to give:

$$2.1 \quad 1.1 \quad -5.6 \quad 2.7 \quad -1.7 \quad 4.6 \quad -4.7 \quad -1.0 \quad 1.7 \quad 0.9$$

and set $d_{10} = 0$. Back calculation using eq. (3.5) yields the following d_i $(1 < i < 10)$:

$$1.60 \quad 0.04 \quad -5.15 \quad 2.40 \quad -1.53 \quad 3.29 \quad -4.88 \quad -0.91 \quad 1.25 \quad d_{10} = 0$$

Now set $d_0 = d_{-1} = \cdots = 0$, and use eq. (3.5) to estimate $\tilde{y}_i = y_i - m$ for $i < 1$. Then

$$\tilde{y}_0 = 0.5\tilde{y}_1 - 0.75d_1 = -0.15, \quad \tilde{y}_{-1} = 0.5\tilde{y}_0 = -0.075, \quad \text{etc.}$$

The $\tilde{y}_t$'s used for the forward calculations are then:

$$-0.01 \quad -0.02 \quad -0.04 \quad -0.075 \quad -0.15 \quad 2.1 \quad 1.1 \quad -5.6 \quad 2.7 \quad -1.7$$
$$4.6 \quad -4.7 \quad -1.0 \quad 1.7 \quad 0.9$$

The first five numbers are estimates. The associated e_t's obtained using eq. (3.4) are:

$$-0.01 \quad -0.02 \quad -0.05 \quad -0.09 \quad -0.18 \quad 2.04 \quad 1.58 \quad -4.97 \quad 1.78 \quad -1.71$$
$$4.16 \quad -3.88 \quad -1.56 \quad 1.03 \quad 0.82$$

Thus with $m = 10.8$, $b = 0.5$, and $h = 0.75$

$$S(m, b, h) = \sum_{t=-5}^{10} e_t^2 = 74.02$$

Next it is necessary to calculate approximations to the derivatives dS/dm, dS/db, and dS/dh. In this instance an adequate approximation can be obtained by taking

$$\frac{dS}{db} = \frac{S(m, b + 0.01, h) - S(m, b, h)}{0.01}$$

$$= \frac{74.02 - 74.58}{0.01} = -5.6$$

If $+0.01$ is replaced by -0.01, the estimate obtained for the derivative is -6.6, which agrees well enough with the figure of -5.6. One or another nonlinear least squares method that requires only an estimate of the first derivatives may then be used. Thus linear least squares may be used to choose Δm, Δb, and Δh to minimize

$$\left(S + \frac{dS}{dm}\Delta m + \frac{dS}{db}\Delta b + \frac{dS}{dh}\Delta h\right)^2$$

This then leads to new estimates $m + \Delta m$, $b + \Delta b$, and $h + \Delta h$, and the cycle of calculations is repeated. There are sufficient potential problems and subtleties that any nonexpert attempting these calculations will be advised to make use of existing programs such as are provided in Box and Jenkins (1976), or of statistical packages such as **Minitab** (Ryan et al., 1981) or GENSTAT (Nelder et al., 1981).

Time Series Residuals

A variant on an autoregressive model involves the fitting of a model such as

$$y_t = b_0 + b_1 t + z_t, \quad (t = 1, \ldots, n) \tag{3.6}$$

where $z_t = e_t + hz_{t-1}$ and the e_t are estimates of independently and identically distributed residuals with mean 0 and variance σ^2. The following approach to estimation may be used:

1. Use ordinary least squares regression to obtain starting values of b_0 and b_1. The residuals from this regression provide starting values of the z_t $(t = 1, \ldots, n)$.
2. Fit $z_t = hz_{t-1} + e_t$.
3. Now consider $y_t - hy_{t-1} = b_0(1 - h) + hb_1 + b_1(1 - h)t + e_t$. Taking the current value of h, improved least squares estimates of b_0 and b_1 may be obtained by regressing $y_t - hy_{t-1}$ upon t.

4. If estimates have converged, stop. Otherwise calculate $z_t = y_t - b_0 - b_1 t$ $(t = 1, \ldots, n)$, and return to Step 2.

The fitting of a variety of models of this type is discussed in Fuller (1976).

8.4. THE FREQUENCY DOMAIN—THE DISCRETE FOURIER TRANSFORM

The main topic of this section is the discrete Fourier transform. Any series of data points may be represented as a sum of periodic (i.e., sine and cosine) components. Such a representation is known as a *frequency domain representation*. Certain operations with the series (e.g., the calculation of the convolution probabilities of two probability series) may sometimes be carried out more easily or more economically in the frequency domain. In time series work the frequency domain representation is used as an aid to finding a model that may have generated the series. In principle it would seem easiest to recognize models in which there are a small number of cyclic components. However, it is not easy to distinguish genuine cyclic components from terms present as part of the frequency domain representation of a model which does not include cyclic terms. A simple regression model with cyclic components at known frequencies will be considered first; such a model involves no new ideas. This provides a convenient lead in to the discussion of the discrete Fourier transform.

Cyclic Phenomena

A cyclic phenomenon of known period $2\pi/\omega$ may be modeled, using

$$\begin{aligned} y_t &= \mu + \rho\cos(\omega t + \phi) + \varepsilon_t \\ &= \mu + \alpha\cos(\omega t) + \beta\sin(\omega t) + \varepsilon_t \end{aligned} \tag{4.1}$$

where it is left to the reader to work out how α and β should be expressed in terms of ρ and ϕ. Least squares methods may then be used to determine estimates of μ, α, and β. Assuming equally spaced time points, the use of standard trigonometric formulae allows some simplification of the calculations. These will not be discussed here; for details, see Bloomfield (1976). Equation (4.1) is easily extended to allow for cycles of two or more different periods, such as weekly as well as annual cycles.

Suppose, however, the frequencies of periodic components which may be present must be deduced from the data itself. Whittaker and Robinson

(1944) give the magnitude of a variable star on 600 successive midnights. In this instance the main periodic component can be determined, with fair accuracy, from visual examination of the data. There are 21 maxima, so that the period is approximately $600/21 = 28.6$ days, and $\omega = 2\pi/28.6 = 0.22$. Some further calculations are needed to demonstrate that superimposed on this there is a second cycle with a period of approximately 24 days.

In general, it may not be easy to recognize such components from the original series. One useful recourse is then examination of the *periodogram* (defined later) which is associated with the Fourier frequency domain representation. A skilled practitioner will be able to work from this to find possible models which need not include cyclic components.

The Discrete Fourier Transform

It will be assumed that observations run from y_0 to y_{n-1}. Define

$$f_j = n^{-1} \sum_{t=0}^{n-1} y_t \exp\left(- \frac{2\pi ijt}{n} \right) \tag{4.2}$$

where $i^2 = -1$. Suppose that y_t is regressed separately on $\cos(2\pi jt/n)$ and $\sin(2\pi jt/n)$. Then the real part of f_j is the coefficient in the first of these regressions, whereas the imaginary part of f_j is the coefficient in the second of these regressions. The *Fourier series* representation of the $\{y_t\}$ is

$$y_t = \sum f_j \exp\left(\frac{2\pi ijt}{n} \right) \tag{4.3}$$

where the summation may be for either of the ranges $-n/2 < j \leq n/2$ or $0 \leq j < n$. In general, y_t may be complex. It is easy to show that, if y_t is real, then f_{n-j} is the complex conjugate of f_j, that is, only half of the f_j's are free to vary independently.

Now write

$$f_j = R_j \exp(i\phi_j)$$

where R_j is known as the modulus and ϕ_j as the argument. The *periodogram*, used in screening for evident frequency components (usually after suitable *smoothing*), is

$$I(\omega_j) = \frac{n}{2\pi} R_j^2$$

where $\omega_j = 2\pi j/n$, for $j = 0, 1, \ldots, n-1$. The definition of $I(\omega_j)$ varies, by the choice of a constant scale factor, from one account to another, depending in part on the purpose for which the periodogram is to be calculated.

An Application of the Fourier Transform to Probability Series

Rather than calculate directly the *convolution* probabilities

$$d_k = p_0 q_k + p_1 q_{k-1} + \cdots + p_k q_0$$

of the two discrete distributions $\{p_i | i = 0, 1, \ldots, \nu\}$ and $\{q_i | i = 0, 1, \ldots, \nu\}$, it will for long series be preferable to multiply the two Fourier transforms to give the Fourier transform of $\{d_k\}$. A minor complication is that each of the original series must first be padded out with $\nu + 1$ trailing zeros to a total length of $n = 2(\nu + 1)$ terms. Then each sum in the convolution determined by the Fourier transform will be a sum to n terms:

$$d_k = p_0 q_k + p_1 q_{k-1} + \cdots + p_n q_{k-n}$$

where the convention is that $q_{j-n} = q_j$. Points in the series may, for purposes of the Fourier transform, suitably be represented on the circumference of a circle. The nth point is identified with the zeroth point. Terms with negative subscripts are zero and do not contribute to the convolution sum.

Identification of Frequency Components

If the aim is to identify frequency components, some authors advocate *tapering* the series before calculating the Fourier transform. *Tapering* modifies data values at the two ends of the series (usually between 10 and 50% of the data) so that they taper off to zero. This reduces the tendency of strong periodic components to leak into other periodic components.

A smoothed form of the periodogram should be used to reduce the influence of random components in the data. The simplest way of smoothing is to replace each point of the periodogram by some linear combination of that point and neighbouring points. See Bloomfield (1976) or Chatfield (1980) for a discussion.

The Fast Fourier Transform

The amount of calculation required to carry out the Fourier transform directly using eq. (4.2) becomes substantial for long series. Let

$$w = \exp\left(-\frac{2\pi i}{n}\right)$$

Then $w^n = 1$, and eq. (4.2) may be written

$$f_j = n^{-1} \sum_{t=0}^{n-1} y_t w^{jt}$$

Some economy can be achieved by tabulating w^t for $t = 0, 1, \ldots, n-1$, which give all the possible distinct values. Subsequent calculations then require n^2 complex multiplications and complex additions, if eq. (4.2) is used directly. This can be reduced by a factor of 2 by noting that f_j is the complex conjugate of f_{n-j}.

The fast Fourier transform is able to improve on this if n is a composite number, that is, $n = n_1 n_2$ for integers n_1 and n_2 both greater than one. Changes in the order in which operations are performed make it possible to identify groups of terms whose sum appears repeatedly in the course of the calculations. The *fast Fourier transform* evaluates each such term once only. Savings are most worthwhile, and the algorithm is simplest, if $n = 2^k$ for some integer k. This is the case that will be discussed here. The restriction to series of length 2^k is not too serious because any series being analyzed can be padded out with zeros until it is the required length. This means that the transform is that of a different (and longer) series. However, the qualitative properties of the transform are preserved so that frequency components apparent in the one case will still be apparent at very nearly the same frequencies in the other case. Tapering becomes more than ever desirable and precedes padding the series out with zeros. See Bloomfield (1976) for details.

Some Details of the Fast Fourier Transform

Assume then that

$$f_j = n^{-1} \sum_{t=0}^{n-1} y_t w^{jt}$$

where $n = 2^k$. This may be rewritten

$$nf_j = \sum_{m=0}^{n/2-1} y_{2m} w^{2mj} + w^j \sum_{m=0}^{n/2-1} y_{2m+1} w^{2mj}$$

where for terms in the first sum $t = 2m$ and for terms in the second sum $t = 2m + 1$. The even- and odd-numbered terms in the sum are handled separately. Now write $j = 2^{k-1} j_1 + j_2$. Then as $w^{2j} = w^{nj_1} w^{2j_2} = w^{2j_2}$, it follows that

$$nf_j = \sum_{m=0}^{n/2-1} y_{2m}(w^2)^{j_2 m} + w^j \sum_{m=0}^{n/2-1} y_{2m+1}(w^2)^{j_2 m}$$

$$= u(j_2) + w^j v(j_2) \tag{4.4}$$

Each of $u(j_2)$ and $v(j_2)$ need be evaluated only for the values $j_2 = 0, 1, \ldots, 2^{k-1}$. In fact $u(j_2)$, for $j_2 = 0, 1, \ldots, 2^{k-1}$, is the Fourier transform of the series consisting of just the even-numbered terms in the original series, and $v(j_2)$ is the transform for the series consisting of just the odd-numbered terms. The original problem, which required calculation of the Fourier transform for one Fourier series of length 2^k, has been replaced by that of handling the same calculations for two series each of length 2^{k-1}. Now consider the number of complex multiplications (together with a corresponding number of complex additions) needed to form the Fourier transform of the original series from $u(j_2)$ and $v(j_2)$. Clearly n multiplications are required; one for each term f_j to be formed. (This number might be further reduced by noting that f_j is the complex conjugate of f_{n-j}.)

Each of the two transform series $u(j_2)$ and $v(j_2)$ may themselves be evaluated as the transforms of two series, each of length 2^{k-2}. Evaluation of the series $u(j_2)$ and $v(j_2)$ from transforms of the respective subseries again requires n complex multiplications. The whole process can be repeated until at the kth such split the problem reduces to that of evaluating the Fourier transforms of 2^k subseries, each of length one! This requires no multiplications or additions at all! However, it will take k sets of n complex multiplications (and additions) to recover then the Fourier transform of the original series. Ignoring other possibilities for reducing the amount of arithmetic, the total number of complex multiplications is $nk = n\log_2 n$. If n is large (around several hundred), the improvement this offers on the n^2 such operations required by the simpleminded use of eq. (4.2) is highly worthwhile.

8.5. FURTHER READING AND REFERENCES

De Boor (1978) gives an elegant account of methods for handling calculations with polynomial splines. Huber (1981) is an authoritative source of information on robust statistical analysis. On elementary aspects of the analysis of time series, see Chatfield (1980) and Bloomfield (1976). Fuller (1976) gives examples of the fitting of a wide variety of time series models.

Ansley (1979) gives an algorithm that may be used for exact maximum likelihood estimation of the parameters of a mixed autoregressive moving average process. See also Newton (1981). Damsleth (1983) makes interesting comparisons between alternative methods for estimating parameters in moving average and other ARMA models.

8.6. EXERCISES

1. Consider the set of (x, y) values:

x	y	x	y	x	y
1	1.7	5	8.2	10	4.5
2	4.5	6	7.5	11	4.0
3	6.8	7	6.8	12	3.2
4	9.2	8	6.0	13	2.8
		9	5.0		

A linear spline (broken stick), with interior joins at $x = 4.5$ and 9.5, is to be fitted by least squares to this set of points. Determine the model matrix $\boldsymbol{X}$. Using a multiple regression computer program, determine the linear spline with the given joins that gives the best least squares fit.

2. Find the robust location estimator following eq. (2.2) for the set of numbers:

6	8	21	6	6	2	3	4	6	1
0	2	8	−19	4	11	0	0	2	

Use the method of eq. (2.10) to test the hypothesis that the location parameter whose estimate has been obtained is zero.

3. Consider the autoregressive moving average model

$$y_t - \mu = \beta(y_{t-1} - \mu) + \varepsilon_t - \eta\varepsilon_{t-1}$$

where the ε_t are independent with common variance σ^2. Show that, if $\beta = \eta$, this is equivalent to the model

$$y_t = \mu + \varepsilon_t$$

4. Consider the model that in its fitted form is

$$y_t = b_0 + b_1 t + z_t, \quad (t = 1, \dots, n)$$

where $z_t = e_t + hz_{t-1}$ and the e_t are estimates of independently and identically distributed residuals with mean 0 and variance σ^2. Determine the weighting matrix $\boldsymbol{W}$ for the following alternative to the method given in eq. (3.6). (The first two steps are unchanged from Section 8.3.)

(i) Use ordinary least squares regression to obtain starting values of b_0 and b_1. The residuals from this regression provide starting values of the z_t $(t = 1,\ldots,n)$.

(ii) Fit $z_t = hz_{t-1} + e_t$.

(iii) Fit $\boldsymbol{y} = \boldsymbol{Xb} + \boldsymbol{z}$ using weighted least squares with weighting matrix $\boldsymbol{W}$.

(iv) If estimates have converged, stop. Otherwise return to Step (ii).

5. Compare the detailed sequence of calculations for the Fourier transform [eq. (4.2)] with that for the fast Fourier transform when the series has length $n = 8$.

CHAPTER 9

Additional Topics

The automatic generation of pseudorandom numbers is the main topic of this chapter. Also included are simple methods for the approximate calculation of the percentage points of standard distributions, the use of recurrences, and sorting and ranking.

9.1. THE AUTOMATIC GENERATION OF PSEUDORANDOM NUMBERS

Everyone is familiar with the use of dice to generate digits uniformly distributed on the set of integers $\{1,\ldots,6\}$. Children's games sometimes use a device patterned on a roulette wheel to generate random decimal digits. War-gamers and other strategists use decimal dice. A decimal die is in the form of a regular icosahedron, with each of the digits 0 to 9 assigned to 2 of the 20 faces. Two throws of the die will generate the two digits of a random number which, assuming an unbiased die, is uniformly distributed in the range 0 to 99. Suppose that the action of firing at an enemy tank destroys it only with probability 0.20, whereas there is probability of 0.35 that it will be put out of action temporarily, and a probability of 0.45 that it will suffer no damage. The war-gamer associates the first outcome with the digits 00 to 19, the second with the digits 20 to 54, and the third with the digits 55 to 99. Or 100 marbles may be labeled 0 to 99, and the number on a marble drawn from a bag used to decide the outcome. In associating plots randomly with treatments in setting up an agricultural field experiment, it may be helpful to associate marbles directly with treatments. Thus each of five treatments may be assigned two marbles. Upon coming to the next plot, the mark on the next marble taken from a bag, without replacement, decides the treatment for that plot.

Various direct electronic equivalents of a die are available. One method uses the points in time at which an electronic noise process rises above or falls below some critical level. A zero or a one is then recorded depending on whether the threshold is crossed during the first or during the second half of an electronic clock cycle whose period (a small fraction of a second) is short relative to the mean time between thresholds. This generates a random sequence of binary digits. The New Zealand post office has a device called Elsie that operates in much this way to decide which of the holders of post office bonus bonds will be the favored recipients of the next issue of bonus cash prizes. In this application it would seem essential to use a method that gives numbers that are in principle unpredictable. Such methods are, however, cumbersome and need careful calibration so that small biases are avoided.

By contrast, pseudorandom numbers are generated according to a precise rule, so in a strict sense the numbers obtained are not random. The intention is that it should be impossible for anyone who does not know the rule to distinguish the sequence from one that has been generated by a random mechanism. If this can be achieved, the sequence can be used—just as if it had been obtained by the use of a probabilistic mechanism—for repeated simulations designed to test the likely outcomes of a certain style of play in a war-gaming contest. (Such simulations are of more than academic interest. Modern nations will increasingly use simulations as they weigh the cost of going to war. It may be that realistic assessment of the costs will make war seem less attractive as a way to settle differences.)

Lancaster (1974) used the digits of π to form a table of random numbers. For this purpose they appear to behave well, though so far it has not been possible to prove that the long-run frequency of all digits is the same. A fatal objection to their use, when the computer is required to generate random numbers, is the complication and intricacy of the algorithm needed.

Very long sequences of random numbers are required when *Monte Carlo* methods are used to determine the percentage points of distributions that are difficult or impossible to handle analytically. Where an individual statistic is to be tested for significance, a very much shorter sequence may suffice; see Section 9.2. Often methods of this type must be used because the statistic is based on a sample that is so small that asymptotic distribution theory can no longer be safely used. Attention should not necessarily be limited to comparison of the statistic value to be tested for significance with its distribution for data that have been randomly generated to follow the model. It is desirable to examine, perhaps graphically, the agreement in detail of the observed data with the results of such a simulation. This allows a more adequate check than is possible from examination of a single percentage point.

Congruential Pseudorandom Number Generators

In practice most methods now either use as it stands or elaborate on the sequence of numbers $\{x_n\}$ generated by taking a linear congruential recursion

$$x_n \equiv ax_{n-1} + c \quad (\text{modulo } m) \tag{1.1}$$

for given values of a, c, and m. When $c = 0$ the generator is *multiplicative congruential*; otherwise ($c \neq 0$), it is *mixed congruential*. If $c \neq 0$, a number in the range $0 \leq x < 1$ is then obtained by dividing x_n by m. (This statement needs to be varied slightly when $c = 0$.) The choice of a and m is crucial. Usually $m = 2^k$ or $m = 2^k - 1$, where k is an integer, in order to make remaindering simple. The choice $m = 2^k - 1$ is attractive if m is a prime, as when $k = 31$. (The number $2^{31} - 1$ has a certain claim to fame; it belongs to Mersenne's list of primes of the form $2^k - 1$, where k is itself prime. Several Mersenne *primes* have since been shown to be composite.) The idea of using generators based on eq. (1.1) has generally been attributed to D. H. Lehmer. The numbers obtained can be shown to fall in a regular lattice pattern, which will be apparent on examining the fine detail of, for example, a plot of x_i upon x_{i-1} from a sequence that extends for a substantial fraction of the period of the generator. It appears possible, in principle, to characterize those lattice structures likely to lead to unsatisfactory generators. See Atkinson (1980). The statistical properties of the sequence obtained can usually be improved by shuffling the numbers in blocks that are large enough that the correlation between successive elements is destroyed.

Consider first the use of the modulus $m = 2^k$. Atkinson (1980) considers $k = 30, 32, 33, 35$. It is known that in such generators the low-order bits have short cycles; this is sometimes undesirable. If the constant c is odd and the multiplier $a \equiv 1 \pmod 4$, then the generator will be full period, that is, all m values $0, 1, \ldots, m - 1$ will in due course appear. If $c = 0$, then the maximum length of the sequence is $m/4$, which is attained when the multiplier $a \equiv 3$ or $5 \pmod 8$ and the seed x_0 is odd. Three multipliers suggested for generators of this type in Marsaglia (1972) and discussed briefly in Atkinson (1980) are 69 069, 71 365, and 100 485.

The choice $m = 2^{31} - 1$, usually with $c = 0$, is attractive on those computers that allow the use of 32 binary bits for integer numbers. The multiplier a is chosen so that the sequence of numbers obtained before recurrence is as long as possible, running over all numbers in the range $0 < x_n < m - 1$. This gives only a small number of generators from which to choose. Fishman and Moore (1982) consider the statistical properties of 16 possible multipliers. Good choices appear to be $a = 630\,360\,016$ $(\equiv 14^{29})$,

$a = 397\ 204\ 094$, $a = 1\ 203\ 248\ 318$, and $a = 764\ 261\ 123$. Fishman and Moore present evidence that the multiplier $a = 16807$ ($= 7^5$), which is used in the portable FORTRAN algorithm of Schrage (1979), may not be altogether statistically satisfactory. Schrage's FORTRAN code illustrates the devices that may be necessary to achieve an efficient implementation. It is not clear whether testing such as that carried out by Fishman and Moore has much relevance to the properties of the sequence which results after shuffling.

Wichmann and Hill (1982) give a portable random number generator based around three simple multiplicative congruential generators with the prime moduli $m_1 = 30\ 269$, $m_2 = 30\ 307$, and $m_3 = 30\ 323$. The multipliers, which are chosen in each case so that the generator is full period, are $a_1 = 171$, $a_2 = 172$, and $a_3 = 170$. The three numbers are scaled to lie between 0 and 1, are added, and the fractional part taken. This has a longer period than Schrage's algorithm (2.78×10^{13} as opposed to 2.15×10^9) and in tests on a PDP-11 computer was several times faster. Wichmann and Hill's algorithm seems unlikely to run into problems with dependence between successive elements.

Kennedy and Gentle (1980) discuss various other methods for improving the sequence of numbers obtained from congruential generators. There are as yet no generators that stand out as clear winners. It has become evident that some generators widely used in the past do not perform satisfactorily. See the comments in Forsythe et al. (1977) on the generator RANDU (whose $m = 2^{31}$, $c = 0$, and $a = 65\ 539$) which appeared in the IBM 360 SSP scientific subroutine library. There is a high correlation between any three successive numbers generated by RANDU.

Ordered Sequences of Uniform Random Numbers

Better methods are available than to generate a sequence of uniform random numbers and sort them. The simplest approach uses the two results:

1. If U is uniformly distributed on $[0,1]$, then $Z = -\log U$ has an exponential distribution.
2. If $Z_1, Z_2, \ldots, Z_{n+1}$ are independent exponentially distributed random variates then the variates

$$X_{(j)} = \frac{\sum_{i=1}^{j} Z_i}{\sum_{i=1}^{n+1} Z_i}, \quad j = 1, \ldots, n,$$

are distributed as the order statistics of a random sample of size n from a uniform distribution on $[0,1]$. For these results see Johnson and Kotz (1970, ch. 25).

The foregoing method requires two passes through the generated list of random numbers. A one-pass method is discussed in Bentley and Saxe (1980). Their method uses the result that the probability distribution of the maximum of n independent uniform random numbers on the interval $[0,1]$ is the same as that as the nth root of a single number from the uniform distribution on $[0,1]$. Moreover, given the largest number $x_{(n)}$, the next largest is distributed as $x_{(n)}U^{1/(n-1)}$, where U is uniformly distributed on $[0,1]$. See Bentley and Saxe (1980) for further discussion.

*9.2. BARNARD'S MONTE CARLO TEST

The method of testing now discussed was suggested by Barnard (1963). Suppose it is desired to test a statistic $\tilde{z}$ for significance at the $100\alpha\%$ level, where large values of $\tilde{z}$ lead to rejection of the null hypothesis H_0. A table of critical values is not available. An exact test at the $100\alpha\%$ level can be carried out on the basis of quite a small number of simulated values of the statistic, at the cost of some loss of power. The critical value is subject to random fluctuation but in such a way that the significance level is unchanged.

Suppose that $N-1$ Monte Carlo determinations, assuming H_0, yield values for the test statistic

$$z_1, z_2, \ldots, z_{N-1}$$

To these $\tilde{z}$ is added, and all values are ranked in order, to give

$$z_{(1)} \leq z_{(2)} \leq \cdots \leq z_{(N)} \tag{2.1}$$

Any ambiguity regarding the placing of $\tilde{z}$ is resolved by placing it as early in the sequence as possible. Under the null hypothesis the probability that $\tilde{z}$ will fall among the m largest values in the sequence is m/N. The type I error for a test that rejects H_0 if $\tilde{z}$ is among the m largest values will thus be $\alpha = m/N$. To fix ideas, take $N = 80$ and $a = 0.05$; then H_0 will be rejected if $\tilde{z}$ lies among the 4 ($= 80 \times 0.05$) largest values in the ordered sequence (2.1).

The Blurring of Boundaries

Marriott (1979) discusses a method for comparing Barnard's Monte Carlo test with the use of a conventional test. Let γ be the probability that a statistic z from the distribution based on H_0 is greater than $\tilde{z}$. A conventional test at the $100\alpha\%$ level would reject H_0 if $\gamma < \alpha$. Barnard's test will do so with a probability that can be calculated as a function of γ. Thus with $N = 80$ and $\alpha = 0.05$, the probability that three or fewer of the values obtained from the Monte Carlo determination will be larger than $\tilde{z}$, leading to rejection of H_0 at the 5% level, is

$$Q(80, 0.05, \gamma) = \sum_{i=0}^{3} \binom{79}{i} \gamma^i (1 - \gamma)^{79-i}$$

The table that follows shows how $Q(80, 0.05, \gamma)$ and $Q(500, 0.05, \gamma)$ vary with γ. Remember that a conventional test at the 5% level would reject H_0 if $\tilde{z}$ is such that $\gamma = 0.05$.

γ	$Q(80, 0.05, \gamma)$	$Q(400, 0.05, \gamma)$
0.08	0.115	0.0035
0.07	0.19	0.029
0.06	0.30	0.15
0.05	0.44	0.48
0.04	0.61	0.85
0.03	0.79	0.990
0.02	0.93	0.99996

*9.3. RANDOM SAMPLES FROM DISTRIBUTIONS OTHER THAN UNIFORM

Suppose that X has cumulative probability distribution function $F(x)$, that is, $\Pr\{X \leq x\} = F(x)$. Suppose that u is a random sample from the uniform distribution on the interval $[0, 1]$, and let $x = F^{-1}(u)$. Then x is a random sample from the distribution of X. For

$$\begin{aligned} \Pr\{X \leq x\} &= \Pr\{F(X) \leq F(x)\} \\ &= \Pr\{U \leq F(x)\} \end{aligned}$$

as U is uniformly distributed on $[0, 1]$. An easy example is provided by the exponential distribution, for which $F(x) = 1 - \exp(-x/\theta)$. Setting $u = F(x)$, the inverse function is $x = -\theta\log(1 - u)$. As $1 - U$ has the same distribution as U, we may equally well write $x = -\theta\log u$. The method is as simple as one could want. For other common continuous distributions, such as the normal, much faster methods are available than can be obtained from inversion of the cumulative distribution.

Normal Random Variates

The best simple method for generating normal random variates is Marsaglia and Bray's improvement on the Box-Muller method (conveniently called Marsaglia and Bray's polar method). For this take u_1 and u_2 independently from a uniform distribution on $[-1, 1]$. (Note that it is not $[0, 1]$, as elsewhere.) Let $w = u_1^2 + u_2^2$. If $w > 1$, it is necessary to select a new pair of uniform random digits before proceeding. (This means that digit pairs are rejected with a probability of 0.21.) Otherwise, set

$$v = \left(\frac{-2\log w}{w}\right)^{1/2}$$

and take $x_1 = u_1 v$, $x_2 = u_2 v$. (Logarithms are, as always, to the base e.) See Atkinson and Pearce (1976) for an outline of the theory underlying this method.

Atkinson and Pearce (1976) and the discussion following their paper devote a great deal of attention to methods for generating normal variates. A recurring theme is that provision of fast reliable random number generators, for whatever distribution, is a task for the expert. When Monte Carlo methods are used, and in simulations, the speed of the random number generator really will, in many instances, limit what can be done. This is a case where it is worthwhile to have the major part of the algorithm coded in a language close to the computer's machine code.

Random Samples from the Gamma and Beta Distributions

The probability density for a random variate with the *gamma* distribution with parameters α and θ is

$$f(x; \theta, \alpha) = \frac{(x/\theta)^{\alpha-1}\exp(-x/\theta)}{\theta\Gamma(\alpha)}$$

The only case that need be considered is that where the scale factor θ is 1. If

$\theta \neq 1$, the gamma variate that is generated must be multiplied by θ. If $2\alpha = \nu$ is an integer, the distribution is the chi-square with ν degrees of freedom (and chi-square scale parameter $\sigma^2 = \theta/2$). A chi-square variate with $\nu = 1$ may be obtained as the square of a normal variate. For $\alpha = 1$ ($\nu = 2$), the distribution becomes the exponential, for which a method was given at the beginning of this section. A simple method that may be used for $\alpha > 1$ is

1. Set $a = (2\alpha - 1)^{-1/2}$, $b = \alpha - \log 4$, and $c = \alpha + a^{-1}$.
2. Generate a pair of numbers u_1 and u_2 independently from a uniform distribution on $[0, 1]$.
3. Set $v = a \log[u_1/(1 - u_1)]$ and $x = \alpha \exp(v)$.
4. If $b + cv - x \geq \log(u_1^2 u_2)$, accept x as the required gamma variate. Otherwise, go to Step 2.

The method can be improved in various ways. See Cheng (1977) for details. However, where efficient calculation is important the method given in Ahrens and Dieter (1982a) is to be preferred.

An algorithm due to Johnk gives an easy way to generate *beta* random variates (from which F-distributed variates can readily be obtained). In addition it caters for gamma random variates with $\alpha < 1$. Details are taken from Atkinson and Pearce (1976). A random variate y is required from a beta distribution with parameters α and β:

$$f(y; \alpha, \beta) = \frac{y^{\alpha-1}(1-y)^{\beta-1}}{B(\alpha, \beta)}, \quad 0 \leq y \leq 1$$

where $B(\alpha, \beta) = \Gamma(\alpha)\Gamma(\beta)/\Gamma(\alpha + \beta)$. Taking u_1 and u_2 independently from the uniform distribution on $[0, 1]$, the steps are

1. Let $v_1 = u_1^{1/\alpha}$, $v_2 = u_2^{1/\beta}$; $\alpha, \beta > 0$.
2. If $w = v_1 + v_2 \leq 1$ put $x = v_1/w$. Otherwise, take new u_1 and u_2 and go to **1**.

If a gamma variate is required, first generate a beta variate with parameters α and $1 - \alpha$. Let z be an independent exponential variate. Then $x = yz$ has a gamma distribution. Again it must be emphasized that these are not the most efficient algorithms available.

For the generation of random variates from a Poisson distribution, see Ahrens and Dieter (1982b). Ahrens and Kohrt (1981) give a general method for efficient sampling from largely arbitrary distributions.

9.4. THE USE OF RECURRENCE RELATIONS

This section provides simple examples of the use of recurrence relations. If a recurrence relation is available that leads to an algorithm with good numerical properties, it is unnecessary to worry whether a function can be expressed in closed algebraic form. For the theory of recurrence relations the reader must look elsewhere. Reingold et al. (1977, ch. 3) give a brief introduction. Fox and Mayers (1968) have a chapter on Computations with Recurrence Relations.

Binomial Probabilities

Let $B(x; n, \pi)$ be the binomial probability that an event occurs exactly x times in a sequence of n Bernoulli trials. Defining $B(x; n, \pi) = 0$ for $x < 0$ or $x > n$, the calculation may, in cases where more accurate results are required than are provided by the normal approximation, be based on any of the three equations

$$B(x; n, \pi) = \binom{n}{x}\pi^x(1 - \pi)^{n-x} \tag{4.1}$$

$$B(x; n, \pi) = \pi B(x - 1; n - 1, \pi) + (1 - \pi)B(x; n - 1, \pi), \quad x = 0, \ldots, n \tag{4.2}$$

$$B(x + 1; n; \pi) = B(x; n; \pi)f_x, \quad x = 1, 2, \ldots, n \tag{4.3}$$

where in eq. (4.3) $f_x = (n - x)\pi/[(x + 1)(1 - \pi)]$.

Suppose that all probabilities $B(x; n, \pi)$ which are greater than some small number ε are required. The direct use of eq. (4.1) becomes unsatisfactory when one of the factors is too large or too small for floating point representation in the computer. Equation (4.2), which is a recurrence relation in n, avoids such problems of overflow or underflow but requires far more computation than an algorithm based on careful use of eq. (4.3), which is a recurrence relation in x. Let $\tilde{x}$ be the value of x that maximizes $B(x; n, \pi)$, and let $\tilde{p} = B(\tilde{x}; n, \pi)$. Then it is readily shown that $\tilde{x}$ is the smallest integer greater than or equal to $(n + 1)\pi - 1$. Other choices of $\tilde{x}$ near the mode of the distribution would serve equally well. Equation (4.3) allows calculation of $B(x; n, \pi)/\tilde{p}$ successively for $x = \tilde{x} - 1, \tilde{x} - 2, \ldots$ and for $x = \tilde{x} + 1, \tilde{x} + 2, \ldots$. In each case calculations continue until $B(x; n, \pi)/\tilde{p} \simeq 0$. Finally $\tilde{p}$ is determined by using the requirement that the probabilities sum to one.

The Classical Occupancy Problem, and Extensions

Balls are placed, in turn and at random, in one of h boxes. Let $P(x; n)$ be the probability that, at the stage when n balls have been placed in boxes, x boxes remain empty. Then

$$P(x; n) = \frac{x+1}{h} P(x+1; n-1) + \frac{h-x}{h} P(x; n-1) \qquad (4.4)$$

The nth ball may either go into one of the $x + 1$ empty boxes, thus reducing the number of empty boxes by one, or it may go into one of the $h - x$ boxes which already contains a ball, when the number of empty boxes is unchanged. Calculations start with $P(h - 1; 1) = 1$, $P(x; 1) = 0$ for $x \neq h - 1$.

Feller (1968, sec. IV.2) gives a combinatorial formula for $P(x; n)$. This requires summation of terms that are alternatively positive and negative, and in any except the simplest cases it is difficult to ensure acceptable precision. Equation (4.4) is much preferable because at each step positive quantities are added. (For some problems the asymptotic Poisson approximation given in Feller, 1968, offers a way around difficulties with the combinatorial formulae.)

Equation (4.4) is capable of generalizations in which the factors $(x + 1)/h$ and $(n - x)/h$ are replaced by more general factors ρ_{x+1} and $1 - \rho_x$, so that

$$P(x; n) = \rho_{x+1} P(x+1; n-1) + (1 - \rho_x) P(x; n-1) \qquad (4.5)$$

Conditional on there being x empty boxes, ρ_x is the probability that the next ball is placed in an empty box. In eq. (4.4), $\rho_x = x/n$. Harkness (1978) considers a generalization of eq. (4.4) in which boxes are successively chosen at random and balls then placed in them with probability τ only, so that in eq. (4.5) $\rho_x = \tau x/h$.

A problem in describing the behavior of tiny wasps that parasitize housefly pupae led D. J. Daley, in work described in Daley and Maindonald (1982) to consider other choices for ρ_x. Conceptually, the housefly pupae were the *boxes*, and the eggs that the wasps lay in the pupae, the *balls*. Consider an experiment in which a female wasp has a choice between a small number of pupae. There is evidence that female wasps examine any pupa encountered before laying an egg. If the pupa is already parasitized the female may move on to look for another pupa. As a plausible approximation, assume that pupae are encountered at random. If the pupa is unparasitized, an egg is laid. If it is already parasitized, then an egg is laid with probability δ only, where $0 < \delta \leq 1$. This leads to

$$\rho_x = \frac{x}{h} + \left(1 - \frac{x}{h}\right)(1 - \delta)\rho_x \qquad (4.6)$$

that is,

$$\rho_x = \frac{x}{\delta(h - x) + x}$$

This is a finite population version of an infinite population model which is discussed, for example, in Griffiths (1977). An advantage of eqs. (4.5) and (4.6) is that it is easy to vary the model by varying the definition of ρ_x.

9.5. SORTING AND RANKING

Methods for sorting and searching are widely discussed in the computing literature. See especially Knuth (1973). If fewer than around 50 items are to be sorted, a simple algorithm such as the *Bubblesort* will perform acceptably. The method is fast for data that are already nearly ordered. For randomly ordered data the expected number of comparisons is of the order of $n^2/2$. Another good method for data that are nearly ordered is the *Insertion* Sort, which is generally much faster than a Bubblesort. See Knuth (1973) for details. For general use with randomly ordered data that are to be sorted in place, a good implementation of Hoare's *Quicksort* algorithm is recommended. This has expected number of comparisons equal to around 1.4 $n \log(n)$. See, for example, Sedgewick (1978). Singleton's CACM algorithm 347 implements Quicksort; FORTRAN and Algol versions are given. Alternating FORTRAN versions of Singleton's Quicksort, with useful improvements, appear in Loeser (1976). Motzkin (1983) proposes further changes leading to her *Meansort* algorithm. BASIC programs for various sorting methods, including Bubblesort and Quicksort, are given in Cotton (1981).

Very large sorting tasks, where most of the data must be held on backup storage (disk or tape), require a specialized program written by experts who are familiar with the particular computer system. Any substantial computer system designed for large-scale commercial use should have such a program.

Calculations Involving Ranks

In statistical applications it is often necessary to determine the sum of the ranks associated with a subset of the data values. Any such sum can be found directly, though inefficiently, using indicator functions in the manner that will now be described. Let $K = \{i_1, i_2, \ldots, i_k\}$ be a subset of the complete set $\{1,\ldots,n\}$ of indexes. Define

$$I(x_i > x_j) = 1 \quad \text{if } x_i > x_j$$
$$= 0, \quad \text{otherwise}$$
$$I(x_i = x_j) = 1, \quad \text{if } x_i = x_j$$
$$= 0, \text{ otherwise}$$

The convention will be adopted that, if two or more data values are equal in magnitude, each takes the average of the ranks that would have been assigned to them if they had been different. The smallest element has rank one. With this convention the sum of the ranks of the elements indexed by K is

$$T_K = \tfrac{1}{2}k(k+1) + \sum_{i \in K,\, j \notin K} \left[I(x_i > x_j) + \tfrac{1}{2}I(x_i = x_j)\right]$$

The verification is straightforward when $k = 1$, and the proof is generally by induction. Use of this result leads to very simple algorithms for calculating the test statistics for the Wilcoxon one- and two-sample tests, and for the Kruskal-Wallis several sample test. Thus the sum of the positive ranks used in the Wilcoxon single-sample test is

$$T_+ = \sum_{1 \le i \le j < n} \left[I(x_i + x_j > 0) + \tfrac{1}{2}I(x_i + x_j = 0)\right]$$

The Wilcoxon two-sample test (also called the Mann-Whitney test) assumes two sets of observations $\{x_i,\, i = 1,\ldots,m\}$ and $\{y_j,\, j = 1,\ldots,n\}$. The test may be based on the sum T_x of the ranks of the x's when the x's and the y's are ranked in a common order. Then if there are m x's and n y's, it may be shown that

$$T_x = \sum \left[I(x_i > y_j) + \tfrac{1}{2}I(x_i = y_j)\right] + \tfrac{1}{2}m(m+1)$$

where the sum is for $1 \le i \le m$ and for $1 \le j \le n$.

A similar formula may be calculated for the Kruskal-Wallis one-way analysis of variance statistic. See Kummer (1981). Use of these formulae is unsatisfactory for large samples—the inefficiencies will be similar to those involved in the use of the Bubblesort.

The Two-Sample Location Problem Again

The location shift estimator $\hat{\theta}$ which corresponds to the two-sample Wilcoxon test is the median of the ordered set of differences $\{y_j - x_i; i = 1,\ldots,m; j = 1,\ldots,n\}$. Let $U(\theta)$ be the number of differences that are greater than or equal to θ. The problem is that of solving the equation $U(\theta) = mn/2$. The function $U(\cdot)$ is monotonic, and hence this equation may in principle be solved using one of the standard iterative methods for solving nonlinear equations. The reader may wish to consider (1) the choice of a suitable starting value for determining $\hat{\theta}$ and (2) how $U(\theta)$ can best be evaluated. (It is desirable to begin by sorting the x's and y's separately.)

The calculation of other order statistics needed for determining confidence intervals may be handled similarly. McKean and Ryan (1977) describe a FORTRAN program (CACM Algorithm 516) based on these ideas. For large data sets the economy achieved (by comparison with methods that require all mn differences to be sorted) is substantial.

Partial sorting, discussed in Chambers (1977, sec. 3.d) offers an alternative approach to the calculation of order statistics.

9.6. SOME USEFUL APPROXIMATIONS

Except in special cases the percentage points, or deviates that correspond to given percentage points, are not available in closed form for such common distributions as the *normal*, *t*, *chi-square*, and *F*. In principle series expansions may be used to give results to any required precision. It is desirable to choose a series expansion that gives rapid convergence for the particular values of the arguments used. It may be necessary to switch between alternative forms of expression. The use, when available, of a closed algebraic expression that gives adequate accuracy will usually be preferred. It may turn out that a closed form of expression can be obtained by taking the first few terms of a rapidly converging series expansion. Some of the best recent approximations (e.g., the Beasley and Springer approximation in Fig. 9.5) use a *rational fraction* (= polynomial/polynomial) approximation to a carefully chosen function of the argument. Modern computing facilities make it easy to investigate alternative forms of approximation, and improved approximations regularly appear in the literature. *Applied Statistics*, (the name by which series C of the Journal of the Royal Statistical Society (JRSSC) is usually known), is a good place to look.

In the sequel BASIC programs are used to document some useful approximations. There are a few instances where several BASIC statements appear on the one line, with the backslash (\) used as a statement separator.

```
5  REM FIG. 9.1
10 PRINT "ENTER STANDARDIZED NORMAL DEVIATE"
15 INPUT Z
20 GOSUB 5010
25 PRINT "CUMULATIVE NORMAL PROBABILITY = "; P3
30 STOP
5000 REM * EASY APPROXIMATION TO CUMULATIVE NORMAL *
5001 REM HAMAKER(1978): APP. STAT. 27, 76-77.
5002 REM THE FIRST TWO DECIMAL DIGITS ARE ALWAYS CORRECT.
5003 REM THE ERROR IN THE THIRD DIGIT IS AT MOST ± 1.
5004 REM
5005 REM TAIL PROBABILITIES OF 0.00001  OR LESS HAVE
5006 REM THE FIRST TWO SIGNIFICANT DIGITS CORRECT.
5007 REM
5010 T3=0.806*ABS(Z)*(1-0.018*ABS(Z))
5015 P3=0.5+0.5*SGN(Z)*SQR(1-EXP(-T3^2))
5020 RETURN
9999 END
```

Figure 9.1. Hamaker's approximation to the cumulative normal.

On machines other than the PDP-11 the separator is usually a colon (:). User defined functions, not available on some versions of BASIC, appear in Fig. 9.12. Otherwise, there should be little problem in implementing the programs. All variable names consist either of a single letter or of a letter followed by a digit. Most microcomputer versions of BASIC allow two-character variable names. Use of two-character names in a program that incorporates any of these routines will avoid any possible clash.

Comments on the BASIC programs will be brief. Consider first the *normal* and *inverse normal*. The two Hamaker approximations, Figs. 9.1 and 9.2, give modest accuracy, but adequate for many practical purposes. A highly accurate approximation to the cumulative normal, remarkable for the simplicity of the code, appears in Fig. 9.3. This approximation requires the

```
5  REM FIG 9.2
10 PRINT "ENTER CUMULATIVE PROBABILITY"
15 INPUT P2
20 IF P2*(1-P2)<=0 THEN 40
25 GOSUB 5035
30 PRINT "STANDARD NORMAL DEVIATE = ";Z
35 STOP
40 PRINT P2;" DOES NOT LIE BETWEEN 0 AND 1."
45 STOP
5030 REM * INVERSE OF CUMULATIVE NORMAL DISTN. *
5031 REM APPROX 2 SIGNIFICANT DIGITS CORRECT FOR
5032 REM DEVIATES LESS THAN 4.
5033 REM HAMAKER(1978): APP. STAT. 27, 76-77.
5035 T2=SQR(-LOG(4*P2*(1-P2)))
5040 Z=SGN(P2-.5)*(1.238*T2*(1+.0262*T2))
5045 RETURN
9999 END
```

Figure 9.2. Hamaker's approximation to the inverse normal.

```
5  REM FIG. 9.3
10 PRINT "ENTER NORMAL DEVIATE"
15 INPUT Z1
20 GOSUB 5505
25 PRINT "UPPER TAIL PROBABILITY =";Q1
30 STOP
5500 REM ** CALCULATE PROBABILITY IN NORMAL TAIL **
5501 REM SEE MORAN(1980), BIOMETRIKA 67, 675-676.
5502 REM DEPENDING ON MACHINE PRECISION, GIVES UP TO
5503 REM 9 DEC. PLACE ACCURACY  FOR DEVIATES < 7.
5505 S9=0
5510 C2=SQR(2)/3*Z1
5515 FOR I=0 TO 12
5520 I5=I+0.5
5525 S9=S9+SIN(I5*C2)*EXP(-I5^2/9)/I5
5530 NEXT I
5535 Q1=0.5-S9/3.141593
5540 RETURN
9999 END
```

Figure 9.3. Moran's approximation to the cumulative normal.

computer to work harder than may be desirable if the calculation is to be repeated a large number of times. Figure 9.4 gives a faster approximation to the cumulative normal which nevertheless gives fair accuracy. Figure 9.5 gives a highly accurate approximation to the inverse of the cumulative normal.

Figure 9.5 rounds to nine decimal digits constants which in Beasley and Springer (1977) are given to 12 or 13 decimal digits. The error in an exact

```
5  REM FIG 9.4
10 PRINT "ENTER NORMAL DEVIATE";
15 INPUT Z
20 GOSUB 5545
25 IF Z<0 THEN PRINT "LOWER";
30 IF Z>0 THEN PRINT "UPPER";
35 PRINT " TAIL PROBABILITY =";Q1
40 STOP
5545 REM * CUMULATIVE NORMAL. *
5546 REM ZELEN & SEVERO MODIFICATION OF HASTINGS APPROX.
5547 REM ABRAMOWITZ & STEGUN (1964), P.932, EQ. 26.2.16
5548 REM ABSOLUTE ERROR < 0.00001
5550 Z1=ABS(Z)
5555 IF Z1>1.9 THEN 5580
5560 T9=1/(1+0.33267*Z1)
5565 A1=0.4361836 \ A2=-0.1201676 \ A3=0.937298
5570 Q1=T9*(A1+A2*T9+A3*T9^2)/EXP(Z1^2/2)/SQR(2*3.141593)
5575 RETURN
5580 REM PEIZER & PRATT (1968) APPROX FOR USE IN TAIL
5581 REM RELATIVE ERROR ALWAYS <0.0005
5585 M1=1-1/(Z1^2+3-1/(.22*(Z1^2+3.2)))
5590 Q1=M1/Z1/10^(0.21714724*Z1^2+0.39909)
5595 RETURN
9999 END
```

Figure 9.4. A further approximation to the cumulative normal.

calculation based on the approximation as given in Fig. 9.5 is never more than would result from changing the cumulative probability by 0.3×10^{-8}. In practice numerical error may increase this on a computer that retains h binary digits in the result of any arithmetical calculation and carries out its arithmetic in binary (rather than hexadecimal or octal), to as much as 3×2^{-h}. (If hexadecimal arithmetic is used, this becomes 20×2^{-h}.) Another and simpler approximation to the inverse of the cumulative normal that may sometimes be useful is that of Curtis (1966, pp. 139–142). For all deviates this gives a maximum relative error of 6.5×10^{-5} in the calculated tail probability.

```
5  REM FIG. 9.5
1Ø PRINT "ENTER CUMULATIVE NORMAL PROBABILITY"
15 INPUT P
2Ø GOSUB 561Ø
25 PRINT "NORMAL DEVIATE = ";XØ
3Ø STOP
56ØØ REM <<<<< GIVES NORMAL DEVIATE, XØ, SUCH THAT *****
56Ø1 REM ***** LOWER TAIL AREA IS P ( Ø < P < 1 ). *****
56Ø2 REM HIGH ACCURACY; SEE TEXT FOR DETAILS.
56Ø3 REM
56Ø4 REM ADAPTED BY  RUSSELL MILLAR AND J.H.M. FROM
56Ø5 REM BEASLEY AND SPRINGER (1977), ALGORITHM AS111,
56Ø6 REM APP. STAT. 26, 118-121.
56Ø7 REM
56Ø8 REM LINE 5621 GIVES HASH SUMS FOR USE IN CHECKING THAT
56Ø9 REM NUMBERS ARE CORRECTLY ENTERED.  EACH IS THE SUM OF
561Ø REM THE ABSOLUTE VALUES OF THE COEFFICIENTS IN THE
5611 REM COLUMN OF FIGURES ABOVE IT.
5612 REM
5615 READ  AØ,A1,A2,A3, B1,B2,B3,B4, CØ,C1,C2,C3, D1,D2
5616 DATA  2.5Ø66282, -18.615ØØØ6,  41.3911977, -25.441Ø6Ø5
5617 DATA -8.47351Ø9,  23.Ø833674, -21.Ø62241Ø,   3.13Ø8291
5618 DATA -2.7871893,  -2.2979648,   4.85Ø1413,   2.3212128
5619 DATA  3.5438892,   1.637Ø678
562Ø REM
5621 REM  17.3112176   45.6334ØØ6   67.3Ø358ØØ   3Ø.8931Ø24
5622 REM HASH CHECKSUMS (ONE FOR EACH COLUMN) ON LINE 5621
5625 Q = P - Ø.5
563Ø IF ABS(Q) > Ø.42 THEN 5655
5635 R = Q * Q
564Ø X = Q * (((A3 * R + A2) * R + A1) * R +AØ)
5645 XØ = X / ((((B4 * R + B3) * R + B2) * R + B1) * R + 1)
565Ø RETURN
5655 R = P
566Ø IF Q > Ø THEN R = 1 - P
5665 IF R =< Ø THEN 5695
567Ø R = SQR( -LOG(R))
5675 X = (((C3 * R + C2) * R + C1) * R + CØ)
568Ø XØ = X / ((D2 * R + D1) * R + 1)
5685 IF Q < Ø THEN XØ = -XØ
569Ø RETURN
5695 PRINT P;" DOES NOT LIE BETWEEN Ø AND 1."
57ØØ RETURN
9999 END
```

Figure 9.5. Beasley and Springer's approximation to the inverse normal.

Figure 9.6 gives an approximation to the cumulative *chi-square* distribution, which may be used for nonintegral as well as for integral *degrees of freedom*. An asymptotic series expansion is used for d.f. (degrees of freedom) that are less than 11. For integral d.f. there is a finite series expansion whose use may substantially reduce the computation required. See Ling

```
5 REM FIG. 9.6
10 PRINT "ENTER CHI-SQUARE STATISTIC, D.F."
15 INPUT X2,D2
20 IF D2<11 THEN GOSUB 5050
30 IF D2<11 THEN 50
35 GOSUB 5110
40 PRINT "EQUIVALENT NORMAL DEVIATE IS ";Z1
45 GOSUB 5505
50 PRINT "UPPER TAIL PROBABILITY = ";Q1
55 STOP
5050 REM LINES 5065-5085 USE EXACT SERIES EXPANSION.
5051 REM SEE APPLIED STATISTICS 29 (1980), PP.113-114.
5052 REM THE APPROXIMATION TO THE COMPLETE GAMMA FUNCTION USED
5053 REM BELOW GIVES AT LEAST 5 DECIMAL PLACE PRECISION.
5055 DEF FNL(A2)=1/(12*A2)*(1-1/A2*(1/30-1/A2*(1/105-1/(140*A2))))
5056 REM USE FNL IN STIRLING'S APPROXIMATION TO THE LOGARITHM OF
5057 REM THE COMPLETE GAMMA FUNCTION.  SEE ABRAHAMOWITZ & STEGUM:
5058 REM HANDBOOK OF MATHEMATICAL FUNCTIONS, SECTION 6.1.41 .
5060 Z=X2/2 \ Z2=Z*Z
5065 C=1 \ G=1 \ D=D2/2 \ A=D \ D3=D+2
5070 A=A+1 \ C=C*Z/A
5075 G=G+C
5080 IF C/G > 0.5E-6 THEN 5070
5085 G=G*EXP(D*LOG(Z)-D3*FNL(D3*D3)-(D3-0.5)*LOG(D3)+D3-Z)*(D+1)
5090 Q1=1.0-G/SQR(2*3.14159)
5095 RETURN
5100 REM * UPPER TAIL OF CHI-SQUARE DISTRIBUTION *
5101 REM CALCULATES EQUIVALENT NORMAL DEVIATE.
5102 REM PEIZER & PRATT(1968), JASA 63, 1416-1456.
5103 REM MAX ERRORS IN CALCULATED PROBABILITY ARE
5104 REM   4 D.F.:0.00058
5105 REM 11 D.F.:0.000046
5106 REM 30 D.F.:0.0000050
5107 REM SEE ALSO MOHAMED EL LOZY (1976),
5108 REM ACM TRANS ON MATH SOFTWARE 2, 393-395.
5109 REM
5110 D1=D2-1
5115 T2=D1/X2
5120 D3=X2-D2+2/3-.08/D2
5125 REM NOW CALCULATE G
5130 GOSUB 5710
5135 Z1=D3*SQR((1+G)/(2*X2))
5140 RETURN
5141 REM ***** END CALC. OF EQUIV. NORMAL DEVIATE >>>>>

include lines 5500 - 5540 from figure 9.3

include lines 5700 - 5765 from figure 9.9

9999 END
```

Figure 9.6. Peizer and Pratt's approximation to the cumulative chi-square.

(1978). The Peizer-Pratt approximation, used earlier for d.f. that are 11 or more, appears superior to anything else available when d.f. are large, unless very high accuracy is required. The chi-square distribution with d.f. which are not necessarily integral is often used to approximate other less tractable distributions. An example is the distribution of the likelihood ratio statistic used in a test for clustering in Scott and Knott (1974 and 1976). The earlier paper gives a less precise version of the approximation.

Figures 9.7 and 9.8 give approximations to the *t-distribution*. A simpler and less precise approximation appears in Bailey (1980). The approximation

```
5 REM FIG. 9.7
10 PRINT "ENTER T-STATISTIC, D.F."
15 INPUT T1,D1
20 IF D1>4 THEN 55
25 REM USE APPROXN AS FOR D1>4 FOR ANY NON-INTEGRAL D1
30 IF D1>INT(D1) THEN 55
40 GOSUB 5155
45 PRINT "PROBABILITY IN UPPER TAIL = ";Q1
50 STOP
55 GOSUB 5200
60 PRINT "EQUIVALENT NORMAL DEVIATE = ";Z1
65 REM NOW FIND TAIL PROBABILITY FOR THIS NORMAL DEVIATE
70 GOSUB 5505
75 PRINT "PROBABILITY IN UPPER TAIL = ";Q1
80 STOP
5150 REM * NORMAL DEVIATE CORRESPONDING TO GIVEN T-STATISTIC *
5151 REM SEE G.W.HILL: ALG.396, C.A.C.M.13, 617-619
5152 REM AND M. EL LOZY(1982), J.STATIST.COMPUT.SIMUL.14,179-189.
5153 REM
5155 IF D1=1 THEN Q1=ATN(T1)/3.14159265
5160 IF D1=2 THEN Q1=.5*T1/SQR(T1^2+2)
5165 IF D1=3 THEN Q1=(ATN(T1/SQR(3))+T1*SQR(3)/(T1^2+3))/3.14159265
5170 IF D1=4 THEN Q1=.5*T1*(1+2/(T1^2+4))/SQR(T1^2+4)
5175 Q1=0.5-Q1
5180 RETURN
5185 REM
5190 REM
5195 REM
5200 IF D1>4 THEN 5210
5205 PRINT "RESULTS ARE INACCURATE FOR SMALL"
5206 PRINT "NON-INTEGRAL DEGREES OF FREEDOM."
5207 REM EQUIVALENT ERROR IN PROB. < 0.0000005 FOR D.F.>=5
5210 A9=D1-0.5 \ B9=48*A9^2 \ T9=T1^2/D1
5215 IF T9>=0.04 THEN Z8=A9*LOG(1+T9)
5220 IF T9<0.04 THEN Z8=A9*(((-T9*0.75+1.0)*T9/3.0-0.5)*T9+1)*T9
5225 P7=((0.4*Z8+3.3)*Z8+24)*Z8+85.5
5230 B7=0.8*Z8^2+100+B9
5235 Z1=((-P7/B7+Z8+3)/B9+1.0)*SQR(Z8)
5240 RETURN
5241 REM

include lines 5500 - 5540 from figure 9.3

9999 END
```

Figure 9.7. G. W. Hill's approximation to the cumulative *t*-distribution.

```
5 REM FIG. 9.8
10 PRINT "GIVE LOWER TAIL PROBABILITY, D.F. FOR T-DISTN."
15 INPUT P,D5
20 GOSUB 5250
25 PRINT "T-DEVIATE =";T1
30 STOP
5250 REM <<<<< GOLDBERG AND LEVINE MODIFICATION *****
5251 REM *****       OF PEISER'S EXPANSION         *****
5252 REM GOLDBERG AND LEVINE (1946), ANNALS OF
5253 REM MATHEMATICAL STATISTICS 17, 216-225.
5254 REM FOR 0.001 < P < 0.999 GIVES AT LEAST
5255 REM 2 D.P. ACCURACY FOR D.F.=10, 3 D.P. FOR D.F.=20.
5260 P4=P*3.141593
5265 IF D5=1 THEN T1=-COS(P4)/SIN(P4)
5270 IF D5=2 THEN T1=SQR(1/(2*P*(1-P))-2)*SGN(P-0.5)
5271 REM EXACT FORMULA FOR D.F. = 1 OR 2.
5275 IF (D5-1)*(D5-2)=0 THEN RETURN
5280 GOSUB 5600
5285 T1=X0*(1+(1+X0^2)/(4*D5)+(3+16*X0^2+5*X0^4)/(96*D5^2))
5290 RETURN

include lines 5600 - 5700 from figure 9.5

9999 END
```

Figure 9.8. Fisher's approximation to the inverse t-distribution.

in Fig. 9.8 to the inverse of the cumulative t-distribution is imprecise for extreme probabilities and small degrees of freedom. Hill (1970) gives an algorithm that will give the deviate accurately for any reasonable probability and any degrees of freedom. Accurate polynomial approximations to particular percentage points of the t and chi-square distributions are given in Bukac and Burstein (1980).

Figure 9.9 gives an approximation to the *F-distribution*, and Fig. 9.10 a rougher approximation to the *noncentral F*. For fast computation of accurate approximations to the F-distribution, it is necessary to move between two alternative forms of series expression, depending on whether or not both d.f. are odd. See Ling (1978).

Figure 9.11 makes use of an approximation to the inverse of the F-distribution in order to give a confidence interval for a *binomial* probability.

Figure 9.12 gives approximate confidence limits for the *correlation* coefficient in a bivariate normal population.

9.7. COMPUTER ARITHMETIC

Consider the integer $39 = 100111_2$. If stored as an integer, most machines will treat it in the binary form 100111_2. Notice, however, that this number

```
5 REM FIG. 9.9
10 PRINT "ENTER F-STATISTIC, NUMERATOR DF, DENOMINATOR DF"
15 INPUT F2,D1,D2
20 IF D1+D2<=2 THEN 70
25 IF D1<2 THEN GOSUB 60
30 IF D2<2 THEN GOSUB 60
35 GOSUB 5310
40 PRINT "EQUIV. NORMAL DEVIATE = ";Z1
45 GOSUB 5505
50 PRINT "TAIL PROBABILITY = ";Q1
55 STOP
60 PRINT "PRECISION MAY BE LESS THAN 2 DECIMAL PLACES."
65 RETURN
70 PRINT "APPROXIMATION CANNOT BE USED FOR D.F. GIVEN."
75 STOP
5300 REM <<<<< CUMULATIVE F-DISTRIBUTION *****
5301 REM PEIZER & PRATT(1968): JASA 63, 1416-1456.
5302 REM MAX ABS ERROR FOR D.F. D1, D2 IS
5303 REM 0.0008 IF D1,D2 >= 4;  0.005 IF D1,D2 >=2.
5304 REM SEE LING (1978): JASA 73, 274-283.
5310 P8=D2/(D1*F2+D2)
5315 Q8=1-P8
5320 N8=(D1+D2-2)/2
5325 S8=(D2-1)/2
5330 T8=(D1-1)/2
5335 D8=S8+1/6-(N8+1/3)*P8+.04*(Q8/D2-P8/D1+(Q8-.5)/(D1+D2))
5340 T2=S8/N8/P8
5345 GOSUB 5710
5350 G1=G
5355 T2=T8/N8/Q8
5360 GOSUB 5710
5365 G2=G
5370 Z1=D8*SQR((1+Q8*G1+P8*G2)/(N8+1/6)/P8/Q8)
5375 RETURN
5380 REM ***** END PEIZER & PRATT APPROX TO CUM F >>>>>

include lines 5500 - 5540 from figure 9.3

5700 REM <<<<< CALCULATE G FUNCTION *****
5701 REM USED IN PEIZER & PRATT APPROXIMATIONS TO
5702 REM TO CUMULATIVE F AND CHI-SQUARED DISTNS.
5703 REM LINES 5730-5745 ARE AN APPROXIMATION TO G
5704 REM WHICH ON A 7 DEC. DIGIT MACHINE IS LESS
5705 REM SUBJECT TO NUMERICAL ERROR WHEN T2>0.9.
5710 G=1
5715 IF T2=0 THEN RETURN
5720 REM
5725 IF ABS(1-T2)>.1 THEN 5755
5730 G=0
5735 FOR J=1 TO 5
5740 G=G+2*(1-T2)^J/((J+1)*(J+2))
5745 NEXT J
5750 RETURN
5755 G=(1-T2^2+2*T2*LOG(T2))/(1-T2)^2
5760 RETURN
5765 REM ***** END CALCULATION OF G FUNCTION >>>>>
9999 END
```

Figure 9.9. Peizer and Pratt's approximation to the cumulative *F*-distribution. (See Section 9.6 for discussion)

```
5 REM FIG. 9.10
10 PRINT "GIVE F,D1,D2,C"
15 PRINT "WHERE F = F-STATISTIC,"
20 PRINT "      D1 = NUMERATOR D.F., D2 = DENOMINATOR D.F.,"
25 PRINT "      C = NON-CENTRALITY PARAMETER."
30 INPUT F2,D1,D2,C1
35 GOSUB 5410
40 PRINT "EQUIVALENT NORMAL DEVIATE = ";Z1
45 GOSUB 5505
50 PRINT "UPPER TAIL PROBABILITY = ";Q1
55 STOP
5400 REM << SEVERO & ZELEN (1960) APPROXN. TO NON-CENTRAL F**
5401 REM BIOMETRIKA 47, 411-416.
5402 REM GENERALIZES PAULSON APPROXIMATION TO CUMULATIVE F.
5403 REM ERROR USUALLY NO MORE THAN 0.01 FOR D1,D2 > 2.
5404 REM
5405 REM SEE TIKU (1966), BIOMETRIKA 53, 606-610;
5406 REM WHERE SEVERAL APPROXNS. TO THE NON-CENTRAL F
5407 REM ARE COMPARED.
5408 REM
5409 REM IF THE NUMERATOR FOR THE F-STATISTIC EQUALS
5410 REM SUM [Z(I)+A(I)]^2, WHERE Z(I) (I=1,... , D1)
5411 REM ARE INDEP. NORMAL VARIATES WITH MEAN=0 AND VAR.=1,
5412 REM THEN NONCENTRALITY  C1 = SUM A(I)^2
5413 REM
5420 F9=(D1*F2/(D1+C1))^(1/3)
5425 A1=2/9*(D1+2*C1)/(D1+C1)^2
5430 A2=2/(9*D2)
5435 Z1=((1-A2)*F9+(A1-1))/SQR(A1+A2*F9^2)
5440 RETURN
5445 REM ***** END SEVERO & ZELEN:  NON-CENTRAL F >>>>>
5450 REM

include lines 5500 - 5535 from figure 9.3

9999 END
```

Figure 9.10. Severo and Zelen's approximation to noncentral *F*. (See Section 9.6 for discussion)

might alternatively be taken as

$$10\ 01\ 11 = 213_4, \quad \text{or } 100\ 111 = 47_8, \quad \text{or } 10\ 0111 = 27_{16}$$

With integers it is a matter of convenience whether the number is regarded as base 2; or by taking k bits at a time from the right, for $k > 1$, as base 2^k. Several alternative methods are available for representing negative integers. The most significant digit of the binary number is the sign bit, equal to zero for positive numbers and equal to one for negative numbers. Assuming there are n bits available for representing the number two common representations for $-x$ are as follows:

1. Twos complement, that is, the number that occupies the full n bits has the bit settings of the positive binary number $2^n - x$.

```
5 REM FIG. 9.11
10 REM Z AND P9 (LINE 15) ARE SET FOR CALCULATING
11 REM 1-SIDED 97.5% LIMITS, OR 2-SIDED 95% LIMITS
15 Z=1.96 \ P9=97.5
20 PRINT "CALCULATE 1-SIDED ";P9;"% CONFIDENCE LIMIT FOR"
21 PRINT "FOR POPULATION PROPORTION - BINOMIAL SAMPLING."
22 PRINT
23 PRINT "FOR 2-SIDED ";100-2*(100-P9);"% LIMITS, CALCULATE"
24 PRINT "UPPER & LOWER 1-SIDED ";P9;"% LIMITS."
30 PRINT  \ PRINT "ENTER X,N"
35 INPUT X1,N
40 IF X1*(N-X1)<0 THEN 125
45 PRINT "ENTER L IF LOWER LIMIT REQUIRED"
46 PRINT "       U IF UPPER LIMIT REQUIRED"
50 INPUT A$
55 IF A$="L" THEN 95
60 IF A$="U" THEN 70
65 PRINT A$;"IS NOT A LEGAL RESPONSE." \ GO TO 45
70 X=X1 \ IF X=N THEN 125
75 GOSUB 5815
80 P2=P8
85 PRINT "1-SIDED ";P9;"% UPPER LIMIT IS: P < ";P2
90 GO TO 135
95 REM GET LOWER LIMIT P1 BY REPLACING X1 BY N-X1 AND
96 REM CARRY OUT CALCULATION AS BEFORE ; THEN P1=1-P8.
100 X=N-X1 \ IF X=N THEN 125
105 GOSUB 5815
110 P1=1-P8
115 PRINT \ PRINT "1-SIDED ";P9;"% LOWER LIMIT IS: ";P1;" < P."
120 GO TO 135
125 PRINT "VALUE GIVEN FOR X AND/OR N IS ILLEGAL."
130 REM
135 PRINT "DO YOU WISH TO DO FURTHER SIMILAR CALCULATIONS"
140 INPUT A$
145 IF A$="Y" THEN 30
150 STOP
155 REM
5800 REM <<<<< APPROXIMATE BINOMIAL CONFIDENCE LIMITS. *****
5801 REM USE THE PAULSON-TAKEUCHI APPROXIMATION DESCRIBED IN:
5802 REM YORITAKE FUJINO: APPROXIMATE BINOMIAL CONFIDENCE LIMITS,
5803 REM BIOMETRIKA(1980)67,677-681.
5804 REM LINE 5830 CALCULATES THE UPPER BINOMIAL PROBABILITY LIMIT,
5805 REM WHERE F2 IS AN APPROXIMATION TO THE REQUISITE UPPER
5806 REM PERCENTAGE POINT OF AN F-STATISTIC WITH D.F. 2*(X+1) AND
5807 REM 2*(N-X).
5808 REM THE APPROXIMATION TO THE INVERSE OF THE CUMULATIVE F IS
5809 REM DERIVED FROM PAULSON'S APPROXIMATION TO THE CUMULATIVE F.
5810 REM IF DESIRED THE USER MAY INPUT THE EXACT F PERCENTAGE POINT
5811 REM PRIOR TO LINE 5830.
5812 REM
5815 A=1/(9*(X+1)) \ A1=1-A
5820 B=1/(9*(N-X)) \ B1=1-B
5825 F2=((A1*B1+Z*SQR(A1^2*B+A*B1^2-A*B*Z^2))/(B1^2-B*Z^2))^3
5830 P8=(X+1)*F2/((N-X)+(X+1)*F2)
5835 RETURN
5840 REM ***** END BINOMIAL CONFIDENCE LIMITS >>>>>
9999 END
```

Figure 9.11. Approximate binomial confidence limits. (See Section 9.6 for discussion)

```
5 REM FIG. 9.12
10 PRINT "CALCULATE 2-SIDED CONF. LIMITS FOR THE POPULATION"
11 PRINT "CORRELATION, ASSUMING SAMPLE POINTS ARE TAKEN"
12 PRINT "INDEPENDENTLY FROM A BIVARIATE NORMAL DISTRIBUTION."
13 REM ASSUMPTION IS THAT POINTS ARE CHOSEN AT RANDOM, THEN
14 REM (X,Y) IS OBSERVED.  ASSUMPTIONS BREAK DOWN IF E.G.
15 REM POINTS ARE CHOSEN TO CORRESPOND TO SPECIFIED X-VALUES.
19 REM SET Z0=1.96 TO OBTAIN 95% CONFIDENCE INTERVAL.
20 Z0=1.96
25 PRINT "ENTER R, NO. OF OBS."
30 INPUT R,N2
35 IF ABS(R)>1.0 THEN 55
40 GOSUB 5410
45 PRINT "95% CONFIDENCE LIMITS ARE: ";R1;" TO ";R2
50 STOP
51 REM
55 PRINT "R =";R;" DOES NOT LIE BETWEEN -1 AND +1"
5400 REM CONF. LIMITS FOR THE POPULATION CORRELATION
5401 REM WINTERBOTTOM (1980): ASYMPTOTIC EXPANSIONS FOR THE
5402 REM BIVARIATE NORMAL CORRELATION.  COMMUNICATIONS IN
5403 REM STATISTICAL SIMULATION AND COMPUTATION B9, 599-609.
5404 REM APPROX. 3-DECIMAL DIGIT ACCURACY FOR SAMPLES OF 10;
5405 REM NEARLY 4-DIGIT ACCURACY FOR SAMPLES OF 25 OR MORE.
5410 V=N2-1
5415 Z=.5*LOG((1+R)/(1-R))
5420 R3=60*R^4-30*R^2+20
5425 R4=165*R^4+30*R^2+15
5430 DEF FNA(X)=Z+X/SQR(V)-R/(2*V)+X*(X^2+3*(1+R^2))/(12*V*SQR(V))
5435 DEF FNB(X)=R*(4*(R*X)^2+5*R^2+9)/(24*V^2)
5440 DEF FNC(X)=X*(X^4+R3*X^2+R4)/(480*V^2*SQR(V))
5445 DEF FNT(X)=FNA(X)-FNB(X)+FNC(X)
5450 DEF FNR(Z)=(EXP(2*Z)-1)/(EXP(2*Z)+1)
5455 X=-Z0
5460 Z1=FNT(X)
5465 X=Z0
5470 Z2=FNT(X)
5475 R1=FNR(Z1)
5480 R2=FNR(Z2)
5485 RETURN
9999 END
```

Figure 9.12. Confidence limits for the product-moment correlation. (See Section 9.6 for discussion)

2. Ones complement, in which case the value would be $(2^n - 1) - x$.

A floating point number is converted into the form $m \times b^c$. The mantissa m gives the most significant digits of the number, b (equal to a power of two) is the assumed base, and c is the characteristic. A common practice is to take m as a number between 0 and 1, scaled so that its leading digit (in the base b representation) is nonzero. Thus assume a base $b = 8$. Then 39 would appear as $0.47_8 \times 8^2$. In effect the number is written as $(2, 0.47)$. In addition provision must be made for the signs both of the mantissa and of the characteristic. The format used on the PDP-11 range of machines will be described by way of illustration. On these machines eight bits are available for the characteristic. Binary exponents from -128 to 127 are represented by the binary equivalents of 0 through to 255. The initial bit of the 24-bit mantissa is not stored; as the number is normalized, it can be assumed that this initial bit is one. An initial sign bit makes up 32 bits in all, requiring two PDP-11 (16-bit) words of storage. The range of numbers that can be represented is from approximately 0.29×10^{-38} to 0.17×10^{39}.

Section 3.4 mentioned the machine constant ε, defined to be the smallest positive number ε such that, as calculated, $1 + \varepsilon > 1$. Consider now the representation of the floating point number 1.0. This will appear as $0.1_b \times b$, where b $(= 10_b)$ is the base. The initial digit (which is the only nonzero digit) occupies a position with place value $b^0 = 1$, whereas the final digit occupies a position with place value b^{-n+1}. Now take any smaller number and slide its mantissa past that of the number 1.0 until bits with the same place value match. Assume that the result of the addition, accurate to $n + 1$ bits (or more) is truncated to n bits. In such a case $\varepsilon = b^{-n+1}$. The details of the way the arithmetic is done, and hence ε, vary from one machine to another. Lines 140 to 180 of the BASIC program in Section 10.7 demonstrate how a rough approximation to ε may be calculated, without detailed knowledge of the form of numeric representation used. Note incidentally that a direct test whether $1 + \varepsilon = 1$ may not work; a code optimizer may take the view that this is never true! The importance of ε is that it is either an upper bound or close to an upper bound for the maximum relative error that results from truncating or rounding a result to machine precision.

The usual forms of representation for strings of characters takes one byte (i.e., eight bits) for each character; however, the commonly used ASCII (American Standard Code for Information Interchange) code makes use of seven only out of the eight available bits. The EBCDIC (Electronic Binary Coded Decimal Interchange Code) requires all eight bits. Numbers which are represented (perhaps for input or output) as strings of characters must in most computer languages be converted to a suitable numeric form before carrying out arithmetic calculations on them. In ASCII "7" (i.e., the character seven) appears as 0110111, whereas in EBCDIC it appears as 10110111.

9.8. FURTHER READING AND REFERENCES

A recent and readable text on simulation and Monte Carlo methods is that of Rubenstein (1981). It needs supplementing by reference to recent journal articles. On methods for the generation of pseudorandom numbers, Knuth (1969) is still well worth consulting. See also chapter 6 of Kennedy and Gentle (1980), where statistical applications of the Monte Carlo method are discussed. Also in Kennedy and Gentle are extensive discussions of computer arithmetic and approximations to probability distributions, including some more accurate than those given in this chapter.

Cheng (1982) has an interesting discussion of methods for reducing the variability of estimators in computer simulation experiments.

Kahan (1983) and Klema (1983) give the background to the IEEE standard for floating point arithmetic that should soon reach an agreed final form. Use of floating point arithmetic processors (*chips*) which conform to the new standard will greatly simplify the writing of programs which can be guaranteed to give arithmetically correct results. The first of these chips to be widely used is the INTEL 8087. The Kahan and Klema papers will appear in the 1983 *Proceedings of the Statistical Computing section of the ASA*.

9.9. EXERCISES

1. Consider the linear congruential generator

$$x_n = ax_{n-1} + c \quad (\text{modulo } m)$$

(i) If $m = 32$ and $c = 0$, which choices of a give the maximum length of period?

(ii) If $m = 32$ and c is odd, which generators are full cycle?

In both (i) and (ii) take one of the generators and plot x_{i+1} against x_i.

2. Given normal variates with mean 0 and variance 1, show how to generate random variates from a bivariate normal distribution with variances σ_x^2 and σ_y^2 and correlation ρ. Generalize to give a method for generating variates from a multivariate normal distribution with correlation matrix $\boldsymbol{C}$. (Hint: Take $\boldsymbol{T}$ upper triangular such that $\boldsymbol{T}'\boldsymbol{T} = \boldsymbol{C}$.)

3. In the recurrence

$$P(x; n) = \rho_{x+1}P(x+1; n-1) + (1 - \rho_x)P(x; n-1)$$

[eq. (4.5)] ρ is a function of δ. Derive a recurrence relation for $dP(x; n)/d\delta$.

4. Consider the calculation of $\log(1 + x)$, where $x < 0.002$, on a (fictitious) machine that stores the 7 most significant decimal digits (after truncation) in any result. Three alternatives that might be considered are:

(i) Direct evaluation of $\log(1 + x)$.

(ii) Use of the series expansion

$$\log(1 + x) = x - \frac{x^2}{2} + \frac{x^3}{3} - \cdots$$

$$= x\left\{1 - x\left[\frac{1}{2} - x\left(\frac{1}{3} - \cdots\right)\right]\right\}$$

(iii) Set $z = x/(2 + x)$, and use the series expansion

$$\log(1 + x) = 2\left(z + \frac{z^3}{3} + \frac{z^5}{5} + \cdots\right)$$

Which of these methods gives best precision? Consider $x = 0.001000005$, $x = 0.0001000005$. Explain.

5. What is the advantage of calculating

$$\frac{a}{\sqrt{(a^2 + b^2)}}$$

as

$$\frac{1}{\sqrt{\left[1 + (b/a)^2\right]}}$$

What checks are needed on the value of b/a when the second of these expressions is used? Explain the use of the constant E2 (calculated in lines 140 to 180) in lines 3340 and 3350 of Fig. 10.8.

CHAPTER 10

Computers and Computer Programs

But now it's really too easy; you can go to the computer and with practically no knowledge of what you are doing, you can produce sense or nonsense at a truly astonishing rate.

G. E. P. Box

Earlier chapters have described algorithms that are useful for statistical calculations. For simple practical problems it may be feasible to carry out the calculations on a hand calculator. More often it is desirable to use a computer that has an existing statistical program or package that implements the algorithm. This allows the statistical analyses to be carried out quickly and effectively, with attention centered on statistical aspects of the problem. This chapter comments briefly on modern digital computing equipment, on its use for statistical analysis, and on some of the programs and packages available for the statistician to use. Included in this chapter are programs—in the BASIC language—for multiple regression, for the singular value decomposition, and for the solution of a nonlinear equation in a single unknown.

10.1. CHOOSING COMPUTING FACILITIES

Until recently, most computer users have found it necessary to accept the machine provided by the installation belonging to their own organization.

This chapter opening quote is taken from Box, G. E. P. (1969). In Milton and Nelder, Eds., *Statistical Computation*, Academic Press, New York, p. 6.

This is changing. It is now common to find a choice between a small computer designed for individual use and terminal access to a medium or large-scale computer or to a network. The small computer may itself become the terminal when access is required to another machine. It may be possible to meet computing requirements by buying several small machines for tasks such as data and program preparation and editing, and hiring time and services from outside for other work. As the machines and the programs which they run continue to change, so must approaches to deploying them.

The Operating System

A computer is at best as good as the programs it runs. Always present, initiating the tasks required from the system and exercising overall control, is the program known as the *operating system*. The operating system command language (job control language or work flow language) allows instructions to be passed to the operating system. Such tasks as the creation and copying of files, editing, sorting, and the running of programs which have special relevance to the particular user, may be initiated directly from the operating system. Operating system command languages vary widely from one system to another and determine whether a system will be flexible or inflexible or easy or hard to use and master. It is difficult for any program or package, however well written, to protect its users from all aspects of an unsatisfactory operating system. For a really good operating system and associated utilities it is necessary to have a machine that is sufficiently large and well designed that there is little need to worry about limitations on the size or number of programs or data sets. On the other hand, a large machine capacity is not a guarantee of a good operating system. Some operating systems in current use have changed little in style since the late 1960s and remain difficult to use and ill-adapted to interactive computing.

What Facilities Do Statisticians Need?

Common forms of statistical analysis require, if they are to be handled efficiently, the use of statistical packages and/or subroutine libraries. It makes no sense for an individual to attempt to repeat the work of the highly competent specialists in statistical and scientific program writing who have contributed to reputable packages and libraries. The point will be argued at greater length in Section 10.2. Most standard packages and subroutines will run on any except the smallest of the commercially available multiuser computer systems. But ease of use and the quality and accessibility of the utilities to assist in data and job preparation and the further processing of results vary widely.

Large Machines, or Small?

In the past it has been common for large organizations to provide users with widely varying requirements with access to a single large system. In many cases computing power equivalent to that of a large system may now be better provided by a well-managed network of smaller machines. The cost of maintaining extra copies of some programs must be offset against the cost of running a large centralized installation, all too often with poor liaison between users and managers. A further advantage of a network is that each individual machine can be managed and equipped so that it caters for the special requirements of a particular subgroup of users.

The terminals used for access into a central computer or network may themselves be personal computers, able to handle text and data entry and editing, and to run small programs. A communications program gives access into the central computer or network when its power or special facilities are required.

Microcomputers

Most of the computers designed for individual use are, technically, microcomputers. Microcomputers are distinguished because their central processing unit (where program control and logical functions are handled) is a mass-produced general purpose control device, built on one or perhaps on several silicon chips. Operating system commands are usually provided as part of a BASIC interpreter. Microcomputers built for the popular or hobbyist market have, to date, almost all used 8-bit microprocessors. This means that data are processed 8 bits (= 1 byte) at a time. On earlier microcomputers primary memory was usually limited to somewhat less than 64K (where 1K = 1024) bytes or characters; a limitation to 128K or 256K bytes is now common. Those microcomputers designed partly or primarily for home entertainment are among the cheapest available. They, along with more expensive counterparts designed for business use, are also remarkable for the ease with which the screen display can be manipulated. The VISICALC electronic spreadsheet package, developed first for the Apple, uses this ability to remarkable effect to allow easy movement between the results of calculations and the formulae used. (VISICALC is a trademark of Software Arts Inc.) VISICALC has found its main use in financial calculations but has equal potential usefulness in mathematical or scientific calculations based on formulae that may need to be modified as calculations proceed. Packages similar to VISICALC have only recently become available for conventional computers.

A remarkable range of devices—joysticks, light pens, voice and music synthesizers, sensors with a limited voice recognition ability, and graphics tablets—are available quite cheaply for plugging directly into popular brands of microcomputers. These devices appear connected to the microcomputers available from the local department store long before most of us get to see them used on conventional larger machines. They add to the attraction of inexpensive equipment which supports a highly interactive style of computing. Thisted (1981) speculates on the potential of these features for use in statistical analysis. It is possible, with imagination and effort, to carry across to programs and systems designed for serious use many of the features that give microcomputer games an appeal which some users find addictive. Carroll (1982) has interesting comments on the contrast between using a text editor and playing the well-known and popular game of Adventure.

The spectacular effects of miniaturization are most obvious in the small hand-held machines, with an alphabetic keyboard and BASIC interpreter, which are now appearing on the market. These very small machines will prove invaluable for the field capture of data and for some immediate processing.

It is well to warn against pressing any computer system, large or small, beyond its proper limits of usefulness. Limitations may arise from the limited capacity of the system, or from the lack of programs and utilities suitable for the task undertaken. The dodges that must be used to adapt a system for a task for which it is not well designed make for waste of time and effort.

The *Scientific American* article by Toong and Gupta (1982) is a useful introduction for anyone wanting to understand how microcomputers work and what they can do.

Statistical Packages for Microcomputers

Statistical packages have been developed for popular microcomputers, and facilities are often comparable to a subset of SPSS and/or **Minitab** (discussed in the next section). Direct implementation on 8-bit machines of packages (e.g., BMDP, PSTAT, or SPSS) designed for conventional computers has been difficult. The FORTRAN compilers and associated operating systems have not had the necessary facilities for making a large program run on a machine with (by current standards) limited primary memory. Now that implementations of large statistical packages are beginning to appear on the more sophisticated 16-bit microcomputers (perhaps better called *supermicros*), there may not be much point in considering implementations on 8-bit machines. Scott, Olsen, and Bryce (1983) describe their experience in adapting several large statistical packages to run on a microcomputer using the Motorola MC68000 chip. Advances in 8-bit machines are however now removing many of their memory and other limitations.

Morganstein and Carpenter (1983) catalog and compare statistical packages for microcomputers, though without looking in any detail at the statistical adequacy of their analyses.

10.2. CHOOSING A PROGRAM

A first requirement for any computer program for practical statistical analysis is that it should give an analysis that conforms with good statistical practice and that results should be correct. Also important are ease of use (for users with the appropriate statistical competence) and the presentation of results in a manner that allows ready assimilation. The writing of programs with these qualities requires both a thorough understanding of the analysis to be performed and a high level of computing skill. The necessary combination of skills is most likely to be found when groups of individuals with a high level of statistical and computing expertise cooperate to write programs intended for wide use. Individuals writing programs for their own use cannot so easily draw on resources of skill other than their own and have less incentive to be sensitive to the criticisms a wider audience would attract. The complications of implementing the analysis in a computer language such as FORTRAN diverts effort from the statistical problem. When it is necessary to write one's own programs for serious analyses, it is desirable that they be tested and criticized prior to use by others competent to judge.

Statistical packages now provide the most popular way of handling statistical analyses. A package is distinguished from a library of subroutines because it is complete in itself—it has its own command language for specifying the form of data input, the manipulations required on the data, and the analysis or analyses required. An ideal is that the command language should allow the statistician to describe the problem in appropriate statistical terms, leaving to the computer package the task of interpreting and acting upon this description. A good example is the use that GENSTAT (Alvey et al., 1982) and P-STAT (Buhler and Buhler, 1979, and Heiberger, 1982) make of the Wilkinson and Rogers (1973) syntax to specify both the block (or error) and treatment structure in a suitably balanced experimental design.

GENSTAT is remarkable because its command language may be used as a powerful general purpose computer language, in which programs may be written for analyses that cannot be handled directly. It may be described as a general purpose data manipulation language with powerful statistical analysis features embedded. The language "S" (Becker and Chambers, 1980) is a tidier general purpose language than GENSTAT but has less powerful statistical analysis features. Other packages with extensive general language features are SAS (SAS Institute, 1979), **Minitab** (Ryan et al., 1981), P-STAT (Buhler and Buhler, 1979), GLIM (Baker and Nelder, 1978), and SPSS (Nie et al., 1975). The P-STAT, SAS, GLIM and "S" systems have commands powerful enough to handle most of the data manipulation and data rearrangement likely to be required prior to an analysis. In addition P-STAT and SAS have extensive file manipulation facilities. Information on a wide variety of packages may be found in Francis (1979).

BMDP (Dixon et al., 1981) functions as a collection of separate analysis programs, but with a command language which is similar across all programs. A wide range of output options are in most cases available, and output is admirably well laid out. Input facilities are less flexible than in the other packages mentioned. If extensive manipulation of data is required prior to an analysis, a separate program or package must be used. BMDP is well suited to use in an interactive computing environment where BMDP programs form part of a collection of utilities for data manipulation and editing, with easy means for using the output from one utility as input to another. A good modern language at the level of the operating system is essential in order to initiate and control the use of the various utilities and programs. SAS users have access to the procedure BMDP in SAS to call a nominated BMDP program. P-STAT has facilities for reading BMDP (or SPSS) save files.

For less standard problems it is necessary to use subroutines from such sources as the algorithms section of *Applied Statistics* (JRSSC) and the Association for Computing Machinery publication *Transactions on Mathematical Software*. Anyone with access to either the IMSL (*International Mathematical and Statistical Libraries*) or the NAG (*Numerical Algorithms Group*) collection should consult these first. The subroutines used must be linked together and supplemented by a main program or programs written by or for the user. Easy means for using the output file from one utility as input to another are even more desirable than when using a program collection such as BMDP. Rather than write one large program to handle the whole analysis, it will often be easiest to split the analysis into a sequence of separate tasks, handled by separate programs. If the programs so formed are likely to be used again, it will be worthwhile to incorporate them into a local program library. Again a good interactive language at the operating system level will make it easy to *package* the separate programs in a manner that allows easy access to the particular program units required.

Skilled users can easily adapt or modify such a loosely knit *package* in accordance with individual requirements.

Reliance on a particular operating system makes it difficult to carry programs across to a different range of machine. Most operating systems are specific to one manufacturer. The Bell Laboratories UNIX system, popular because of its power and adaptability, is exceptional because of the variety of machines on which it has been implemented. A point of criticism is the barrier its often cryptic commands present to casual users. UNIX has been a point of reference in much recent discussion of operating systems; see, for example, Denning (1982). Its success gives hope that operating system languages will in due course be standardized.

Prospects for Packages

Such progress has been made in providing statisticians with packages and languages for statistical analysis that it is easy to underestimate what has still to be achieved. Most packages give little attention to automatic checks that data values are reasonable or to providing plots that assist their users in making such checks. Coarse checks that data satisfies minimal conditions for an analysis are, to date, uncommon. Thus it makes sense, when the usual product-moment correlation coefficient is calculated, to check for linearity against the alternative of a monotonic nonlinear relationship. Calculation of the Spearman rank correlation provides one of several alternative possibilities for making such a check. If it is substantially greater than the product-moment correlation, a nonlinear relation is indicated. Computers are ideally suited to the mechanical checking and flagging of a wide range of conditions that would make an intended analysis inappropriate. Humans find it bothersome to carry out more than a few crude checks; computers will, if rightly instructed, go on as long as necessary. The user must be allowed to interact with what the computer may reveal, asking for further checks or alternative analyses when these seem desirable.

The use of matrix manipulations to handle calculations not otherwise available is, in current packages, awkward and inefficient in the use of machine resources. Good facilities for forming and working with upper triangular matrices, or with band matrices are lacking. Instead the emphasis is on facilities for matrix inversion. It is not possible to take advantage of the special form of band matrices.

It is not easy to present computer output so that it will convey its message effectively. Ehrenburg (1981) gives rules for the presentation of numerical information that merit attention. The display of more than two or three decimal digits in each number may make it difficult to discern patterns which are present. Presentation of results to any larger number of digits is usually best delayed to the final stage of the analysis.

The handling of a statistical analysis should be an educative experience, challenging understanding and imagination.

Beyond Packages—Expert Systems

A common complaint against statistical packages has been that they encourage mindless use by individuals who lack statistical expertise. This may be unfair; statistical methods were misused before the computer appeared. Preece (1982) is unhappy with many of the examples of the use of the t-test which appear in statistical textbooks. Several of his examples predate statistical packages. In no case would a computer have been needed to handle the calculations. The problem lies with a careless or mechanical use of the particular method of analysis.

Expert statistical systems take very seriously the desirability for expert guidance as the analysis proceeds. Statistical analysis becomes a process of interaction between user and system. The system becomes an adviser, giving guidance on experimental design, on the collection of data, and on analyses as they are carried out. Guidance would extend to engaging in dialogue with the user to determine preliminary analyses and/or desirable graphical inspections, with a similar scrutiny of results at each subsequent stage. Such a system would closely model the manner in which a good statistical consultant works with clients. Ideally the system should have the ability to learn, in dialogue with expert statisticians. Already there has been some limited success in developing expert systems that give access to other areas of professional expertise. See Chambers (1981), Gale and Pregibon (1983), and Hahn (1983). The successful development of expert statistical systems would make professional statistical expertise far more widely available than at present.

10.3. LANGUAGES FOR WRITING COMPUTER PROGRAMS

To date the language most commonly used for statistical programs which are intended for wide use has been FORTRAN (FORmula TRANslation). The current official definition of FORTRAN is that published in the 1977 standard. The new standard defines a far more versatile language than the 1966 standard which it replaced but retains many unsatisfactory features which discourage good programming practices. For an introduction to FORTRAN 77 which compares it with FORTRAN 66, see Balfour and Marwick (1979).

The most widely used language for simple interactive tasks is BASIC. In one or other of its versions it has become standard on all microcomputers. As a language in which to write carefully structured programs it is inferior even to FORTRAN 66. But it does have pleasant string handling features, such as became available only in FORTRAN 77. A further attraction is that most (but not all) implementations of BASIC interpret and execute BASIC programs line by line as they appear, so that users can be allowed to move freely between the running of a program and correction or editing or the checking of intermediate results. This is a boon when programs are developed or tested interactively at a terminal. (When FORTRAN and allied languages are used, the program must usually be compiled, or translated whole into lower-level codes, before any of it is executed. The compiled program should run very much faster than a program that is interpreted line by line.) The ANSI (American National Standards Institute) X3.60-1978 standard for minimal BASIC has very restricted features, and program writers and computer manufacturers have taken little notice of it. The BASIC standards committee (X3J2) has now (1982) released a proposal for a powerful version of BASIC, which is expected to become a new ANSI standard. Included are some features (e.g., named subprograms) not available in most present microcomputer versions of BASIC. The new standard describes a language that in most respects compares favorably with FORTRAN 77. Provisional information on the new standard may be found in Kurtz (1982).

The writers of subroutine libraries have to date favored FORTRAN because of the wide availability of FORTRAN compilers. Thus IMSL subroutines are all in FORTRAN; NAG gives a choice of FORTRAN or **Algol60**. The large investment in these libraries need not be a hindrance to moving to a more modern language such as **Pascal** or perhaps PL/I when new programs or subroutines are written. On a modern computing system the user who writes in **Pascal** (or even BASIC) can be allowed the freedom of calling previously compiled FORTRAN (or other) subroutines.

The language **Ada**, for which commercial compilers should shortly be available, can be expected to come into wide use within the next few years. The **Ada** designers took **Pascal** as their starting point; **Pascal** was itself developed from **Algol60**. **Ada**'s sponsorship by the U.S. Department of Defense seems certain to ensure that it will slowly replace FORTRAN. Serious misgivings have been expressed regarding the size and complexity of the final language definition. See Hoare (1981), in a published lecture which is remarkable for its wit and wisdom.

The interactive language APL (A Programming Language) has gained an enthusiastic following among some statisticians. It is easy to embed within

APL abilities similar to those of a statistical package. This approach to handling statistical calculations is well described in Anscombe (1981).

Programming Style

There are now widely agreed principles that ought to guide the writing of computer programs. Kernighan and Plauger (1978) is a readable popular exposition of such principles. Much of what is said is relevant to any nontrivial use of the computer, where a sequence of commands must be put together in order to carry out the desired tasks. See also Turner (1980).

Technical aids now in prospect, which will assist and change the programmer's art, are discussed in Wasserman and Gutz (1982).

Particular Computer Packages

The next two sections of this chapter give brief accounts of the packages **Minitab** and GLIM. Later sections of this chapter give listings of BASIC computer programs that implement a selection of the algorithms described in earlier chapters. The BASIC programs are intended as a starting point for readers of this book who have access to a microcomputer or a terminal that runs BASIC and wish to experiment with the algorithms.

10.4. USING A PACKAGE FOR STATISTICAL ANALYSIS—MINITAB

This section will discuss, briefly, a few of the features of the statistical package **Minitab**. Section 10.5 will discuss the package GLIM. These two packages have been chosen for discussion because the author is familiar with them, and because they can be implemented on quite small computers. Both have been implemented on the Digital Equipment LSI11, which is suitably described as a microcomputer version of a PDP11. **Minitab** is aimed in the first place at the novice, with an attempt made to check on particular features of the data which may make the analysis invalid or inappropriate. It is thus very attractive also to statistical professionals, who comprise the main audience at which GLIM is aimed.

Minitab is patterned in part on the OMNITAB statistical system (Hogben et al., 1971), which keeps a *worksheet* where data are laid out in columns. In **Minitab** a column is just as appropriately viewed as a vector of values of one variate. Data that are to be placed in the *worksheet* may be included, preceded by a READ or SET command which gives details of where the data are to be placed, among the **Minitab** commands. Or READ *filename* (followed by a list of columns) or SET *filename* (followed by a column) may be used to read data onto the worksheet from a specified file. The system keeps track of the number of the final row, in each column, which has been used to store data. Columns are referred to as C1, C2, . . . , up to some fixed limit (usually in the range 25 to 200). Alternatively, a column may be referred to using a name (in single quotes) which has been assigned using the optional NAME command. In addition to columns of data **Minitab** allows the use of *stored constants* (scalars) K1, K2, . . . and matrices M1, M2, ...

Commands appear one to a line, though there is a device for extending a command over from one line to the next. The arguments (numbers, columns, matrices, or stored constants) that follow any command word except LET may have explanatory text interspersed; such text is ignored by **Minitab**. Any argument must have the right type for the position it occupies (in sequence) following the command word. The LET command gives a powerful facility for transforming data values. Thus in the example in Fig. 10.1 below the statement

$$\text{LET C13} = \text{LOG(C3} + 100)$$

adds 100 to each of the values in C3 (column 3), takes the logarithm, and stores the result in C13. It is easy to see how any BASIC or FORTRAN arithmetic assignment can be changed into the **Minitab** form. Calculations involving columns are carried out for all rows of the specified columns.

A sequence of **Minitab** commands may be stored together, under a single name, as a macro. EXEC, followed by the macro name, then causes execution of the whole sequence of commands.

```
SET CROSS-SECTIONAL DIAMETER (MM) IN C1
76 70 73 74 77 74 72 71 69 73 73 72 73 77 70
72 70 75 71 71 69 74 72 73 72
SET LENGTH THROUGH CORE (MM) IN C2
68 64 65 73 69 69 70 70 64 69 76 69 65 65 65
67 68 64 65 65 65 70 66 69 67
SET WEIGHT (GM) LESS ONE HUNDRED IN C3
68 50 42 80 64 64 62 60 45 63 74 52 58 70 50
63 56 56 49 51 40 66 55 55 58
LET C11=LOG(C1)
LET C12=LOG(C2)
LET C13=LOG(C3+100)
REGRESS Y IN C13 ON 2 PREDICTORS IN C11 AND C12, C15,C16
NOTE THE TWO EXTRA COLUMNS ARE FOR RESIDUALS (C15)
NOTE AND FITTED VALUES (C16)
PLOT C15 VS C16
STOP
```

Figure 10.1. Minitab example.

Now consider, as shown in Fig. 10.1, the use of **Minitab** to carry out a regression of log(apple weight) upon log(cross-sectional diameter) and log(distance through core), for data from a study of apple shapes. This gives the output shown in Fig. 10.2. If two further columns, say C15 and C16, are specified in the REGRESS statement, they will be used to store, respectively, the standardized residuals and the fitted values. The plot obtained by using the command

PLOT RESIDUALS IN C15 VS FITTED VALUES IN C16

is then a desirable adjunct to the analysis in Fig. 10.2. The worksheet, or some of its columns, may be stored on an output file, making it easy to take

```
THE REGRESSION EQUATION IS
Y = -  2.66 +  1.01 X1  + 0.807 X2

                                             ST. DEV.      T-RATIO =
        COLUMN       COEFFICIENT             OF COEF.      COEF/S.D.
        --             -2.662                 1.003          -2.66
X1      C11            1.0095                0.2253          4.48
X2      C12            0.8072                0.1572          5.14

THE ST. DEV. OF Y ABOUT REGRESSION LINE IS
S = 0.03222
WITH (  25- 3) =  22 DEGREES OF FREEDOM

R-SQUARED = 74.7 PERCENT
R-SQUARED = 72.4 PERCENT, ADJUSTED FOR D.F.

ANALYSIS OF VARIANCE

 DUE TO       DF           SS          MS=SS/DF
REGRESSION     2      0.067316         0.033658
RESIDUAL      22      0.022842         0.001038
TOTAL         24      0.090158

FURTHER ANALYSIS OF VARIANCE
SS EXPLAINED BY EACH VARIABLE WHEN ENTERED IN THE ORDER GIVEN

 DUE TO       DF           SS
REGRESSION     2      0.067316
C11            1      0.039929
C12            1      0.027388

          X1           Y      PRED. Y     ST.DEV.
ROW       C11         C13      VALUE      PRED. Y     RESIDUAL     ST.RES.
  3      4.29     4.95583    5.03868     0.00907     -0.08285      -2.68R
 11      4.29     5.15906    5.16488     0.01954     -0.00583      -0.23 X
 14      4.34     5.13580    5.09253     0.01744      0.04327       1.60 X

R DENOTES AN OBS. WITH A LARGE ST. RES.
X DENOTES AN OBS. WHOSE X VALUE GIVES IT LARGE INFLUENCE.

DURBIN-WATSON STATISTIC = 2.58
```

Figure 10.2. Output from the **Minitab** example (Fig. 10.1).

up the analysis from this point at a later terminal session. Program output can be similarly stored.

Version 82.1 of **Minitab** will handle two-way analysis of variance with an equal number of observations in each cell, with provision for the storage of both fitted values and residuals. Checks on residuals are not, as in regression, provided automatically; the user must attend to this. Also included (from Version 81.1 onward) are a number of the analyses suggested by J. W. Tukey under the name EDA (Exploratory Data Analysis). The EDA commands are based on Velleman and Hoaglin (1981). **Minitab** provides commands that are very easy to use for fitting simple Box-Jenkins models to time series data. Estimates are obtained by the unconditional least squares method outlined in Section 8.3. A final point is that **Minitab** has extensive data manipulation features that go well beyond the use of the LET command.

10.5. USING A PACKAGE FOR STATISTICAL ANALYSIS—GLIM

GLIM (Generalized Linear Interactive Modelling) has been developed in the U.K. by a Royal Statistical Society working party. Baker and Nelder (1978) describe version 3. The project has had the benefit of the experience Nelder and Baker and other workers at Rothmasted Experimental Station had gained with GENSTAT. With the release of version 4, GLIM will become one of several modules in a new system, provisionally called PRISM. Other PRISM modules will cater for analysis of variance in the style of GENSTAT (see Section 5.5) and for graphics.

Statistical analysis features are set in the context of a language that handles calculations and manipulations both on columns of numbers (variates or factors) and on scalars. The general language features extend to looping (but not to a GOTO), and the use of macros. A *macro* is a set of statements that may be invoked by using the macro name and giving any necessary arguments (parameters). GLIM is suited for handling the calculations of normal theory linear models, and this is as far as many users will wish to take their use of GLIM. It may be used for calculations in a balanced analysis of variance model, but for this purpose it gives a form of output that is far from ideal. For this reason PRISM will have a separate module to cater for the analysis of data from designed experiments. But the particular strength of GLIM is its ability to handle *generalized linear models* as defined in Nelder and Wedderburn (1972). Thus GLIM will handle logit and probit analysis, and the analysis of multiway tables with multinomial errors.

The names of variates or factors in GLIM may be of arbitrary length, but characters after the fourth are ignored. Directive names (or command words) have $ (or whatever else is used as the currency symbol) as their first character. A directive is terminated with the appearance of the directive name that begins the next directive. These comments should suffice to make sense of the set of GLIM statements in Fig. 10.3 (with output in Fig. 10.4) which carry out the same regression calculations as in the **Minitab** example in Section 10.4. Diagnostic information is not printed out automatically, as when **Minitab** is used. Instead the user has access to a wide variety of system vectors and system scalars and is expected to use these to provide whatever diagnostic information is thought appropriate to the analysis in hand. In Fig. 10.3 use is made of the fitted values stored in the system vector %FV.

Finally, consider two simple examples (Figs. 10.5 and 10.6) that illustrate the use of GLIM for analyzing multiway tables. The examples are taken

```
$UNITS 25
$C the next two directives read in the
cross-sectional diameters of the 25 apples
$DATA DIAM  $READ
76 70 73 74 77 74 72 71 69 73 73 72 73 77 70
72 70 75 71 71 69 74 72 73 72
$C the next two directives read in the
lengths through the core of the 25 apples
$DATA LENC  $READ
68 64 65 73 69 69 70 70 64 69 76 69 65 65 65
67 68 64 65 65 65 70 66 69 67
$C the next two directives read in apple
weights, less 100
$DATA W100  $READ
68 50 42 80 64 64 62 60 45 63 74 52 58 70 50
63 56 56 49 51 40 66 55 55 58
$CALC LDIAM=%LOG(DIAM)
$CALC LLEN=%LOG(LENC)
$CALC LWT=%LOG(W100+100)
$PRINT 'regress log(weight) on log(length)'
$PRINT '                    and log(diameter)'
$PRINT $
$YVAR LWT  $FIT LDIAM+LLEN  $
$C print out parameter estimates and their
standard errors. (there are a variety of other
possibilities.)
$DIS E $
$PRINT
$PRINT 'plot of residuals versus fitted values'
$PRINT
$CALC RES=LWT-%FV  $PLOT RES %FV $
$STOP
```

Figure 10.3. GLIM example.

```
regress log(weight) on log(length)
                    and log(diameter)

 CYCLE  DEVIANCE      DF
    1   0.2284E-01    22

        ESTIMATE       S.E.       PARAMETER
    1   -2.662        1.003       %GM
    2    1.010        0.2253      LDIA
    3   0.8072        0.1572      LLEN
   SCALE PARAMETER TAKEN AS    0.1038E-02


plot of residuals versus fitted values

 0.640E-01 *
 0.560E-01 *
 0.480E-01 *                   R                  R
 0.400E-01 *                            R
 0.320E-01 *
 0.240E-01 *      R          R
 0.160E-01 *        R       R  R
 0.800E-02 *   R       R    R      R  R     R
 0.000E+00 *                    R     R  R R
-0.800E-02 *           R                              R
-0.160E-01 *
-0.240E-01 *
-0.320E-01 *
-0.400E-01 *     R                    R          R
-0.480E-01 *                        R
-0.560E-01 *
-0.640E-01 *
-0.720E-01 *
-0.800E-01 *                 R
-0.880E-01 *
-0.960E-01 *
-0.104     *
  .........*.........*.........*.........*.........*.........*
         4.95      5.00      5.05      5.10      5.15      5.20
```

Figure 10.4. Output from the GLIM example (Fig. 10.3).

from Section 7.6. In Fig. 10.5 the output appears along with the GLIM directives.

10.6. BASIC PROGRAMS FOR STATISTICAL CALCULATIONS

The programs given in the next two sections show how a program for regression calculations may be built around the algorithms described in Chapters 1 through 4. Complication has been kept to a minimum by leaving out checks for some of the more obvious problems (e.g., a variate whose values are all the same) and by omitting any diagnostic information beyond the variance-covariance matrix and calculation of residuals. Readers will

```
$C example from R.O.MURRAY 'AGONIES OF THE ASPIRING ATHLETE'
   NEW SCIENTIST, VOL.55, P.434, AUG.31 1972.
$UNITS 6
$FACTOR SCHOOL 3   HIPS 2
$C SCHOOL level 1 - athletic regime  2 - intellectual regime
                3 - control group from industry.
   HIPS   level 1 - damage apparent  2 - no evident damage.
$CALC SCHOOL=%GL(3,1)
$CALC HIPS=%GL(2,3)
$DATA NUMS
$READ
   23  7 12
   71 70 68
$YVAR NUMS   $ERROR P
$FIT SCHOOL+HIPS  $DISPLAY ER  $
        SCALED
 CYCLE  DEVIANCE       DF
    3     7.564         2

        ESTIMATE       S.E.        PARAMETER
    1    2.756       0.1744        %GM
    2 -0.1995        0.1537        SCHO(2)
    3 -0.1613        0.1521        SCHO(3)
    4    1.605       0.1689        HIPS(2)
   SCALE PARAMETER TAKEN AS        1.000

   UNIT  OBSERVED      FITTED       RESIDUAL
     1        23        15.73         1.833
     2         7        12.88        -1.639
     3        12        13.39       -0.3790
     4        71        78.27       -0.8218
     5        70        64.12        0.7349
     6        68        66.61        0.1699

$C E displays estimates and standard errors
$C R displays standardized residuals and fitted values
$CALC FITV=%FV   $C this gives fitted values in each cell.
                    %LP would give log(fitted values).
$CALC SRES=(NUMS-%FV)/%SQRT(%FV)
$LOOK SCHOOL HIPS FITV SRES  $
   1     1.000        1.000        15.73         1.833
   2     2.000        1.000        12.88        -1.639
   3     3.000        1.000        13.39       -0.3790
   4     1.000        2.000        78.27       -0.8218
   5     2.000        2.000        64.12        0.7349
   6     3.000        2.000        66.61        0.1699
$STOP
```

Figure 10.5. The use of the GLIM program in analyzing a two-way table, with output.

find it a useful exercise to consider what changes are needed to improve the programs to a standard where they would be suitable for use by novices.

A secondary aim has been to allow readers to experiment with implementing regression and other calculations on a small computer. The programs have been written in BASIC because BASIC is available in one of its versions on almost all popular brands of microcomputers. The ease with which BASIC programs can be made interactive compensates in part for the

```
$C analysis of a three-way frequency table
   (fictitious data). If sex is ignored and
   the table analysed as a two-way table,
   misleading results are obtained.
$UNITS 8
$FACTOR SEX 2  PORR 2  SLEP 2
$DATA NUMS
$CALC SEX=%GL(2,2)  :PORR=%GL(2,4)  :SLEP=%GL(2,1)
$READ
     405 100       11  42
      47  15      107 415
$YVAR NUMS  $ERROR P
$FIT PORR+SLEP   $C deviance will approximately
                    equal the chi-square statistic
                    for testing for no association
                    in the two-way table.
$C
$DIS E   $
$FIT +SEX+PORR.SEX+SLEP.SEX  $DIS ER  $
$C the deviance is an approximate chi-square
   statistic for testing for no association between
   porridge and sleepwalking, after adjusting for
   association between sex and porridge, and
   between sex and sleepwalking.
$STOP
```

Figure 10.6. The use of GLIM to analyze a three-way table with ficticious data.

absence of features that make it easy to display program structure. A severe disability is that there is no satisfactory means for dividing BASIC programs up into independent modules. It will be necessary to shorten the programs in order to implement them on very small machines (16K bytes of random access memory should be adequate; if comments are left out 8K bytes may be adequate). Several of the hand machines now available are capable of running cut-down versions of programs, such as those given.

Implementations of BASIC vary widely. In the interests of conciseness use has been made of a small number of features which either are not universally available or else are implemented differently in different BASICs. Notes on some of these features now follow.

Nonstandard Features Used in the BASIC Programs

1. *Multiple statements on a line* (\ *used to separate statements*). Most microcomputer versions of BASIC use a colon (:) as a statement separator, rather than the backslash. If there is no provision for multiline statements then the second and subsequent statements must be moved to their own lines, each with their own number. Thus replace

   260 N = 0 \ M8 = N8 \ GOSUB 2000

by

```
260 N = 0
262 N8 = M8
264 GOSUB 2000
```

(Make sure that the first statement on the line retains the original line number.)

2. *IF THEN statement.* For example,

```
4820 IF D1 < > 0 THEN D1 = 1/D1
```

In some BASICs the only form of IF statement allowed is

```
IF condition THEN linenumber
```

Statement 4820 may be replaced by

```
4820 IF D1 = 0 THEN 4830
4822 D1 = 1/D1
```

The logic is now harder to follow.

3. *Use of < > for "is not equal to."* Various symbols are used. Some BASICs use /= or # in place of < > . If all else fails use the technique of note 2 above so that the test becomes one for equality rather than for inequality.

4. *LET has been omitted prior to arithmetic assignments.* Thus E1 = 1 + E0 appears in place of LET E1 = 1 + E0. A few versions of BASIC insist that LET should always appear in such statements. Some versions, although not requiring LET in most contexts, would insist that statement 4820 be modified to

```
4820 IF D1 < > 0 THEN LET D1 = 1.0/D1
```

5. *Use of STR$( ); concatenation of strings using +.* One solution is to omit entirely the lines which use these features. Omission of lines 2030–2050 will make the default name for each variate a null string. Omission of lines 9520–9540 will mean that numbers are displayed to more significant digits than are usually required, and do not appear in tidy columns.

In place of STR$( ), some versions of BASIC use STR( ) or STRING$( ) or STRING( ).

6. *Use of SEG$(,,).* Various alternatives to SEG$(,,) at lines 9680 and 9710 are in use. A simple solution to any problems is to delete lines 9630–9710 and 9770 entirely, and change line 9620 to

```
9620 INPUT X$
```

It will then be necessary to enter each new variate value on a new line.

Microsoft BASIC, which is widely used on microcomputers, uses MID$(L$,K1,L) in order to select the substring consisting of L characters which starts at character K1. Thus SEG$(L$,K1,K2), which selects characters K1 to K2 inclusive, becomes MID$(L$,K1,K2 − K1+1). Lines 9680 and 9710 become

```
9680 K$ = MID$(L$,K,1)
```

and

```
9710 X$ = MID$(L$,K1,K2 − K1+1)
```

7. *Arrays are assumed to start at element* 0. If arrays start at element 1, then change

```
20 I0 = 0
```

to

```
20 I0 = 1
```

Lien (1981) gives comparative information on a variety of microcomputer versions of BASIC.

10.7. MULTIPLE REGRESSION—VIA THE NORMAL EQUATIONS

This first version of a program for multiple regression calculations has fewer BASIC statements and should run more quickly than the program in Section 10.8. Formation of the normal equations as here, rather than direct reduction of the model matrix to upper triangular form using planar rotations as in Section 10.8, will give less accurate results when there are strong dependencies between the explanatory variates. Figure 10.7 shows where the program differs from the listing in Fig. 10.8 in the next section. At each point the program tells the user what to do next. When prompted to enter a row of data you may instead, by entering DEL and following the prompt, request deletion of an earlier row of data.

```
5  REM FIG. 10.7
10 REM ****************************************************
11 REM * REGRESSION CALCULATIONS - FORM CSSP MATRIX AND    *
12 REM * FROM IT CHOLESKY DECOMPOSITION TO UPPER TRI FORM *
13 REM ****************************************************
(replace lines 5-13 in fig. 10.8)

200 REM SET TOLERANCE
210 T0=1.0E-3
(insert these lines)

370 REM DATA IS IN PLACE AND CSSP MATRIX HAS BEEN FORMED
371 REM * * * * * * * * * * * * * * * * * * * * * * * * * *
372 PRINT "* CSSP MATRIX *"
380 U3=0 \ GOSUB 8350
390 REM FORM FIRST ROW OF CHOLESKY UPPER TRIANGLE MATRIX
400 D0=U(I0,I0) \ D=SQR(D0) \ U(I0,I0)=D
410 FOR J=I0+1 TO Q7 \ U(I0,J)=U(I0,J)*D \ X(J)=U(J,J) \ NEXT J
420 REM NOW FORM REMAINING ROWS OF CHOLESKY DECOMPOSITION
430 GOSUB 3450
440 REM NEXT PRINT MEANS, S.D.'S, & R-SQUARED'S
450 L0=I0+1 \ L9=Q7 \ GOSUB 8010
460 GOSUB 8110
(replace existing lines 370-460)

omit line 530
omit line 590

2690 REM
2700 L9=Q7 \ GOSUB 3250
2710 REM
2720 REM
(replace existing lines 2690-2720)

3250 REM { { { Form CSSP Matrix
3260 W5=1 \ IF W9=1 THEN W5=X(Q9)
3270 W5=W5*D9
3280 REM D9=1 => ADD ROW. D9=-1 => DELETE ROW
3290 W=U(I0,I0)+W5
3300 U(I0,I0)=W
3310 FOR J=I0+1 TO Q7
3320 D=(X(J)-U(I0,J))*W5
3330 U(I0,J)=U(I0,J)+D/W
3340 X(J)=X(J)-U(I0,J)
3350 FOR I=I0+1 TO J
3360 U(I,J)=U(I,J)+X(I)*D
3370 NEXT I
3380 NEXT J
3390 RETURN \ REM end Form CSSP Matrix } } }
```

Figure 10.7. A BASIC program for multiple regression via the normal equations.

```
3400 REM
3450 REM { { { Chol. Decomp. of Pos. Def. Symm. Matrix
3460 F=0 \ REM F WILL COUNT NO. OF SINGULARITIES
3470 REM NEGATIVE VALUE WILL INDICATE AN ERROR
3480 FOR J=L0 TO Q7
3490 D=U(J,J) \ E1=D*E0*J \ T1=D*T0
3500 J1=J-1
3510 FOR K=J TO Q7
3520 W=U(J,K)
3530 IF J=L0 THEN 3570
3540 W0=0
3550 FOR I=L0 TO J1 \ W0=W0+U(I,J)*U(I,K) \ NEXT I
3560 W=W-W0
3570 IF K>J THEN 3650
3580 D=0
3590 IF W>T1 THEN 3640
3600 PRINT "DIAG. EL. NO.";J;" REDUCES FROM";U(J,J);" TO";W
3610 IF (W+E1)<0 THEN 3690
3620 PRINT "FAILS TOLERANCE TEST; ELEMENT IS SET TO ZERO."
3630 F=F+1 \ GO TO 3650
3640 D=1/SQR(W)
3650 U(J,K)=W*D
3660 NEXT K
3670 NEXT J
3680 RETURN
3690 F=-J
3700 PRINT "*** MATRIX IS NOT OF FORM  X' X ***"
3701 PRINT "**(OR ELSE SUBROUTINE HAS FAILED)**"
3710 STOP
3720 REM end Chol. Decomp. } } }
(replace existing lines 3250-3680)

omit 3762

8180 W=X(J)
8190 P5=SQR(W/(N-1)) \ PRINT TAB((J-I0)*9);
(replace existing lines 8180-8190)

9999 END
```

Figure 10.7. (*Continued*)

A demonstration run is given in Section 2.6. When the program is first entered, do a test run, using the same data as in Section 2.6 and checking results at each point against those given in that section. If simplification is necessary in order to accommodate the program, then provision for interchanging columns of the model matrix may be deleted. Delete lines 510–520, 580, 620–710, and 4250–4680.

Unless memory is very limited, the number (N8) of rows of data which will be stored, and the number of columns allowed (controlled by Q8), should be increased by making the appropriate changes in lines 30–100. Line 3590 (c.f. lines 210, 3470) checks whether, to within a specified *tolerance*, each new variate is a linear combination of earlier variates. Variates which fail the test will not be used as explanatory variates. To change the tolerance, make the required change in line 210.

Weights w_i, one for each row, may be specified. The matrix of weights is then taken as $W = \text{diag}[w_i]$. It is assumed that $\text{var}[y_i] = cw_i^{-1}$. The error mean square estimates c.

The leverage associated with each observation is available from the calculations, but the program as it stands does not print out this information. See the remark in line 9141.

10.8. MULTIPLE REGRESSION—USING PLANAR ROTATIONS

This program (Fig. 10.8) uses planar rotations to reduce the model matrix to upper triangular form. Instructions for running it are essentially the same as for the version of the program which is described in Section 10.7. Figure 10.9 lists the variable and array names used, and gives details of sections of the code which are called as subroutines.

Only if a diagonal element of the Cholesky upper triangular matrix happens to be exactly zero will automatic action be taken to exclude from use an explanatory variate which is exactly or nearly a linear combination of earlier variates. In other cases it is necessary to use the output information (squared multiple correlations, diagonal elements of the Cholesky upper triangular matrix, and diagonal elements of the variance-covariance matrix) to check for this. Variates can then be omitted or transposed as necessary.

Bear in mind that it is possible to specify an explanatory variate as the current dependent variate, and in this way get detailed information on its dependence on earlier explanatory variates.

10.9. THE SINGULAR VALUE DECOMPOSITION

Given a real matrix Z, the singular value decomposition determines orthogonal matrices U and V and a diagonal matrix D, chosen to have all diagonal elements positive, such that $U'ZV = D$. The diagonal elements d_i of D are the singular values. The columns of V are eigenvectors of $Z'Z$ and the squares of the singular values d_i are eigenvalues of $Z'Z$. If Z is symmetric, then the eigenvectors are also those of Z, and the singular values d_i are the absolute values of the eigenvalues of Z itself.

The algorithm implemented in Fig. 10.10 is chosen on account of its simplicity. It is suitable for matrices Z with no more than about 15 rows. The algorithm works with Z'; one has $V'Z'U = D$. Calculations are carried out in place, and Z' is overwritten with those columns of V that correspond to nonzero singular values. The algorithm is designed for applications (such as those in Chapter 6) where it is not necessary to form U explicitly.

```
5 REM FIG. 1Ø.8
1Ø REM ***************************************************
11 REM * REGRESSION CALCULATIONS - USE PLANAR ROTATIONS *
12 REM * TO REDUCE DATA MATRIX TO UPPER TRIANGULAR FORM.*
13 REM ***************************************************
2Ø IØ=Ø  \ REM ARRAY ELEMENTS ARE STORED STARTING AT IØ.
3Ø N8=2Ø \ REM AT MOST N8 ROWS OF DATA WILL BE STORED.
4Ø Q8=1Ø \ REM ARRAY ELEMENTS RUN FROM IØ TO AT MOST Q8.
5Ø DIM XØ(2Ø,1Ø) \ REM DATA MATRIX
51 REM INSTEAD OF STORING DATA IN XØ, ROWS OF DATA MAY
52 REM ALTERNATIVELY BE WRITTEN OFF TO SECONDARY STORAGE,
53 REM AND RECOVERED FROM THERE WHEN REQUIRED FOR
54 REM CALCULATION OF RESIDUALS.
55 REM SEE LINES 241Ø-247Ø.
6Ø DIM X(1Ø)     \ REM TEMPORARY STORAGE OF ROW OF DATA
7Ø DIM U(1Ø,1Ø) \ REM UPPER TRIANGULAR MATRIX
8Ø DIM B(1Ø)     \ REM REGRESSION COEFFICIENTS
9Ø DIM V2(1Ø)    \ REM DIAGONAL ELEMENTS OF INVERSE OF X'X
1ØØ DIM N$(1Ø)   \ REM VARIATE NAMES
11Ø REM ****************************************************
111 REM THE CONSTANT TERM, IF PRESENT, IS TAKEN AS VARIATE Ø
112 REM IN COLUMN IØ.  IN THIS CASE MØ=IØ.
113 REM IF THE CONSTANT TERM IS OMITTED, THEN MØ=IØ-1.
12Ø REM ****************************************************
121 REM THE I TH VARIATE OCCUPIES COLUMN MØ+I.
122 REM THE FINAL VARIATE IS THE Q TH, OCCUPYING COLUMN Q7.
123 REM Q9=Q7+W9, WHERE W9=Ø (EQUAL WTS), OR W9=1 (UNEQUAL WTS).
124 REM AN IMPLEMENTATION RESTRICTION IS Q9<=Q8 (SEE LINE 4Ø).
125 REM COLUMN FOR DEPENDENT VARIATE IS QØ.
126 REM COLUMNS FOR EXPLANATORY VARIATES ARE IØ, ... , PØ.
127 REM ****************************************************
13Ø REM CALCULATE MACHINE EPSILON
14Ø EØ=1
15Ø EØ=EØ/2
16Ø E1=1+EØ
17Ø IF E1>1 THEN 15Ø
18Ø E2=SQR(EØ) \ EØ=EØ*2
19Ø REM
25Ø REM ****************************
251 REM * PREPARE FOR ENTRY OF DATA *
252 REM ****************************
26Ø N=Ø \ M8=Ø \ MØ=IØ \ GOSUB 2ØØØ
27Ø REM ****************************
28Ø FOR J=IØ TO Q7
29Ø FOR I=IØ TO J \ U(I,J)=Ø \ NEXT I
3ØØ NEXT J
31Ø REM ************************************************
32Ø PRINT "ENTER VALUES ROW BY ROW, IN THE ORDER:"
33Ø LØ=MØ+1 \ L9=Q9 \ GOSUB 8Ø1Ø
34Ø PRINT \ PRINT "FINISH WITH:  EOD" \ PRINT
35Ø GOSUB 225Ø
```

Figure 10.8. A BASIC program for multiple regression, with planar rotations used to reduce the model matrix to upper triangular form. (See Section 10.8 for discussion)

```
36Ø PRINT
37Ø REM * * * * * * * * * * * * * * * * * * * * * * * * *
38Ø REM
39Ø LØ=IØ+1 \ L9=Q7
4ØØ GOSUB 8Ø1Ø \ REM PRINT VARIATE NAMES
41Ø REM NEXT PRINT MEANS, S.D.'S, & UNADJUSTED R-SQUARED'S
411 REM R-SQUARED'S GIVE A ROUGH CHECK ON LINEAR DEPENDENCE
42Ø DØ=U(IØ,IØ) \ GOSUB 811Ø
43Ø REM
44Ø REM
45Ø REM
46Ø REM * * * * * * * * * * * * * * * * * * * * * * * * *
47Ø PRINT "* CHOLESKY UPPER TRIANGULAR MATRIX *"
48Ø LØ=IØ \ L9=Q7 \ C3=Ø \ GOSUB 8Ø1Ø
49Ø U3=Ø \ GOSUB 835Ø
5ØØ PRINT "ENTER  P  TO PROCEED WITH REGRESSION CALCULATIONS"
51Ø PRINT "OTHERWISE ENTER  D (DELETE A COLUMN)"
52Ø PRINT "               OR  T (TRANSPOSE A COLUMN)"
53Ø PRINT "               OR  A (ADD OR DELETE DATA.)"
54Ø PRINT "               OR  L (LOOK AT DATA)"
55Ø PRINT "               OR  J (JUMP TO NEXT SET OF OPTIONS)."
56Ø INPUT R$ \ PRINT
57Ø IF R$="P" THEN 73Ø
58Ø IF R$="T" THEN 63Ø
59Ø IF R$="A" THEN 32Ø
6ØØ IF R$="L" THEN GOSUB 94ØØ
61Ø IF R$="J" THEN 92Ø
62Ø IF R$<>"D" THEN 5ØØ
63Ø PRINT "MOVE VARIATE NUMBER:  "; \ INPUT L \ L=L+IØ
64Ø M=Q7 \ M3=1 \ IF R$="D" THEN 67Ø
65Ø M3=Ø
66Ø PRINT "TO BECOME VARIATE NO.:"; \ INPUT M \ M=M+IØ
67Ø GOSUB 425Ø
68Ø IF L<=Q7 THEN Q7=Q7-M3
69Ø MØ=IØ-1 \ IF N$(IØ)="CONST." THEN MØ=IØ
7ØØ PRINT "NEW ORDERING AND NUMBERING OF VARIATES IS:"
71Ø LØ=IØ \ L9=Q7 \ GOSUB 8Ø1Ø
72Ø Q9=Q7+W9 \ PRINT \ GOTO 5ØØ
73Ø IF C3=Ø THEN 77Ø
74Ø PRINT "* CURRENT VERSION OF CHOLESKY UPPER TRIANGULAR MATRIX *"
75Ø L9=Q7 \ GOSUB 8Ø1Ø
76Ø U3=Ø \ GOSUB 835Ø
77Ø PRINT "* INVERSE OF UPPER TRIANGULAR MATRIX *"
78Ø GOSUB 475Ø
79Ø U3=1 \ GOSUB 835Ø
8ØØ PRINT "* CALCULATE REGRESSION COEFFICIENTS *"
81Ø PRINT "GIVE DEP. VARIATE - TAKE VAR. NO."; \ INPUT QØ
82Ø QØ=QØ+IØ
83Ø PRINT "INCLUDE EXPLAN. VAR'S UP TO NO."; \ INPUT PØ
84Ø PØ=PØ+IØ \ GOSUB 4ØØØ
85Ø FOR J=IØ TO PØ \ B(J)=X(J) \ NEXT J
86Ø GOSUB 85ØØ
```

Figure 10.8. (*Continued*) (See Section 10.8 for discussion)

```
87Ø PRINT "* VARIANCE-COVARIANCE MATRIX *"
88Ø GOSUB 8Ø1Ø
89Ø GOSUB 88ØØ
9ØØ PRINT "* TABLE OF RESIDUALS *"
91Ø GOSUB 9ØØØ
92Ø PRINT "ENTER  F (FURTHER ANALYSIS ON THE SAME DATA)"
93Ø PRINT "OR     N (NEW PROBLEM)"
94Ø PRINT "OR     Q (QUIT)"
95Ø INPUT R$
96Ø PRINT
97Ø IF R$="F" THEN 5ØØ
98Ø IF R$="N" THEN 26Ø
99Ø IF R$<>"Q" THEN 92Ø
1ØØØ STOP
1Ø1Ø GO TO 92Ø
1Ø2Ø REM
2ØØØ REM { { {  Preliminaries to Entry of Data Matrix
2Ø1Ø PRINT "GIVE NO. OF VAR'S"; \ INPUT Q
2Ø2Ø Q7=Q+IØ \ IF Q7>Q8 THEN 215Ø
2Ø3Ø FOR I=1 TO Q
2Ø4Ø N$(I+IØ)="VØ"+STR$(I)
2Ø5Ø NEXT I
2Ø6Ø PRINT
2Ø7Ø PRINT "DO YOU WISH TO NAME THE VAR'S (Y OR N)"; \ INPUT R$
2Ø8Ø IF R$<>"Y" THEN 214Ø
2Ø9Ø PRINT "GIVE NAMES (UP TO 8 CHAR'S)"
21ØØ FOR I=IØ+1 TO Q7
211Ø PRINT I; \ INPUT V$ \ N$(I)=V$
212Ø NEXT I
213Ø PRINT
214Ø N$(IØ)="CONST." \ PRINT
215Ø PRINT "DO ALL ROWS HAVE EQUAL WEIGHT (Y OR N)"; \ INPUT R$
216Ø W9=Ø \ IF R$<>"Y" THEN W9=1
217Ø Q9=Q7+W9 \ PRINT \ IF W9=1 THEN N$(Q9)="*WEIGHT*"
218Ø IF Q9<=Q8 THEN 222Ø
219Ø PRINT "ARRAY BOUNDS & Q8 AT LINES 4Ø-1ØØ ARE SET TOO SMALL"
22ØØ PRINT "SET Q8 ETC. TO ";Q9
221Ø STOP
222Ø RETURN \ REM end Preliminaries } } }
223Ø REM
225Ø REM { { { Get  Data One Row at a Time & Update Upper
2251 REM Triangle of matrix U.
226Ø I=MØ \ X(IØ)=1 \ D9=1
227Ø IF I>MØ THEN PRINT "..";
228Ø GOSUB 96ØØ \ REM - GET NEW ROW OF DATA
229Ø IF I>MØ THEN 234Ø
23ØØ IF V$="EOD" THEN RETURN
231Ø IF V$="DEL" THEN  GOSUB 251Ø
232Ø IF D9=-1 THEN 226Ø
233Ø PRINT V$;" IS NOT A LEGAL INPUT STRING" \ GO TO 226Ø
234Ø IF I<Q9 THEN 227Ø
235Ø IF M8=N THEN M8=M8+1
```

Figure 10.8. (*Continued*) (See Section 10.8 for discussion)

```
2360 N=N+1
2370 IF N>M8 THEN 2400
2380 IF N=N8+1 THEN GOSUB 2430
2390 FOR I=I0 TO Q9 \ X0(N,I)=X(I) \ NEXT I
2391 REM STORE VALUES IN X0(,) - IF THERE IS ROOM!
2400 GOSUB 3250
2410 GO TO 2260
2420 REM
2430 PRINT "RESIDUALS WILL BE GIVEN FOR ROWS 1 TO";N8;" ONLY."
2440 PRINT "ALL ROWS OF THE ARRAY X0(,) ARE NOW USED."
2450 M8=N8
2460 RETURN
2461 REM end Get Data Row by Row . . . } } }
2470 REM
2500 REM { { { Delete Row of Data
2510 PRINT "LAST ENTERED ROW WAS NO.";N
2520 PRINT "INPUT NO. OF ROW TO BE DELETED" \ INPUT K0
2530 D9=-1
2540 IF (K0-1)*(N-K0)<0 THEN 2750
2550 IF K0<=M8 THEN  GOSUB 2790
2560 IF K0>M8 THEN  GOSUB 2820
2570 PRINT  \ PRINT "ROW OF VALUES TO BE DELETED IS:"
2580 FOR I=M0+1 TO Q9 \ PRINT X(I); " "; \ NEXT I
2590 PRINT  \ PRINT "PROCEED?  RESPOND WITH Y"
2600 PRINT "(N => RESUME DATA ENTRY IMMEDIATELY.)"
2610 INPUT R$ \ IF R$="N" THEN 2760
2620 IF R$<>"Y" THEN 2590
2630 N=N-1
2640 IF K0<=M8 THEN M8=M8-1
2650 IF K0>M8 THEN 2690
2660 FOR I=K0 TO M8
2670 FOR J=I0 TO Q9 \ X0(I,J)=X0(I+1,J) \ NEXT J
2680 NEXT I
2690 IF W9=0 THEN 2710
2700 W5=SQR(X(Q9)) \ FOR J=I0 TO Q7 \ X(J)=W5*X(J) \ NEXT J
2710 L9=Q7 \ GOSUB 3750
2720 GOSUB 3500
2730 PRINT  \ PRINT "NOW RESUME ENTRY OF DATA"
2740 RETURN \ REM } } }
2750 PRINT "ILLEGAL ROW NUMBER"
2760 PRINT "NO ACTION TAKEN - RESUME ENTRY OF DATA"
2770 RETURN \ REM end Delete Row of Data } } }
2780 REM
2790 REM GET DATA FROM X0(,)
2800 FOR J=I0 TO Q9 \ X(J)=X0(K0,J) \ NEXT J
2810 RETURN
2820 PRINT "INPUT ROW OF DATA VALUES FROM KEYBOARD:"
2830 IF W9=1 THEN PRINT "INPUT WEIGHT AS LAST VALUE IN ROW"
2840 IF I>M0 THEN PRINT "..";
2850 GOSUB 9600
2860 IF I<Q9 THEN 2840
2870 RETURN
```

Figure 10.8. (*Continued*) (See Section 10.8 for discussion)

```
288Ø REM
325Ø REM { { { Rotate Row of Data into Upper Tri. Scheme
326Ø IF W9=Ø THEN 33ØØ
327Ø W5=SQR(X(Q9))
328Ø FOR I=IØ TO Q7 \ X(I)=W5*X(I) \ NEXT I
329Ø REM
33ØØ FOR I=IØ TO Q7
331Ø D1=U(I,I) \ D2=X(I)
332Ø DØ=ABS(D1) \ IF DØ<ABS(D2) THEN DØ=ABS(D2)
333Ø IF DØ=Ø THEN 345Ø
334Ø D1=D1/DØ \ D2=D2/DØ \ IF ABS(D2)<E2 THEN 345Ø
335Ø IF ABS(D1)<E2 THEN D1=Ø
336Ø REM E2 IS SUCH THAT 1+E2^2 IS CALCULATED AS 1.Ø
337Ø D=SQR(D1^2+D2^2) \ U(I,I)=D*DØ
338Ø CØ=D1/D \ SØ=D2/D
339Ø IF I=Q7 THEN 345Ø
34ØØ I1=I+1
341Ø FOR J=I1 TO Q7
342Ø D1=U(I,J) \ D2=X(J)
343Ø U(I,J)=D1*CØ+D2*SØ \ X(J)=-D1*SØ+D2*CØ
344Ø NEXT J
345Ø NEXT I
346Ø REM AT THIS POINT D2 IS THE CURRENT GIVENS RESIDUAL
347Ø RETURN \ REM end Rotate Row into Upper Tri. Scheme } } }
348Ø REM
35ØØ REM { { { Rotate Row out of Upper Tri. Scheme
351Ø IF V>=1 THEN W=Ø
3511 REM V MAY SLIGHTLY EXCEED 1.Ø BECAUSE OF MACHINE ERROR.
3512 REM WILL LEAD TO W>1.Ø AT LINE 366Ø .
352Ø IF V<1 THEN W=SQR(1-V)
353Ø FOR I=IØ TO Q7
354Ø M=Q7-I+IØ
355Ø D1=X(M) \ D2=W \ DØ=ABS(D2) \ IF DØ<ABS(D1) THEN DØ=ABS(D1)
356Ø IF DØ=Ø THEN 365Ø
357Ø D1=D1/DØ \ D2=D2/DØ \ IF ABS(D1)<E2 THEN 365Ø
358Ø IF ABS(D2)<E2 THEN D2=Ø
359Ø D=SQR(D1^2+D2^2) \ W=D*DØ
36ØØ CØ=D2/D \ SØ=D1/D
361Ø FOR J=M TO Q7
362Ø D1=U(M,J) \ D2=X(J) \ IF J=M THEN D2=Ø
363Ø U(M,J)=D1*CØ-D2*SØ \ X(J)=D1*SØ+D2*CØ
364Ø NEXT J
365Ø NEXT I
366Ø PRINT "THE FOLLOWING SHOULD BE 1.Ø: ";W
367Ø RETURN
3671 REM end Rotate Row out of Upper Tri. Scheme } } }
```

Figure 10.8. (*Continued*) (See Section 10.8 for discussion)

```
3680 REM
3750 REM { { { Soln. of Lower Tri. System
3760 REM REQUIRED
3761 REM (I) for calc. of S.E.'s of fitted values (line 9130)
3762 REM (II) for row deletion using lines 2500-2740
3770 V=0
3780 FOR L=I0 TO L9
3790 W=0 \ L1=L-1
3800 IF L=I0 THEN 3820
3810 FOR J=I0 TO L1 \ W=W+U(J,L)*X(J) \ NEXT J
3820 D=U(L,L) \ IF D=0 THEN 3840
3830 D=1/D
3840 X(L)=(X(L)-W)*D
3850 V=V+X(L)^2
3860 NEXT L
3870 RETURN \ REM end Soln. of Lower Tri. System } } }
3880 REM
4000 REM { { { Solve Upper Tri. System of Equns.
4010 FOR J=I0 TO Q0 \ X(J)=U(J,Q0) \ NEXT J
4020 FOR L=I0 TO P0
4030 M=P0+I0-L \ M1=M+1 \ W=0
4040 IF M=P0 THEN 4060
4050 FOR J=M1 TO P0 \ W=W+U(M,J)*X(J) \ NEXT J
4060 D=U(M,M) \ IF D=0 THEN 4080
4070 D=1/D
4080 X(M)=(X(M)-W)*D
4090 NEXT L
4100 RETURN \ REM end Solve Upper Tri. System of Equns. } } }
4110 REM
4250 REM { { { transpose col. L to become col. M.
4260 IF L>Q7 THEN 4460
4270 IF M>Q7 THEN 4460
4280 K0=1 \ IF L>M THEN K0=-1
4290 IF L=M THEN 4360
4300 C3=1 \ L1=L \ M1=M
4310 IF L<M THEN M1=M-1
4320 IF L>M THEN L1=L-1
4330 FOR K=L1 TO M1 STEP K0
4340 GOSUB 4500
4350 NEXT K
4360 M2=M+W9*M3 \ M1=M2-K0 \ IF L=M2 THEN 4470
4370 V$=N$(L)
4380 FOR K=L TO M1 STEP K0 \ N$(K)=N$(K+K0) \ NEXT K
4390 N$(M2)=V$
4400 FOR I=1 TO M8
4410 X1=X0(I,L)
4420 FOR K=L TO M1 STEP K0 \ X0(I,K)=X0(I,K+K0) \ NEXT K
4430 X0(I,M2)=X1
4440 NEXT I
4450 RETURN
4460 PRINT "REQUIRE BOTH COLUMN NUMBERS <= ";Q7-I0
4470 PRINT "NO ACTION TAKEN"
4480 RETURN \ REM end transpose col L to become col. M } } }
```

Figure 10.8. (*Continued*) (See Section 10.8 for discussion)

```
4490 REM
4500 REM { { { interchange cols. K, K+1.  Called from 4320
4510 FOR I=I0 TO K
4520 T=U(I,K+1) \ U(I,K+1)=U(I,K) \ U(I,K)=T
4530 NEXT I
4540 D1=T \ D2=U(K+1,K+1)
4550 D0=ABS(D1) \ IF D0<ABS(D2) THEN D0=ABS(D2)
4560 IF D0=0 THEN 4660
4570 D1=D1/D0 \ D2=D2/D0 \ IF ABS(D2)<E2 THEN 4660
4580 IF ABS(D1)<E2 THEN D1=0
4590 U(K+1,K+1)=0
4600 D=SQR(D1*D1+D2*D2) \ U(K,K)=D*D0
4610 C0=D1/D \ S0=D2/D
4620 FOR J=K+1 TO Q7
4630 D1=U(K,J) \ D2=U(K+1,J)
4640 U(K,J)=C0*D1+S0*D2 \ U(K+1,J)=S0*D1-C0*D2
4650 NEXT J
4660 RETURN
4670 REM end interchange cols K, K+1 } } }
4680 REM
4690 REM
4750 REM { { { Invert Upper Tri., Store in Lower Tri.
4760 FOR I=I0 TO Q7 \ IF I=I0 THEN 4870
4770 I1=I-1 \ D=U(I,I) \ IF D=0 THEN 4790
4780 D=1/D
4790 FOR J=I0 TO I1 \ W=0
4800 FOR K=J TO I1
4810 D1=U(K,J) \ IF K>J THEN 4830
4820 IF D1<>0 THEN D1=1/D1
4830 W=W+D1*U(K,I)
4840 NEXT K
4850 U(I,J)=-W*D
4860 NEXT J
4870 NEXT I
4880 RETURN \ REM end Invert Upper Tri. . . . } } }
4890 REM
8000 REM { { { Print Variate Names
8010 PRINT "VAR. NO.";
8020 FOR I=L0 TO L9 \ PRINT TAB((I-L0+1)*9);I-I0; \ NEXT I
8030 PRINT  \ PRINT "    NAME";
8040 FOR I=L0 TO L9
8050 V$=N$(I) \ PRINT TAB((I-L0+1)*9+1);V$;
8060 NEXT I
8070 PRINT
8080 RETURN
8090 REM end Print Variate Names } } }
8091 REM
8100 REM { { { Print Means, S. D.'s, and R-squared values
8110 PRINT "MEANS:  ";
8120 FOR J=I0+1 TO Q7
8130 P5=U(I0,J)/D0 \ PRINT TAB((J-I0)*9); \ GOSUB 9500
8140 NEXT J
8150 PRINT
```

Figure 10.8. (*Continued*) (See Section 10.8 for discussion)

```
8160 PRINT "S.D.'S: ";
8170 FOR J=I0+1 TO Q7
8180 W=0 \ FOR I=I0+1 TO J \ W=W+U(I,J)^2 \ NEXT I
8190 X(J)=W \ P5=SQR(W/(N-1)) \ PRINT TAB((J-I0)*9);
8200 GOSUB 9500
8210 NEXT J
8220 PRINT  \ PRINT
8230 PRINT "R-SQUARED: ----";
8240 FOR J=I0+2 TO Q7
8250 P5=1-U(J,J)^2/X(J) \ PRINT TAB((J-I0)*9); \ GOSUB 9500
8260 NEXT J
8270 PRINT
8280 PRINT "(THIS MEASURES DEPENDENCE ON EARLIER VARIATES)"
8290 PRINT \ PRINT
8300 RETURN
8310 REM end Print Means, S.D.'s, and R-squared values } } }
8320 REM
8350 REM { { { Print Upper (U3=0), Or Lower (U3=1),
8360 REM Triangle of Square Matrix
8370 FOR I=L0 TO L9
8380 PRINT I-I0;
8390 FOR J=I TO L9
8400 D=U(I,J) \ IF U3=0 THEN 8430
8410 D=U(J,I) \ IF I<>J THEN 8430
8420 IF D<>0 THEN D=1/D
8421 REM DIAG ELS OF LOWER TRI ARE INVERTED
8430 P5=D \ PRINT TAB((J-L0+1)*9); \ GOSUB 9500
8440 NEXT J
8450 PRINT
8460 NEXT I
8470 PRINT  \ PRINT
8480 RETURN \ REM end Print Upper or Lower Tri. . . . } } }
8490 REM
8500 REM { { { Print Coeffs & S.E.'s Following Regression
8501 PRINT  \ PRINT "DEPENDENT VAR. IS < ";N$(Q0);" >.";
8510 S2=0 \ FOR I=P0+1 TO Q0 \ S2=S2+U(I,Q0)^2 \ NEXT I
8520 F=N-P0+I0-1 \ IF F<=0 THEN 8540
8530 S2=S2/F \ PRINT "    ERROR VAR. ="; \ P5=S2 \ GOSUB 9500
8540 PRINT " (DF ="; F;")"
8550 PRINT
8560 L0=I0 \ L9=P0 \ GOSUB 8010
8570 PRINT "COEFFS.:";
8580 FOR J=I0 TO P0
8590 P5=B(J) \ PRINT TAB((J-I0+1)*9); \ GOSUB 9500
8600 NEXT J
8610 PRINT \ PRINT "S.E.'S: ";
8620 FOR J=I0 TO P0 \ W=0
8630 FOR I=J TO P0
8640 D=U(I,J) \ IF D=0 THEN 8670
8650 IF I=J THEN D=1.0/D
8660 W=W+D*D
8670 NEXT I
```

Figure 10.8. (*Continued*) (See Section 10.8 for discussion)

```
868Ø V2(J)=W \ IF F<=Ø THEN 87ØØ
869Ø P5=SQR(W*S2) \ PRINT TAB((J-IØ+1)*9); \ GOSUB 95ØØ
87ØØ W=Ø \ FOR I=MØ+1 TO J \ W=W+U(I,J)^2 \ NEXT I
871Ø X(J)=W
872Ø NEXT J
873Ø PRINT
874Ø FOR J=MØ+1 TO PØ
875Ø V=X(J)*V2(J) \ IF V<1ØØØ THEN 877Ø
876Ø PRINT "PLEASE CHECK VARIATE";J;" FOR INTERDEPENDENCE";
8761 PRINT " WITH OTHER EXPLANATORY VARIATES."
877Ø NEXT J
878Ø PRINT \ PRINT
879Ø RETURN \ REM end Print Coeffs. & S.E.'s } } }
88ØØ REM { { { Print Var-Cov Matrix
88Ø1 REM ELEMENTS OF MATRIX ARE NOT STORED
881Ø IF F<=Ø THEN PRINT "  VAR'S & COV'S ARE PROPORTIONAL TO:"
882Ø L9=PØ \ IF F<=Ø THEN S2=1.Ø
883Ø FOR I=IØ TO PØ
884Ø PRINT I-IØ;
885Ø FOR J=I TO PØ
886Ø W=Ø
887Ø FOR K=J TO PØ
888Ø D1=U(K,I) \ IF K=I THEN IF D1<>Ø THEN D1=1/D1
889Ø D2=U(K,J) \ IF K=J THEN IF D2<>Ø THEN D2=1/D2
89ØØ W=W+D1*D2
891Ø NEXT K
892Ø P5=W*S2
893Ø PRINT TAB((J-IØ+1)*9); \ GOSUB 95ØØ
894Ø NEXT J
895Ø PRINT
896Ø NEXT I
897Ø PRINT  \ PRINT
898Ø RETURN \ REM end Print Var-Cov. Matrix } } }
899Ø REM
9ØØØ REM { { { Residuals from Regression Model
9Ø1Ø PRINT "ROW NO.";
9Ø2Ø PRINT "  OBSERVED  RESIDUAL  EXPECTED   SE(EXP)   WEIGHT"
9Ø3Ø IF F<=Ø THEN PRINT TAB(4Ø); "PROP.TO"
9Ø4Ø L9=PØ
9Ø5Ø FOR K=1 TO M8
9Ø6Ø FOR I=IØ TO Q9 \ X(I)=XØ(K,I) \ NEXT I
9Ø7Ø YØ=Ø \ Y=X(QØ)
9Ø8Ø FOR I=IØ TO PØ \ YØ=YØ+X(I)*B(I) \ NEXT I
9Ø9Ø RØ=Y-YØ
91ØØ WØ=1
911Ø IF W9=Ø THEN 913Ø
912Ø WØ=X(Q9)
913Ø GOSUB 375Ø
914Ø SØ=SQR(V*S2)
9141 REM LEVERAGE OF DATA POINT IS  V*WØ
915Ø REM
916Ø PRINT K;TAB(9);Y;
```

Figure 10.8. (*Continued*) (See Section 10.8 for discussion)

```
9170 P5=RØ \ PRINT TAB(19); \ GOSUB 95ØØ
9180 P5=YØ \ PRINT TAB(29); \ GOSUB 95ØØ
9190 P5=SØ \ PRINT TAB(39); \ GOSUB 95ØØ
92ØØ PRINT TAB(49);WØ
921Ø NEXT K
922Ø PRINT
923Ø RETURN
9231 REM end Residuals . . . } } }
924Ø REM
94ØØ REM { { { print back data
941Ø PRINT \ PRINT "GIVE NUMBERS OF FIRST & LAST VARIATES"
942Ø INPUT L,M \ LØ=L+IØ \ L9=M+IØ \ IF L>M THEN 949Ø
943Ø IF (Q9-L9)*(LØ-IØ)<Ø THEN 949Ø
944Ø GOSUB 8Ø1Ø
945Ø FOR I=1 TO M8
946Ø FOR J=LØ TO L9 \ PRINT TAB((J-L+1)*9); XØ(I,J); \ NEXT J
947Ø PRINT
948Ø NEXT I
949Ø PRINT \ RETURN
9491 REM end print back data } } }
9492 REM
95ØØ REM { { { print rounded to five significant digits
951Ø IF ABS(P5)<1E-2Ø THEN P5=Ø
952Ø IF P5=Ø THEN 955Ø
953Ø E$="1E"+STR$(4-INT(Ø.4342945*(LOG(ABS(P5))+.ØØØØØ5)))
954Ø E5=VAL(E$) \ P5=INT(P5*E5+.5)/E5
955Ø PRINT P5;
956Ø RETURN
9571 REM end print rounded to five significant digits } } }
958Ø REM
96ØØ REM { { { Decode L$ into a sequence of nos. *****
961Ø REM        Store in X(IØ+1,...,I).
962Ø L$="" \ V$="" \ INPUT L$
963Ø L1=LEN(L$)+1
964Ø K1=1
965Ø FOR K=1 TO L1
966Ø K2=K-1
967Ø IF K=L1 THEN 97ØØ
968Ø K$=SEG$(L$,K,K)
969Ø IF K$<>" " THEN 978Ø
97ØØ IF K=K1 THEN 977Ø
971Ø V$=SEG$(L$,K1,K2)
972Ø IF I>MØ THEN 975Ø
973Ø IF V$="EOD" THEN RETURN
974Ø IF V$="DEL" THEN RETURN
975Ø I=I+1
976Ø X(I)=VAL(V$)
977Ø K1=K+1
978Ø NEXT K
979Ø RETURN
98ØØ REM end Decode L$ . . . } } }
98Ø1 REM . . . . . . . . . . . . . . . . . . . . . .
9999 END
```

Figure 10.8. (*Continued*) (See Section 10.8 for discussion)

SUMS OF SQUARES AND PRODUCTS OF PARTIAL RESIDUALS

The following additional statements allow the printing of matrices of sums of squares and products of partial residuals. Special cases are
(i) The SSP matrix (hold constant variates up to -1)
(ii) The CSSP matrix (hold constant variates up to 0)

```
535 PRINT "          OR  S (TO PRINT MATRIX OF PARTIAL SSP'S)"

595 IF R$="S" THEN 1800

1800 PRINT "* SSP MATRIX OF PARTIAL RESIDUALS *"
1810 PRINT "HOLD VARIATES CONSTANT UP TO NO.";
1820 INPUT L \ L0=L+I0+1
1830 PRINT "SHOW SSP'S FOR VARIATES UP TO NO.";
1840 INPUT M \ L9=M+I0 \ IF L9>Q7 THEN 1880
1850 IF L9<L0 THEN 1880
1860 GOSUB 8010
1870 GOSUB 7800
1880 PRINT \ GOTO 500

7800 REM { { { Print Matrix of Partial SSP's
7810 FOR I=L0 TO L9
7820 PRINT I-L0;
7830 FOR J=I TO L9
7840 W=0 \ FOR K=L0 TO I \ W=W+U(K,I)*U(K,J) \ NEXT K
7850 PRINT TAB((J-I0)*9); \ P5=W \ GOSUB 9500
7860 NEXT J
7870 PRINT
7880 NEXT I
7890 PRINT  \ PRINT
7900 RETURN \ REM end Print Matrix of Partial SSP's } } }
```

CALCULATION OF PARTIAL RESIDUALS

The following additional statements calculate the partial residuals needed for partial regression plots:

```
65 DIM Y1(10) \ REM USE FOR STORING PARTIAL RESIDUALS

9022 FOR J=I0 TO P0
9024 IF V2(J)<>0 THEN V2(J)=1.0/V2(J)
9026 NEXT J
9028 I1=I0 \ IF N$(I0)="CONST." THEN I1=I0+1

9145 GOSUB 9250
9146 REM AT THIS POINT PRINT OUT, OR STORE
9147 REM X(I),Y1(I) (I=M0+1,... ,P0)

9250 REM { { { Calculation of Partial Residuals
9260 GOSUB 4020
9270 FOR I=I1 TO P0 \ X(I)=X(I)*V2(I) \ Y1(I)=R0+B(I)*X(I)
9280 NEXT I
9290 RETURN
9300 REM end Calculation of Partial Residuals } } }
```

Figure 10.8. (*Continued*) (See Section 10.8 for discussion)

```
>>>>>  FIRST OCCURRENCES OF VARIABLE & ARRAY NAMES  <<<<<

B()- 8Ø     CØ- 338Ø    C3- 48Ø
D- 337Ø     DØ- 42Ø     D1- 331Ø    D2- 331Ø    D9- 226Ø
EØ- 14Ø     E1- 16Ø     E2- 18Ø     E5- 954Ø    E$- 953Ø
F- 852Ø     I- 29Ø      IØ- 2Ø      I1- 34ØØ    J- 28Ø
K- 433Ø     K$- 968Ø    KØ- 252Ø    K1- 964Ø    K2- 966Ø
L- 63Ø      L$- 962Ø    LØ- 39Ø     L1- 379Ø    L9- 39Ø
M- 64Ø      MØ- 26Ø     M1- 4Ø3Ø    M2- 436Ø    M3- 64Ø     M8- 26Ø
N- 26Ø      N$()- 1ØØ   N8- 3Ø      PØ- 82Ø     P5- 813Ø
Q- 2Ø1Ø     QØ- 8ØØ     Q7- 28Ø     Q8- 4Ø      Q9- 71Ø
RØ- 9Ø9Ø    R$- 56Ø     SØ- 338Ø    S2- 851Ø    T- 452Ø
U(,)- 7Ø    U3- 49Ø     V- 351Ø     V2()-9Ø     V$- 211Ø
W- 351Ø     WØ- 91ØØ    W5- 27ØØ    W9- 72Ø
X()- 6Ø     XØ(,)- 5Ø   X1- 441Ø    Y- 9Ø7Ø     YØ- 9Ø7Ø
```

```
>>>>>   SUBROUTINES   <<<<<<
```

START	END	CALLED FROM	PURPOSE, GOSUBs, & VARIABLES OR ARRAYS WHICH MUST BE PASSED AS PARAMETERS
2ØØØ	222Ø	26Ø	Preliminaries to entry of data matrix. Pass as parameters: IØ,N$(),Q7,Q8,Q9,W9
225Ø	241Ø	37Ø	Input data. GOSUBs to 96ØØ, 25ØØ, 243Ø 325Ø. E2,IØ,MØ,M8,N,Q7,Q9,XØ(,),W9,U(,),X()
243Ø	246Ø	238Ø	Print out message when XØ(,) can hold no more data. M8,N8
25ØØ	277Ø	231Ø	Delete row of data. GOSUBs to 279Ø, 282Ø, 375Ø, and 35ØØ. D9,E2,IØ,MØ,M8,N,Q7,Q9,U(,),W9,X(),XØ(,)
279Ø	281Ø	255Ø	Get data from array XØ(,) KØ,IØ,Q9,X(),XØ(,)
282Ø	287Ø	256Ø	Input data from keyboard. GOSUB 96ØØ. MØ,Q9,W9,X()
325Ø	347Ø	24ØØ	Rotate row of data in. E2,IØ,N,Q7,Q9,U(,),W9,X()
35ØØ	367Ø	272Ø	Rotate row of data out. E2,IØ,Q7,U(,),V,X()
375Ø	387Ø	271Ø 913Ø	Solve lower triangular system. IØ,L9,U(,),V,X()

Figure 10.9. Name and subroutine reference for Fig. 10.8: (*a*) variable and array names used, with the line numbers at which they first appear, and (*b*) details of sections of the code called as subroutines, together with any names which are not local.

The method is described in Nash (1975) and modifies a method due to Kaiser. The BASIC program is based in part on a program due to Nash and Lefkovitch.

If Z is a singular symmetric matrix, then the singular value decomposition of Z stops short of finding all the eigenvectors. A suitable device is to find the eigenvectors and eigenvalues of $Z + cI$, where c is some positive

4ØØØ	41ØØ	84Ø	Solve upper triangular system. X(),IØ,PØ,QØ,U(,)
425Ø	449Ø	67Ø	Transpose column L to become column M. GOSUB 45ØØ. E2,IØ,L,M,M3,M8,N$(),Q7,U(,),W9,XØ(,)
45ØØ	466Ø	434Ø	Interchange two columns. E2,IØ,K,Q7,U(,)
475Ø	487Ø	78Ø	Invert upper triangular matrix. IØ,Q7,U(,)
8ØØØ	8Ø8Ø	33Ø, 4ØØ 48Ø, 71Ø 75Ø, 88Ø 856Ø	Print variate names. IØ,LØ,L9,N$()
81ØØ	831Ø	42Ø	Print means, S.D.'s, & R-squared's. GOSUB 95ØØ. DØ,IØ,N,Q7,U(,),X()
835Ø	848Ø	49Ø, 76Ø 79Ø	Print upper or lower triangle of square matrix. GOSUB 95ØØ. LØ,U3,L9,U(,)
85ØØ	879Ø	86Ø	Print coefficients and S.E.'s GOSUBs to 8Ø1Ø, 95ØØ. B(),IØ,MØ,N,N$(),QØ,PØ,S2,U(,),V2()
88ØØ	898Ø	89Ø	Print variance-covariance matrix. GOSUB 95ØØ. IØ,PØ,S2,U(,)
9ØØØ	923Ø	91Ø	Calculate and print residuals. GOSUBs to 375Ø, 95ØØ. B(),IØ,M8,N,PØ,QØ,Q9,S2,U(,),V,W9,X(),XØ(,)
94ØØ	949Ø	6ØØ	IØ,M8,XØ(,)
95ØØ	955Ø	813Ø, 82ØØ 829Ø, 843Ø 853Ø, 859Ø 869Ø, 893Ø 917Ø-919Ø	Print real value, rounded to 5 significant digits. P5
96ØØ	979Ø	228Ø 284Ø	Decode line of input data. MØ,I,X(),V$

Figure 10.9. (*Continued*)

constant. The eigenvectors are the same as those of Z, whereas the eigenvalues are increased, in each case, by an amount c.

10.10. THE SOLUTION OF A NONLINEAR EQUATION IN ONE UNKNOWN

This program (Fig. 10.11) is essentially a BASIC version of the Algol and FORTRAN programs in R. P. Brent's *Algorithms for Minimization without Derivatives* (Prentice-Hall, © 1973). Another FORTRAN version, which was the immediate antecedent of this program, appears in Forsythe et al. (1977).

```
5 REM FIG. 10.10
10 REM Calculation of Singular values and vectors of
11 REM an arbitrary real matrix.
12 REM Nash (1975) - Modifies a method due to Kaiser.
13 REM Based in part on Nash and Lefkovitch (1977):
14 REM  Programs for sequentially updated principal
15 REM  components and regression by singular value
16 REM  decomposition.  Mary Nash Information Services,
17 REM  Ontario, Canada.
18 REM
20 PRINT "SINGULAR VALUE DECOMPOSITION.  THE PROGRAM"
21 PRINT " PROMPTS FOR INPUT OF A REAL MATRIX  Z' ."
22 PRINT
23 PRINT "LET  U  AND  V  TO BE ORTHOGONAL MATRICES, &"
24 PRINT "D  A DIAGONAL MATRIX,  SUCH THAT  U' Z V = D ."
25 PRINT "THE PROGRAM GIVES THE DIAGONAL ELEMENTS OF  D,"
26 PRINT "TOGETHER WITH COLUMNS OF  V  CORRESPONDING TO"
27 PRINT "NON-ZERO DIAGONAL ELEMENTS."
28 PRINT
30 DIM U(6,6),X(6)
40 REM LINES 50-110 CALCULATE MACHINE EPSILON
50 E0=1
60 FOR I=1 TO 50
70 E0=E0/2
80 E=1+E0
90 IF E=1 THEN 110
100 NEXT I
110 E0=E0*2
120 PRINT "MACHINE EPSILON=";E0
130 PRINT \ PRINT "GIVE DIMENSIONS OF ARRAY"; \ INPUT L,M
140 L0=1 \ M0=1
150 L2=L0+L-1 \ M2=M0+M-1
160 E2=L*E0^2
170 PRINT \ PRINT "INPUT DATA MATRIX"
180 GOSUB 9800
190 IF S9=0 THEN 260
200 PRINT "GIVE AMOUNT (>= 0) TO BE ADDED TO MATRIX DIAGONAL"
210 INPUT Z0
220 FOR I=0 TO L-1
230 U(L0+I,M0+I)=U(L0+I,M0+I)+Z0
240 NEXT I
250 REM PROCEED WITH SVD CALCULATIONS
260 GOSUB 7010
270 REM PRINT RESULTS
280 GOSUB 7700
290 STOP
300 GOTO 130
```

Figure 10.10. A BASIC program for the singular value decomposition.

10.11. FURTHER READING AND REFERENCES

Standards for statistical computer programs have improved quite remarkably since around 1975. A major stimulus was the formation in 1974 of the Committee on Evaluation of Statistical Program Packages of the American Statistical Association. See Francis et al. (1975). Yates (1966), in a paper that saw the changes resulting from the use of the computer in statistics as a

```
310 REM
7000 REM { { { SVD calculations
7010 M3=M*(M-1)/2
7020 FOR J=M0 TO M2-1
7030 FOR K=J+1 TO M2
7040 X2=0 \ H=0 \ Y2=0
7050 FOR I=L0 TO L2
7060 X0=U(I,J) \ Y0=U(I,K)
7070 H=H+X0*Y0
7080 X2=X2+X0^2
7090 Y2=Y2+Y0^2
7100 NEXT I
7110 C0=0 \ S0=1
7120 IF X2<Y2 THEN 7200
7130 C2=X2*Y2
7140 IF C2<=E2 THEN 7220
7150 IF H^2<E2*C2 THEN 7220
7160 Q=X2-Y2 \ H2=2*H
7170 R=SQR(Q^2+H2^2)
7180 C0=SQR((R+Q)/(2*R))
7190 S0=H2/(2*R*C0)
7200 GOSUB 7320
7210 GO TO 7230
7220 M3=M3-1
7230 NEXT K
7240 NEXT J
7250 PRINT "SWEEP END - # ROTATIONS = ";M3
7260 IF M3>0 THEN 7010
7270 RETURN
7280 REM end SVD calculations } } }
7290 REM
7300 REM
7310 REM { { { rotation
7320 FOR I=L0 TO L2
7330 D1=U(I,J) \ D2=U(I,K)
7340 U(I,J)=D1*C0+D2*S0
7350 U(I,K)=-D1*S0+D2*C0
7360 NEXT I
7370 RETURN
7380 REM end rotation } } }
```

Figure 10.10. (*Continued*)

revolution, expresses concerns and judgments which are as pertinent as ever. See also Muller (1970) and Milton and Nelder (1969). Francis (ed., 1979) includes a comprehensive (up to 1977) bibliography of papers on the assessment of statistical computer programs and allied topics.

Chambers (1977) discusses program design and structure and program evaluation at greater length than has been attempted here. See also the chapter on *Programming and Statistical Software* in Kennedy and Gentle (1980).

The choice of statistical packages for mention in Section 10.2 reflects the author's own work and experience. An author whose experience was in econometrics, rather than in biological statistics, would undoubtedly make a

```
7390 REM
7700 REM { { { calc. singular values and normalize vectors
7710 FOR J=M0 TO M2
7720 X2=0
7730 FOR I=L0 TO L2
7740 X2=X2+U(I,J)^2
7750 NEXT I
7790 X(J)=SQR(X2)
7800 NEXT J
7810 FOR J=M0 TO M2
7820 IF X(J)=0 THEN 7860
7830 FOR I=L0 TO L2
7840 U(I,J)=U(I,J)/X(J)
7850 NEXT I
7860 NEXT J
7870 PRINT "  SINGULAR VALUES & VECTORS"
7880 PRINT
7890 FOR J=M0 TO M2
7900 PRINT "SINGULAR VALUE (";J;")=";X(J)-Z0
7910 IF X(J)=0 THEN 7960
7920 PRINT "EIGENVECTOR:-"
7930 FOR I=L0 TO L2
7940 PRINT I,U(I,J)
7950 NEXT I
7960 PRINT
7970 NEXT J
7980 RETURN
7990 REM end calc. singular values and normalize vectors } } }
7991 REM
9800 REM { { { input data matrix
9810 S9=0
9820 IF L<>M THEN 9860
9830 PRINT "IS MATRIX SYMMETRIC (Y OR N)"
9840 INPUT A$
9850 IF A$="Y" THEN S9=1
9860 L1=L-1 \ M1=M-1 \ I1=0
9870 FOR I=0 TO L1
9880 IF S9=1 THEN I1=I
9890 FOR J=I1 TO M1
9900 PRINT "EL. AT (";I+1;",";J+1;")=";
9910 INPUT U(L0+I,M0+J)
9920 IF J=I THEN 9940
9930 IF S9=1 THEN U(L0+J,M0+I)=U(L0+I,M0+J)
9940 NEXT J
9950 PRINT
9960 NEXT I
9970 RETURN
9971 REM end input data matrix } } }
9990 END
```

Figure 10.10. (*Continued*)

different selection. An extensive list of statistical packages may be found in the two comparative reviews by Francis (1979 and 1981). The second of these volumes compares developers' assessments of strengths and weaknesses with those of users. But because different packages are designed for and attract very different groups of users, this material does not provide a basis for satisfactory comparisons between packages.

```
5  REM FIG. 10.11
10 DEF FNA(X)=X-(1/1.31)*LOG(1+1.93*X)
20 A0=.33 \ B0=1
30 PRINT "FIND ROOT IN INTERVAL (";A0;",";B0;")"
40 T=1.00000E-03
50 PRINT "INTENDED ACCURACY =";T
60 REM CHANGE LINES 10 - 50 AS REQUIRED IN ANY NEW PROBLEM
70 A=A0 \ B=B0
80 GOSUB 1060
90 PRINT
100 PRINT "ROOT = ";U
110 STOP
1000 REM { { { find root in specified interval
1010 REM BASIC VERSION COURTESY T.R.HOPKINS
1011 REM BASED ON ALGOL & FORTRAN PROGRAMS IN BRENT (1973).
1012 REM SEE ALSO THE FORTRAN PROGRAM IN
1013 REM                     FORSYTHE, MALCOLM & MOLER (1977).
1020 REM
1030 REM
1040 REM ROOT IS ASSUMED TO LIE IN INTERVAL (A,B)
1050 REM TEST IF ROOT CONTAINED IN INTERVAL
1060 F1=FNA(A)
1070 F2=FNA(B)
1080 IF F1*F2<=0 THEN 1120
1090 PRINT "GIVEN INTERVAL DOES NOT CONTAIN ROOT"
1100 PRINT "OR CONTAINS AN EVEN NUMBER OF ROOTS"
1110 STOP
1120 REM T SETS DESIRED ACCURACY OF RESULT
1130 T=ABS(T)
1140 REM COMPUTE RELATIVE MACHINE PRECISION
1150 E1=1
1160 E1=E1/2
1170 IF 1+E1>1 THEN 1160
1180 REM FIRST STEP
1190 C=A
1200 F3=F1
1210 D=B-A
1220 E=D
1230 IF ABS(F3)>=ABS(F2) THEN 1310
1240 A=B
1250 B=C
1260 C=A
1270 F1=F2
1280 F2=F3
1290 F3=F1
```

Figure 10.11. A BASIC version of Brent's algorithm for the solution of a nonlinear equation in a single unknown.

The following are important sources of information on statistical computing:

Proceedings of the Statistical Computing Section of the American Statistical Association (published annually since 1976). (American Statistical Association, 806 15th St, N.W., Washington, D.C. 20005.)

Proceedings of Computer Science and Statistics: The 15th Annual Symposium on the Interface. The 15th annual symposium, held in March

```
1300 REM TEST FOR CONVERGENCE
1310 T1=2*E1*ABS(B)+.5*T
1320 X=.5*(C-B)
1330 IF ABS(X)<=T1 THEN 1740
1340 IF F2=0 THEN 1750
1350 REM NEED BISECTION
1360 IF ABS(E)<T1 THEN 1620
1370 IF ABS(F1)<=ABS(F2) THEN 1620
1380 REM IS QUADRATIC INTERPOLATION POSSIBLE
1390 IF A<>C THEN 1460
1400 REM LINEAR INTERPOLATION
1410 S=F2/F1
1420 P=2*X*S
1430 Q=1-S
1440 GO TO 1520
1450 REM INVERSE QUAD INT
1460 Q=F1/F3
1470 R=F2/F3
1480 S=F2/F1
1490 P=S*(2*X*Q*(Q-R)-(B-A)*(R-1))
1500 Q=(Q-1)*(R-1)*(S-1)
1510 REM ADJUST SIGNS
1520 IF P<=0 THEN 1540
1530 Q=-Q
1540 P=ABS(P)
1550 REM TEST INTERPOLATION ACCEPTABLE
1560 IF 2*P>3*X*Q-ABS(T1*Q) THEN 1620
1570 IF P>=ABS(.5*E*Q) THEN 1620
1580 E=D
1590 D=P/Q
1600 GO TO 1650
1610 REM BISECTION
1620 D=X
1630 E=D
1640 REM COMPLETE STEP
1650 A=B
1660 F1=F2
1670 IF ABS(D)<=T1 THEN 1700
1680 B=B+D
1690 GO TO 1710
1700 B=B+SGN(X)*T1
1710 F2=FNA(B)
1720 IF F2*SGN(F3)>0 THEN 1190
1730 GO TO 1230
1740 U=B
1750 RETURN
1760 REM end find root in specified interval } } }
9999 END
```

Figure 10.11. (*Continued*)

1983, was published (edited by James E. Gentle) by North-Holland, Amsterdam and New York. Details of other proceedings which were available at that time appear in the front of that volume.

Compstat: *Proceedings in Computational Statistics*. (The 4th symposium was held in 1980.) The biennial volumes are published by Physica-Verlag, Vienna.

Texts that describe the languages mentioned in Section 10.3 are the following:

Ada: Wegner (1980)
APL: Anscombe (1981)
BASIC: Alcock (1977)
FORTRAN: Balfour and Marwick (1979)
Pascal: Tiberghien (1981)
PL/I: Barnes (1979)

Epilogue

Apart from material that may be found in elementary texts in numerical analysis, or that standardly forms part of courses in computer science, a few important areas of statistical computing have received little attention. These include the special problems that arise in working with large data sets, data validation, cluster analysis and problems in pattern recognition generally, and statistical graphics.

Advances in graphics and in pattern recognition must now depend heavily on the work of computer scientists in these areas. Unless aided by human intervention at crucial points, computer programs are at present poor at recognizing patterns that, in two or three dimensions, are immediately obvious to the human eye. Success with pattern recognition problems almost certainly requires the use of machines designed to process information in parallel. The area is important to research in artificial intelligence. A successful technique with present machines, which process information serially, has been to use the computer to provide the eye with a variety of two- or three-dimensional perspectives on higher-dimensional data, in the hope that important features of the data will then be made obvious. See the discussion in Donoho et al. (1979) of the successive PRIM computer systems (from PRIM9 through to PRIMH). Substantial progress in automated pattern recognition will have important implications for practical statistical analysis.

Little has been said about such methods for assessing sampling variability as the bootstrap, aptly described by Diaconis and Efron (1983) as "computer-intensive." The implications of research into computer-intensive methods are almost as important for theoretical as for applied statistics. Efron (1979), in a paper that includes a brief account of jacknife and bootstrap methods, comments that the advent of the computer has redefined what should be regarded as simple in mathematical statistics.

APPENDIX

Major Cholesky

André-Louis Cholesky, born October 15, 1875 in Montguyon (Charente-Inférieure), entered l'Ecole Polytechnique at the age of 20 and upon graduation went into the artillery branch. Attached to the Geodesic Section of the Geographic Service in June 1905, he made himself noticed at once by his extraordinary intelligence, his great aptitude for mathematics and his inquiring mind and original ideas, which, though sometimes paradoxical, were always marked by great eloquence of expression.

It was in that same period that the revision of the French triangulation had just been decided. The plan was to continue the revision of the meridian line of Paris which was to be used as the base of a new cadastral triangulation. The problem of adjusting the grid preoccupied many officers of the Geodesic Section. It required fixing methods, in the sense of speed, convenience, and maximal precision, not yet entirely understood. Cholesky tackled this problem, bringing to it, as in everything he did, a marked originality. He invented in his solution of the condition [normal, Tr.] equations in the method of least squares, a very ingenious computational

---

This is a free rendering of a translation by Richard W. Cottle with the assistance of Eduardo Aguado and is used with permission. Professor Cottle first presented it in Operations Research 314—Matrix Analysis and Inequalities—at Stanford University on October 15, 1975. It was first published in the *Bullétin géodésique* (1922) entitled *Union géodésique et géophysique internationale, Première Assemblé Génèrale, Rome, May 1922*, Section de Géodésie, Toulouse, Privat, 1922, no. 8, pp. 159–161.

Translators note. The method was subsequently published by Commandant (i.e., Major) Benôit, Note sur une méthode de résolution des équation normales provenant de l'application de la méthode des moindres carrés á un système d'equations linéaires en nombre inférieure á celui des inconnues. Application de la méthode á la résolution d'un système defini d'equations linéaires. (Procédé de Commandant Cholesky.) *Bullétin géodésique*, vol. 7, no. 1 (1924), pp. 67–77.

procedure that immediately proved extremely useful and most certainly would have great benefits for all geodesists if it were published some day.

In the midst of all this there came up a project for mapping Crete, then occupied by international troops. Following a proposal by Colonel Lubanski, commanding officer of the French troops in Crete, a former geodesist, and then a rapid reconnaissance made in March to April 1906 by Lieutenant Colonel Bourgeois, chief of the Geodesic Section, the French Geographic Service was chosen to undertake the triangulation of the French and British sectors of the island of Crete (departments of San Nicolo and Candie) and the topographical survey of the French sector.

Three officers, including Major Lallemand, chief of the mission, and Lieutenant Cholseky, who had left France in November 1907, proceeded in three months to the fundamental operations: measurement of a base (in the Kavousi plain), determination of an astronomical latitude and azimuth at the southern end. Then Cholesky alone remained to execute the triangulation.

The reconnaissance and the construction of the markers were pursued in midwinter. If one is mindful of the fact that Crete, whose breadth is scarcely 40 kilometers, is covered with mountains whose altitude occasionally exceeds 2400 meters, one appreciates how particularly hard the operations were. At the summit of the Lassithi heights, the water necessary for the detachment was still furnished at the end of May by the melting of enormous drifts of snow. Nevertheless, surmounting all obstacles, Cholesky succeeded in finishing the work on June 15, 1908. Unfortunately, in the end political circumstances did not permit the topographers to go to Crete to execute the surveys.

Afterward, the geodesic activity of Cholesky underwent a two-year interruption (September 1909 to September 1911) because of the obligation he had toward carrying out his tour of duty as battery commander.

Sometime after his return to the Geodesic Section, he was sent to Algiers to take direction of the precision leveling of Algeria and Tunisia, executed by the Geodesic Section on behalf of the Governor General of Algeria and the Regency of Tunis.

To his new functions Cholesky brought the zeal he applied to everything. One detects it from reading the *Report on the operations of the precision levelling of Algeria and Tunisia during the campaigns* 1910–11, 1911–12, 1912–13.*

---

*Cahiers du Service géographique de l'Armée, no. 35, Paris (*Imprimerie du Service géographique de l'Armée*), 1913, 36 p.

†See Rapport sur les travaux exécutés en 1912, Cahiers du Service géographique de l'Armée, no. 36, Paris (*Imprimerie du Service géographique de l'Armée*) 1913, 90 p., 15 pl., pp. 4–17.

The primary Tunisian grid was completed on site in the winter 1913–1914. Right after that all the calculations were collected and reviewed, and the grid was checked and adjusted. The publication which will ensue will be the result of this effort.

Meanwhile, for the problem of leveling in Morocco under rather impossible conditions, Cholesky sought to fit the means to the purpose, and arrived at the methods that would be used.[†]

In May 1913 Cholesky was assigned to the Ministry of Foreign Affairs and put in charge of a vast service, the Topographical Service of the Regency of Tunis. He hardly had time to show the extent of what he would have been able to do there. Indeed, the war broke out shortly thereafter.

During the campaign, he was one of the officers who best understood and developed the role of geodesy and topography in the organization of artillery firing. In October 1916 his technical qualities resulted in his being sent on a mission with the Geographical Service of the Romanian army. He returned in February 1918 having rendered distinguished service there.

Five months later, on August 31, 1918 he died for his country, two months before his comrade Levesque, also the commander of an artillery group. This was a sad and irretrievable loss for geodesy.

References

[Abbreviations: ACM = Association for Computing Machinery, ASA = American Statistical Association, JASA = Journal of the American Statistical Association, JRSS = Journal of the Royal Statistical Society.]

Ahrens, J. H., and U. Dieter (1982a). Generating gamma variates by a modified rejection technique. *Commun. ACM* **25**, 47–54.

Ahrens, J. H., and U. Dieter (1982b). Computer generation of Poisson deviates from modified normal distributions. *ACM Trans. Math. Soft.* **8**, 163–179.

Ahrens, J. H., and K. D. Kohrt (1981). Computer methods for efficient sampling from largely arbitrary distributions. *Computing, Archives for Informatics and Numerical Computation* **26**, 19–31.

Aitkin, M. A. (1974). Simultaneous inference and the choice of variable subsets in multiple regression. *Technometrics* **16**, 221–227.

Alcock, D. (1977). *Illustrating BASIC*. Cambridge University Press, Cambridge, U.K.

Alvey, N. G., P. W. Lane, and N. Galwey (1982). *Introductory Guide to Genstat*. Academic Press, London.

Allen, D. M., and F. B. Cady (1982). *Analysing Experimental Data by Regression*. Wadsworth, Belmont, Calif.

Ansley, C. F. (1979). An algorithm for the exact likelihood of a mixed autoregressive-moving average process. *Biometrika* **66**, 59–65.

Anscombe, F. J. (1967). Topics in the investigation of linear relations fitted by the method of least squares. *JRSS B* **29**, 1–52.

Anscombe, F. J. (1981). *Computing in Statistical Science through APL*. Springer-Verlag, New York.

Atkinson, A. C. (1980). Tests of pseudo-random numbers. *Applied Statistics* **29**, 164–171.

Atkinson, A. C. (1982). Regression diagnostics, transformations, and constructed variates (with discussion). *JRSS B* **44**, 1–36.

Atkinson, A. C., and M. C. Pearce (1976). The computer generation of beta, gamma and normal random variates. *JRSS A* **139**, 431–461.

Baker, R. J. (1981). GLIM-4 et al. *GLIM newsletter, issue 5* (December), p. 17.

Baker, R. J., and J. A. Nelder (1978). *GLIM Manual, Release 3*. Numerical Algorithms Group and Royal Statistical Society, c/o Numerical Algorithms Group, Oxford, U.K.

Bailey, B. J. R. (1980). Accurately normalizing transformations of a Student's t variate. *Applied Statistics* **29**, 304–306.

Balfour, A., and D. H. Marwick (1980). *Programming in Standard Fortran 77*. Heinemann Educational Books, London.

Barnard, G. A. (1963). (In discussion following a paper by Bartlett.) *JRSS B* **25**, 294.

Barnes, R. A. (1979). *PL/I for Programmers*. Prentice-Hall Englewood Cliffs, N.J.

Barnett, V., and T. Lewis (1978). *Outliers in Statistical Data*. Wiley, Chichester, U.K.

Beasley, J. D., and S. G. Springer (1977). The percentage points of the normal distribution (Algorithm AS111). *Applied Statistics* **26**, 118–121.

Becker, R. A., and J. M. Chambers (1980). *S: A Language and System for Data Analysis*. Bell Laboratories, Murray Hill, N.J.

Belsey, D. A., E. Kuh, and R. E. Welsch (1980). *Regression Diagnostics: Identifying Influential Data and Sources of Collinearity*. Wiley-Interscience, New York.

Benôit (1924). Note sur une méthode de résolution des équation normales provenant de l'application de la méthode des moindres carrés á un système d'equations linéaires en nombre inférieure á celui des inconnues—Application de la méthode á la résolution d'un système defini d'equations linéaires. (Procédé du Commandant Cholesky.) *Bullétin Géodésique* **7**, 67–77.

Bentley, J. L, and J. B. Saxe (1980). Generating sorted lists of random numbers. *ACM Trans. Math. Soft.* **6**, 359–364.

Berk, K. N. (1977). Tolerance and condition in regression calculations. *JASA* **72**, 863–866.

Berk, K. N. (1978). Comparing subset regression procedures. *Technometrics* **20**, 1–6.

Bishop, Y. M. M., S. E. Fienberg, and P. W. Holland (1975). *Discrete Multivariate Analysis*. The MIT Press, Cambridge, Mass.

Bloomfield, P. (1976). *Fourier Analysis of Time Series: An Introduction*. Wiley-Interscience, New York.

Box, G. E. P., and G. M. Jenkins, (1976). *Time Series Analysis: Forecasting and Control*, rev. ed. Holden-Day, San Francisco.

Breaux, H. J. (1968). A modification of Efroymson's technique for stepwise regression analysis. *Commun. ACM* **11**, 556–557.

Brent, R. P. (1973). *Algorithms for Minimization without Derivatives*. Prentice-Hall, Englewood Cliffs, N.J.

Brier, S. S. (1980). Analysis of contingency tables under cluster sampling. *Biometrika* **67**, 591–596.

Buhler, S., and R. (1979). *PSTAT 78 User's Manual*. P-Stat Inc., Princeton, N.J.

Bukac, J., and H. Burstein (1980). Approximations to Student's t and chi-square percentage points. *Communications in Statistics—Simulation and Computation B* **9**, 665–672.

Campbell, S. L., and C. D. Meyer (1979). *Generalized Inverses of Linear Transformations*. Pitman, London.

Carroll, J. M. (1982). The adventure of getting to know a computer. *IEEE Computer* **15**, 49–58.

Chambers, J. M. (1977). *Computational Methods for Data Analysis*. Wiley-Interscience, New York.

Chambers, J. M. (1981). Some thoughts on expert software. In W. F. Eddy, ed., *Computer Science and Statistics: 13th Annual Symposium on the Interface*. Springer-Verlag, New York, pp. 36–40.

Chan, T. F. (1982). An improved algorithm for computing the Singular Value Decomposition. *ACM Trans. Math. Soft.* **8**, 72–83 and (for ACM Algorithm 581) 84–88.

Chatfield, C. (1980). *The Analysis of Time Series: An Introduction*, 2nd. ed. Chapman and Hall, London.

Cheng, R. C. H. (1977). The generation of gamma variables with non-integral shape parameters. *Applied Statistics* **26**, 71–74.

Cheng, R. C. H. (1982). The use of antithetic variates in computer simulations. *Journal of the Operational Research Society* **33**, 229–237.

Cochran, W. G., and G. M. Cox (1957). *Experimental Designs*, 2nd ed. Wiley, New York.

Cook, R. D. and Weisberg, S. (1980). Characterizations of an empirical influence function for detecting influential cases in regression. *Technometrics* **22**, 495–508.

Cormack, R. M. (1981). Loglinear models for capture-recapture experiments on open populations. In R. W. Hiorns and D. Cooke, eds., *The Mathematical Theory of the Dynamics of Biological Populations*. Academic Press, London.

Corsten, L. C. A. (1958). Vectors, a tool in statistical regression theory. *Mededelingen van de Landbouwhogeschool te Wageningen* **58**, 1–92.

Cotton, G. (1981). About sorts—part II. *Interface Age* **6** (9), 82–92.

Cox, D. R., and D. V. Hinkley (1974). *Theoretical Statistics*. Chapman and Hall, London.

Cox, D. R., and E. J. Snell (1974). The choice of variables in observational studies. *Applied Statistics* **23**, 51–59.

Curtis, A. R. (1966). Theory and calculation of best rational approximations. In D. C. Handscomb, ed., *Methods of Numerical Approximation*. Pergamon Press, Oxford, U.K., pp. 139–148.

Dahlquist, B., and A. Bjorck (1974). *Numerical Methods*. Prentice-Hall, Englewood Cliffs, N.J.

Daley, D. J., and J. H. Maindonald (1982). A Unified view of models describing the avoidance of super-parasitism. *Technical Report, Statistics Department, Australian National University, Canberra*.

Damsleth, E. (1983). Estimation in moving average models, why does it fail? *Journal of Statistical Computation and Simulation* **16**, 109–128.

Daniel, C. and F. S. Wood (1980). *Fitting Equations to Data*, 2nd ed. Wiley-Interscience, New York.

David, H. A. (1963). *The Method of Paired Comparisons*. Griffin's, London.

Davies, R. B., and B. Hutton (1975). The effect of errors in the independent variables in linear regression. *Biometrika* **62**, 383–392.

De Boor, C. (1978). *A Practical Guide to Splines*. Springer-Verlag, New York.

Dempster, A. P., and M. Gasko-Green (1981). New tools for residual analysis. *Annals of Statistics* **9**, 945–959.

Dempster, A. P., N. M. Laird, and D. B. Rubin (1977). Maximum likelihood estimation via the EM algorithm. *JRSS B* **39**, 1–38.

Denning, P. J. (1982). Are operating systems obselete? *Commun. ACM* **25**, 225–227.

Dennis, J. E., D. M. Gay, and R. E. Welsch (1981). An adaptive nonlinear least squares algorithm. *ACM Trans. Math. Soft.* **7**, 348–368.

Diaconis, P. and B. Efron (1983). Computer-Intensive Methods in Statistics. *Scientific American* **248**, 96–108.

Dixon, W., M. Brown, L. Engleman, J. Frane, M. Hill, R. Jennrich, and J. Toporek, eds. (1981). *BMDP Manual*. University of California Press, Berkeley.

Dongarra, J. J., J. R. Bunch, C. B. Moler, and G. W. Stewart (1979). *Linpack Users' Guide*. Society for Industrial and Applied Mathematics, Philadelphia.

Donoho, D., P. J. Huber, and H.-M. Thoma (1981). The use of kinematic displays to represent high dimensional data. In W. F. Eddy, ed., *Computer Science and Statistics: Proceedings of the 13th Symposium on the Interface*, pp. 275–278.

Doolittle, M. H. (1878). Method employed in the solution of normal equations and the adjustment of a triangulation. *U.S. Coast and Geodetic Survey Report*, pp. 115–120.

Draper, N. R., and H. Smith (1981). *Applied Regression Analysis*, 2nd ed. Wiley-Interscience, New York.

Durand, D. (1956). A note on matrix inversion by the square root method. *JASA* **51**, 288–292.

Dutter, R. (1977). Numerical solution of robust regression problems: Computational aspects, a comparison. *Journal of Statistical Computation and Simulation* **5**, 207–238.

Dwyer, P. S. (1941). The Doolittle technique. *Annals of Mathematical Statistics* **12**, 449–458.

Dwyer, P. S. (1945). The square root method and its use in correlation and regression. *JASA* **40**, 493–503.

Efron, B. (1979). Computers and the theory of statistics: thinking the unthinkable. *SIAM Review* **21**, 460–480.

Ehrenburg, E. H. C. (1981). The problem of numeracy. *American Statistician* **35**, 67–71.

Enslein, K., A. Ralston, and H. S. Wilf, eds. (1977). *Statistical Methods for Digital Computers* (vol. 3 of *Mathematical Methods for Digital Computers*). Wiley-Interscience, New York.

Ezekial, M. (1924). A method of handling curvilinear correlation for any number of variables. *JASA* **19**, 431–453.

Feller, W., (1968). *An Introduction to Probability Theory and its Applications*, 3rd. ed. Wiley, New York.

Fienberg, S. E. (1979). Log linear representation for paired comparison models with ties and within-pair order effects. *Biometrics* **35**, 479–481.

Fienberg, S.E., (1980). *The Analysis of Cross-classified Categorical Data, 2nd ed. The MIT Press, Cambridge, Mass.*

Fishman, G. S., and L. R. Moore (1982). A statistical evaluation of multiplicate congruential random number generators with modulus $2^{31} - 1$. *JASA* **77**, 129–136.

Forsythe, G. E., M. A. Malcolm and C. B. Moler (1977). *Computer Methods for Mathematical Computations*. Prentice-Hall, Englewood Cliffs, N.J.

Fox, L., and D. F. Mayers (1968). *Computing Methods for Scientists and Engineers*. Clarendon Press, Oxford, U.K.

Francis, I., ed. (1979). *A Comparative Review of Statistical Software*. International Association for Statistical Computing, 428 Prinses Beatrixlaan, 2270 AZ Voorburg, Netherlands.

Francis, I. (1981). *Statistical Software: A Comparative Review*. North-Holland, New York.

Francis, I., R. M. Heiberger, and P. F. Velleman (1975). Criteria and considerations in the evaluation of statistical program packages. *American Statistics* **29**, 52–56.

Fujino, Y. (1980). Approximate binomial confidence limits. *Biometrika* **67**, 677–681.

Fuller, W. A. (1976). *Introduction to Statistical Time Series*. Wiley, New York.

Gale, W. A., and D. Pregibon (1983). Using expert systems for developing statistical strategy. *1983 Proceedings of the Statistical Computing Section of the ASA*, forthcoming.

Gentleman, J. F., and M. B. Wilk (1975). Detecting Outliers. II. Supplementing the direct analysis of residuals. *Biometrics* **31**, 387–410.

Gilchrist, R., ed. (1982). *GLIM 82: Proceedings of the International Conference on Generalized Linear Models*. Springer-Verlag, Berlin.

Gill, P. E., G. H. Golub, W. Murray, and M. A. Saunders (1974). Methods for modifying matrix factorizations. *Mathematics of Computation* **28**, 505–535.

Gnanadesikan, R. (1977). *Methods for Statistical Data Analysis of Multivariate Observations*. Wiley-Interscience, New York.

Golub, G. H. (1969). Matrix decompositions and statistical calculations. In R. C. Milton and J. A. Nelder, eds., *Statistical Computation*. Academic Press, New York.

Golub, G. H., and G. P. H. Styan (1973). Numerical computations for univariate models. *Journal of Statistical Computation and Simulation* **2**, 253–274.

Gordon, A. D. (1981). *Classification*. Chapman and Hall, London.

Graybill, F. A. (1982). *Introduction to Matrices with Applications in Statistics*, 2nd ed. Wadsworth, Belmont, Calif.

Griffiths, D. A. (1977). Avoidance-modified generalized distributions and their application to studies of superparasitism. *Biometrics* **33**, 102–112.

Haberman, S. (1974). *The Analysis of Frequency Data*. University of Chicago Press, Chicago.

Hahn, G. J. (1983). Expert systems for statistical consulting. *1983 Proceedings of the Statistical Computing Section of the ASA*, forthcoming.

Hammarling, S. (1974). A note on modifications to the Givens plane rotation. *Journal of the Institute of Mathematics and its Applications* **13**, 215–218.

Hampel, F. R. (1974). The influence curve and its role in robust estimation. *JASA* **69**, 383–393.

Harkness, W. L. (1978). The Classical Occupancy Problem Revisted. In V. F. Kolchin, A. V. Sevast'yanov, and V. P. Chistyakov, eds., transl. by A. V. Balakrishnan, *Random Allocations* (Scripta Series in Mathematics). Wiley, New York.

Harris, R. J. (1975). *A Primer of Multivariate Statistics*. Academic Press, London.

Harville, D. A. (1977). Maximum likelihood approaches to variance component estimation and related topics. *JASA* **72**, 320–340.

Hawkins, D. M., and W. J. R. Eplett (1982). The Cholesky factorization of the inverse correlation or covariance matrix in multiple regression. *Technometrics* **24**, 191–198.

Heiberger, R. (1982). *P-STAT ANOVA Command*. P-STAT Inc., Princeton, N.J.

Hemmerle, W. J. (1967). *Statistical Computations of a Digital Computer*. Blaisdell, Waltham, Mass.

Hemmerle, W. J. (1982). Algorithm 591: A comprehensive matrix-free algorithm for analysis of variance. *ACM Trans. Math. Soft.* **8**, 383–401.

Hill, G. W. (1970). Algorithm, 395: Student's *t*-quantiles. *Commun. ACM* **13**, 619–620.

Hoare, C. A. R. (1981). The Emperor's old clothes. *Commun. ACM* **24**, 75–83. Reprinted in *Byte* **6** (1981), 414–425.

Hoaglin, D. C. and R. E. Welch (1978). The Hat Matrix in Regression and ANOVA. *American Statistician* **32**, 17–22 and *Corrigenda* **32**, 146.

Hocking, R. R. (1976). The analysis and selection of variables in linear regression. *Biometrics* **32**, 1–44.

Hocking, R. R. (1983). Developments in Linear Regression Methodology: 1959–1982, with discussion. *Technometrics* **25**, 219–249.

Hofstadter, D. R. (1979). *Godel, Escher, Bach: Eternal Golden Braid*. Basic Books, New York.

Hogben, D., S. T. Peavey, and R. N. Varner (1971). *OMNITAB II User's Reference Manual*. Technical Note 551, National Bureau of Standards, Washington, D.C.

Holt, D., A. J. Scott, and P. G. Ewings (1980). Chi-square tests with survey data. *JRSS A* **143**, 303–320.

Houthakker, H. S. (1951). Some calculations on electricity consumption in Great Britain. *JRSS A* **114**, 359–371.

Huber, P. J. (1981). *Robust Statistics*. Wiley, New York.

Hughes, D. T., and J. G. Saw (1972). Approximating the percentage points of Hotelling's generalized T^2 statistic. *Biometrika* **59**, 224–226.

IMSL (1982). *IMSL Library Reference Manual*. IMSL Inc., Houston, Tex.

Jarrett, R. G. (1978). The analysis of designed experiments with missing observations. *Applied Statistics* **27**, 38–46.

Jennrich, R. I. (1977). Stepwise Regression. In Enslein, K., A. Ralston, and H. S. Wilf, eds. (1977). *Statistical Methods for Digital Computers* (q.v.) Wiley-Interscience, New York.

Johnson, N. L., and S. Kotz (1970). *Continuous Univariate Distributions*. Vol. 2. Houghton Mifflin, Boston.

Jowett, G. H. (1963). Application of Jordan's procedure for matrix inversion in multiple regression and multivariate distance analysis. *JRSS B* **25**, 352–357.

Kempthorne, O. (1980). The term *Design* Matrix. (Letter to the Editor.) *American Statistician* **34**, 249.

Kempton, R. A. (1978). Statistical consultancy at an agricultural research institute. *Mathematical Scientist* **3**, 9–21.

Kennedy, W. J., Jr., and J. E. Gentle (1980). *Statistical Computing*. Marcel Dekker, New York.

Kernighan, B. W., and P. J. Plauger (1978). *The Elements of Programming Style*, 2nd ed. McGraw-Hill, New York.

Knuth, D. E. (1969). *The Art of Computer Programming. Vol. 2 / Seminumerical Algorithms*. Addison-Wesley, Reading, Mass.

Knuth, D. E. (1973). *The Art of Computer Programming. Vol. 3 / Sorting and Searching*. Addison-Wesley, Reading, Mass.

Kotz, S., N. L. Johnson, and C. B. Read (1982–). *Encyclopedia of Statistical Sciences. Wiley, New York*.

Kuki, H., and W. J. Cody (1973). A study of the accuracy of floating point number systems. *Commun. ACM* **16**, 223–230.

Kummer, G. (1981). Formulas for the computation of the Wilcoxon test and other rank statistics. *Biometrical Journal* **23**, 237–243.

Kurtz, T. E. (1982). On the way to standard BASIC. *Byte* **7**, 182–218.

Lamagna, E. A. (1982). Fast computer algebra. *IEEE Computer* **15**, 43–58.

Lancaster, H. O. (1974). *An Introduction to Medical Statistics*. Wiley, New York.

Lawal, H. B. (1980). Tables of percentage points of Pearson's goodness of fit statistic for use with small expectations. *Applied Statistics* **29**, 292–298.

Lawal, H. B., and G. J. G. Upton (1980). An approximation to the distribution of the χ^2 goodness-of-fit statistic for use with small expectations. *Biometrika* **67**, 447–453.

Lawson, C. L., and R. J. Hanson (1974). *Solving Least Squares Problems*. Prentice-Hall, Englewood Cliffs, N.J.

Lien, D. A. (1981). *The Basic Handbook*, 2nd ed. Compusoft Publishing, San Diego, Calif.

Ling, R. F. (1974). Comparison of several algorithms for computing sample means and variances. *JASA* **69**, 859–866.

Ling, R. F. (1978). A study of the accuracy of some approximations for t, χ^2, and F tail probabilities. *JASA* **73**, 274–283.

Loeser, R. (1976). Survey on algorithms 347, 426, and Quicksort. *ACM Trans. Math. Soft.* **2**, 290–299.

Longley, J. W. (1981). Modified Gram-Schmidt process vs. classical Gram-Schmidt. *Communications in Statistics—Simulation and Computation B* **10**, 517–528.

McCullagh, P. (1980). Regression models for ordinal data. *JRSS B* **42**, 109–142.

McCullagh, P. and J. A. Nelder (1983). Generalized Linear Models. Chapman and Hall, London.

McIntosh, A. A. (1982). *Fitting Linear Models: An Application of Conjugate Gradient Algorithms*. Springer-Verlag, Berlin.

McKean, J. W., and T. A. Ryan, Jr. (1977). Algorithm 516: An algorithm for obtaining confidence intervals and point estimates based on ranks in the two-sample location problem [G1]. *ACM Trans. Math. Soft.* **3**, 183–185.

Maindonald, J. H. (1977). Least squares computations based on the Cholesky decomposition of the correlation matrix. *Journal of Statistical Computation and Simulation* **5**, 247–258.

Marquardt, D. A. (1970). Generalized inverses, ridge regression, biased linear estimation, and nonlinear estimation. *Technometrics* **12**, 591–612.

Marriott, F. H. C. (1979). Barnard's Monte Carlo tests: How many simulations? *Applied Statistics* **28**, 75–77.

Marsaglia, G. (1972). The structure of linear congruential sequences. In S. K. Zaremba, ed., *Applications of Number Theory to Numerical Analysis*. Academic Press, New York.

Miller, A. (1976). Cholesky and Cholesky factorisation. *DMS Newsletter (Division of Mathematics and Statistics, CSIRO)* **17**, 3–4.

Miller, A. (1982). More on selecting subsets of regressor variables. *DMS Newsletter (Division of Mathematics and Statistics, CSIRO)* **90**, 1–4.

Miller, A. J. (1984). Selecting subsets of regression variables. JRSS A, forthcoming.

Morganstein, D., and J. F. Carpenter (1983). Statistical Packages available for microcomputers. *1983 Proceedings of the Statistical Computing Section of the ASA*, forthcoming.

Motzkin, D. (1983). Meansort. *Communications of the ACM*, **26** 250–251.

Mosteller, F., and J. Tukey (1977). *Data Analysis and Regression*. Addison-Wesley, Reading, Mass.

Muller, M. E. (1970). Computers as an instrument for data analysis. *Technometrics* **12**, 259–293.

Mullet, G. M., and T. W. Murray (1971). A new method for examining rounding error in least-squares regression computer programs. *JASA* **66**, 496–498.

Murray, R. O. (1972). Agonies of the aspiring athlete. *New Scientist* **55** (issue of Aug. 31), 434–436.

Nai-Kuan Tsao (1975). A note on implementing the Householder transformation. *SIAM Journal of Numerical Analysis* **12**, 53–58.

Nash, J. C. (1975). A one-sided transformation method for the singular value decomposition and algebraic eigenproblem. The Computer Journal **18**, 74–76.

Nash, J. C. (1979). *Compact Numerical Methods for Computers*. Adam Hilger, Bristol, U.K.

Nelder, J. A. (1965). The analysis of randomised experiments with orthogonal block structure (Parts I and II). *Proceedings of the Royal Society of London A* **283**, 147–178.

Nelder, J. A., et al. (1981). *Genstat—A General Statistical Program*. Rothamsted Experimental Station, Harpenden, Herts. AL5 2JQ, U.K.

Nelder, J. A., and R. W. M. Wedderburn (1972). Generalized linear models. *JRSS A* **135**, 370–384.

Newton, H. J. (1981). On some numerical properties of ARMA parameter estimation procedures. In W. F. Eddy, ed., *Computer Science and Statistics: Proceedings of the 13th Symposium on the Interface*, pp. 172–177.

Nie, N. H., C. H. Hull, J. G. Jenkins, K. Steinbrenner, and D. H. Bent, (1975). *SPSS*, 2nd ed. McGraw-Hill. (Note also Hull and Nie: SPSS Update, 1979).

Numerical Algorithms Group (1981). *NAG FORTRAN Mini Manual, Mark 9*. Numerical Algorithms Group, 256 Banbury Rd., Oxford OX2, 7DE.

Paige, C. C. (1979). Computer solution and perturbation analysis of generalized linear least squares problems. *Mathematics of Computation* **33**, 171–183.

Paige, C. C., and M. A. Saunders (1982). LSQR: An algorithm for sparse linear equations and sparse least squares. *ACM Trans. Math. Soft.* **8**, 43–71 and (for ACM Algorithm 583) 195–209.

Palmgren, J. (1981). The Fisher information matrix for log linear models arguing conditionally on observed explanatory variables. *Biometrika* **68**, 563–566.

Parlett, B. N. (1980). *The Symmetric Eigenvalue Problem*. Prentice-Hall, Englewood Cliffs, N.J.

Payne, R. W., and G. N. Wilkinson (1977). A general algorithm for analysis of variance. *Applied Statistics* **26**, 251–260.

Pearce, S. C., T. Calinski, and T. F. de C. Marshall (1974). The basic contrasts of an experimental design with special reference to the analysis of data. *Biometrika* **61**, 449–460.

Pearson, E. S., and H. O. Hartley, eds. (1972). *Biometrika Tables for Statisticians*. Vol. 2. Cambridge University Press, Cambridge, U.K.

Peizer, D. B., and J. W. Pratt (1968). A normal approximation for binomial F, beta, and other common, related tail probabilities, I. *JASA* **63**, 1416–1456.

Plackett, R. L. (1960). *Principles of Regression Analysis*. Clarendon Press, Oxford, U.K.

Plackett, R. L. (1981). *The Analysis of Categorical Data*. Griffin's, London.

Preece, D. A. (1982). *t* is for trouble (and textbooks): a critique of some examples of the paired-samples *t*-test. *The Statistician* **31**, 169–195.

Pregibon, D. (1981). Logistic regression diagnostics. *Annals of Statistics* **9**, 705–724.

Rao, C. R. (1967). Least squares theory using an estimated dispersion matrix and its application to measurement of signals. *Proc. Fifth Berkeley Symposium on Mathematical Statistics and Probability* **1**, 355–372. University of California Press, Berkeley.

Rao, C. R. (1973). *Linear Statistical Inference and Its Applications 2nd. ed.* Wiley, New York.

Reid, J. K. (1971). On the method of conjugate gradients for the solution of large sparse systems of linear equation. In J. K. Reid, ed., *Large Sparse Sets of Linear Equations* (Conference Proceedings). Academic Press, New York.

Reingold, E. M., Jurg Nievergelt, and Narsingh Deo (1977). *Combinatorial Algorithms: Theory and Practice*. Prentice-Hall, Englewood Cliffs, N.J.

Rubinstein, R. Y. (1981). *Simulation and the Monte Carlo Method*. Wiley-Interscience, New York.

Ryan, T. A., B. L. Joiner, and B. F. Ryan (1981). *Minitab Reference Manual*. Minitab Project, Philadelphia.

SAS Institute (1979). *SAS User's Guide, 1979 Edition*. SAS Institute Inc., Raleigh, N.C.

Schrage, L. (1979). A more portable Fortran random number generator. *ACM Trans. Math. Soft.* **5**, 132–138.

Scott, A. J., and M. Knott (1974). A cluster analysis method for grouping means in the analysis of variance. *Biometrics* **30**, 507–512.

Scott, A. J., and M. Knott (1976). An approximate test for use with AID. *Applied Statistics* **25**, 103–106.

Scott, D. T., J. Olsen, and G. R. Bryce (1983). Experiences in converting some statistical programs to an MC68000 based microcomputer. *1983 Proceedings of the Statistical Computing Section of the ASA*, forthcoming.

Searle, S. R. (1978). A summary of recently developed methods of estimating variance components. In A. R. Gallant and T. M. Gerig, eds., *Computer Science and Statistics: 11th Annual Symposium on the Interface*, pp. 64–69.

Searle, S. R., and H. V. Henderson (1983). Faults in a computing algorithm for reparameterizing linear models. *Communications in Statistics, Simulation and Computation* **B12**, 67–76.

Searle, S. R., F. M. Speed, and H. V. Henderson (1981). Some computational and model equivalences in analysis of variance of unequal-subclass-numbers data. *American Statistician* **35**, 16–33.

Searle, S. R., et al. (1978–1980). *Annotated Computer Outputs*: *BMDP-2V*, *GENSTAT* (*ANOVA and REGRESSION*), *RUMMAGE*, *SAS GLM*, *SAS HARVEY*, *SPSS* (*ANOVA*), *MOPT* (*MINITAB*, *OSIRIS*, *P-STAT*, *and TROLL*); and for variance components and mixed models: *BMDP-V* (*P3V*, *P2V and P8V*), *SAS GLM VARCOMP*, *SAS HARVEY*, *SAS RANDOM* (*and NESTED*). Biometrics Unit, Ithaca, 14853.

Seber, G. A. F. (1977). *Linear Regression Analysis*. Wiley-Interscience, New York.

Sedgewick, R. (1978). Implementing Quicksort programs. *Commun. ACM* **21**, 847–857.

Smith, P. J., D. S. Rae, R. W. Manderscheid, and S. Silbergeld (1981). Approximating the moments and distribution of the likelihood ratio statistic for multinomial goodness of fit. *JASA* **76**, 737–740

Spjotvoll, E. (1972). Multiple comparison of regression functions. *Annals of Mathematical Statistics* **43**, 1076–1088.

Stewart, G. W. (1973). *Introduction to Matrix Computations*. Academic Press, New York.

Stewart, G. W. (1979). The effects of rounding error on an algorithm for downdating a Cholesky factorization. *Journal of the Institute of Mathematics and its Applications* **23**, 203–213.

Stoer, J., and R. Bulirsch (1980). *Introduction to Numerical Analysis*. Springer-Verlag, New York.

Thisted, R. A. (1981). The effect of personal computers on statistical practice. In W. F. Eddy, ed., *Computer Science and Statistics: Proceedings of the* 13*th Symposium on the Interface, pp*. 223–244.

Thompson, M. L. (1978). Selection of variables in multiple regression. Part I. A review and evaluation. *International Statistical Review* **46**, 1–19. Part II. Chosen procedures, computations and examples. *International Statistical Review* **46**, 129–146.

Thompson, R., and R. J. Baker (1981). Composite link functions in generalized linear models. *Applied Statistics* **30**, 125–131.

Tiberghien, J. (1981). *The Pascal Handbook*. Sybex, Berkeley, Calif.

Toong, H. D., and A. Gupta (1982). Personal computers. *Scientific American* **247**, 88–99.

Turner, J. (1980). The structure of modular programs. *Commun. ACM* **23**, 272–277.

Turner, J. C., and W. J. Rogers (1980). *STATUS: a statistical computing language*. New Zealand Mathematical Society, Victoria University, Wellington.

Velleman, P. F., and D. C. Hoaglin (1981). *Applications, Basics, and Computing of Exploratory Data Analysis. Duxbury Press, Duxbury, Mass.*

Velleman, P. F., and R. E. Welsch (1981). Efficient computing of reggression diagnostics. *American Statistician* **35**, 234–242.

Wampler, R. H. (1970). A report on the accuracy of some widely used least squares computer programs. *JASA* **65**, 549–565.

Wampler, R. H. (1980). Test procedures and test problems for least squares algorithms. *Journal of Econometrics* **12**, 3–22.

Wasserman, A. I., and S. Gutz (1982). The future of programming. *Commun. ACM* **25**, 196–206.

Wegner, P. (1980). *Programming with Ada: an Introduction by Means of Graded Examples*. Prentice-Hall, Englewood Cliffs, N.J.

Whittaker, E. T. and G. Robinson (1944). *The Calculus of Observations*, 4th ed. Blackie and Son, London. Reprinted by Dover Publications, 1967.

Wichmann, B. A., and I. D. Hill (1982). An efficient and portable pseudo-random number generator. (Algorithm AS183). *Applied Statistics* **31**, 188–190.

Wilkinson, G. N., and C. E. Rogers (1973). Symbolic description of factorial models for analysis of variance. *Applied Statistics* **22**, 392–399.

Wilkinson, J. H. (1963). *Rounding Errors in Algebraic Processes*. Prentice-Hall, Englewood Cliffs, N.J.

Wilkinson, J. H. (1965). *The Algebraic Eigenvalue Problem*. Clarendon Press, Oxford, U.K.

Wilkinson, J. H. (1966). A priori analysis of algebraic processes. *Proceedings of the International Congress of Mathematicians*, Moscow (1966), pp. 629–640.

Wilkinson, J. H. (1974). Error analysis and related topics. The classical error analyses for the solution of linear systems. *Bulletin of the Institute of Mathematics and its Applications* **10**, 175–180.

Wilkinson, J. H., and C. Reinsch, (1971). *Handbook for Automatic Computation, II, Linear Algebra*, F. L. Bauer, et al., eds. Springer-Verlag, New York.

Williams, E. R., D. Ratcliff, and T. P. Speed (1981). Estimating missing values in multi-stratum experiments. *Applied Statistics* **30**, 71–72.

Wood, F. S. (1973). The use of individual effects and residuals in fitting equations to data. *Technometrics* **15**, 677–686.

Worthington, B. A. (1975). General iterative method for analysis of variance when the block structure is orthogonal. *Biometrika* **62**, 113–120.

Yates, F. (1966). Computers, the second revolution in statistics. (The first Fisher memorial lecture.) *Biometrics* **22**, 233–251.

Suppliers

ADDRESSES FOR SUPPLIERS OF PACKAGES AND SUBROUTINE LIBRARIES MENTIONED IN CHAPTER 10

ACM Algorithms Distribution Service, c/o IMSL Inc. (q.v.).

BMDP Statistical Software, 1964 Westwood Blvd., Los Angeles, CA 90025.
(213)475–5700

IMSL Inc., 7500 Bellaire Boulevard, 6th Fl., NBC Bldg., Houston, TX 77036–5085.
(713)772–1927

Minitab, Inc., 215 Pond Laboratory, University Park, PA 16802.
(814)865–1595

Numerical Algorithms Group Ltd, 256 Banbury Road, Oxford OX2 7DE, U.K. (Suppliers of GLIM and GENSTAT, as well as of the NAG library.)
(0865)511245

Address in North America:
1131 Warren Avenue, Downers Grove, IL 60515
(312)971–2337

P-STAT Inc., P.O. Box 285, Princeton, NJ 08540.
(609)924–9100

SAS Institute Inc., P.O. Box 8000, Cary, NC 27511–8000.
(919)467–8000

SPSS Inc., 444 N. Michigan Avenue, Chicago, IL 60611.
(312)329–2400.

Index